D1745137

WITHDRAWN
FROM
UNIVERSITY OF PLYMOUTH
LIBRARY SERVICES

Charles Seale-Hayne Library
University of Plymouth
(01752) 588 588
LibraryandITenquiries@plymouth.ac.uk

SEDIMENTARY PROCESSES: QUANTIFICATION USING RADIONUCLIDES

RADIOACTIVITY IN THE ENVIRONMENT

A companion series to the Journal of Environmental Radioactivity

Series Editor
M.S. Baxter
Ampfield House
Clachan Seil
Argyll, Scotland, UK

SEDIMENTARY PROCESSES: QUANTIFICATION USING RADIONUCLIDES

J. Carroll

Akvaplan-niva AS
Polar Environmental Center
N-9296 Tromsø
NORWAY

and

I. Lerche

Department of Geological Sciences
University of South Carolina
Columbia, SC 29208
USA

ELSEVIER

2003

AMSTERDAM – BOSTON – LONDON – NEW YORK – OXFORD – PARIS
SAN DIEGO – SAN FRANCISCO – SINGAPORE – SYDNEY – TOKYO

ELSEVIER SCIENCE Ltd
The Boulevard, Langford Lane
Kidlington, Oxford OX5 1GB, UK

© 2003 Elsevier Science Ltd. All rights reserved.

This work is protected under copyright by Elsevier Science, and the following terms and conditions apply to its use:

Photocopying
Single photocopies of single chapters may be made for personal use as allowed by national copyright laws. Permission of the Publisher and payment of a fee is required for all other photocopying, including multiple or systematic copying, copying for advertising or promotional purposes, resale, and all forms of document delivery. Special rates are available for educational institutions that wish to make photocopies for non-profit educational classroom use.

Permissions may be sought directly from Elsevier's Science & Technology Rights Department in Oxford, UK: phone: (+44) 1865 843830, fax: (+44) 1865 853333, e-mail: permissions@elsevier.com. You may also complete your request on-line via the Elsevier Science homepage (http://www.elsevier.com), by selecting 'Customer Support' and then 'Obtaining Permissions'.

In the USA, users may clear permissions and make payments through the Copyright Clearance Center, Inc., 222 Rosewood Drive, Danvers, MA 01923, USA; phone: (+1) (978) 7508400, fax: (+1) (978) 7504744, and in the UK through the Copyright Licensing Agency Rapid Clearance Service (CLARCS), 90 Tottenham Court Road, London W1P 0LP, UK; phone: (+44) 207 631 5555; fax: (+44) 207 631 5500. Other countries may have a local reprographic rights agency for payments.

Derivative Works
Tables of contents may be reproduced for internal circulation, but permission of Elsevier Science is required for external resale or distribution of such material.
Permission of the Publisher is required for all other derivative works, including compilations and translations.

Electronic Storage or Usage
Permission of the Publisher is required to store or use electronically any material contained in this work, including any chapter or part of a chapter.

Except as outlined above, no part of this work may be reproduced, stored in a retrieval system or transmitted in any form or by any means, electronic, mechanical, photocopying, recording or otherwise, without prior written permission of the Publisher. Address permissions requests to: Elsevier's Science & Technology Rights Department, at the phone, fax and e-mail addresses noted above.

Notice
No responsibility is assumed by the Publisher for any injury and/or damage to persons or property as a matter of products liability, negligence or otherwise, or from any use or operation of any methods, products, instructions or ideas contained in the material herein. Because of rapid advances in the medical sciences, in particular, independent verification of diagnoses and drug dosages should be made.

First edition 2003

Library of Congress Cataloging in Publication Data
A catalog record from the Library of Congress has been applied for.

British Library Cataloguing in Publication Data
A catalogue record from the British Library has been applied for.

The UNIVERSITY of PLYMOUTH	
Item No.	900 5689876
Date	28-OCT-03 S
Class No.	552.5 CAR
Contl. No.	✓
LIBRARY SERVICES	

ISBN: 0-08-044300-1 ✓

⊗ The paper used in this publication meets the requirements of ANSI/NISO Z39.48-1992 (Permanence of Paper).

Printed in The Netherlands.

Contents

Preface

Without an accurate method of dating sediments there is no way that one can provide a quantitative description of Earth processes. Prior to the advent of radionuclide methods for absolute age dating, there were available relative methods of deciding when a formation was deposited prior to others, so that a paragenetic description could be given of the ordering of events. However, there was no way of deciding on the absolute temporal duration of events, on the age of formations or indeed on any other process in which time played an integral and fundamental role.

The advent of radionuclide dating, with the attendant absolute measure of age, changed the views of geology and processes that can occur in ways that have been of major significance. In addition, the radionuclides allow a direct comparison of absolute age to the prior proxies. Accordingly, not only had one available a true age dating but also a proxy scale that could be tied back directly to the true scale. In this way, what had previously been paragenetic sequences of events could be given duration in time so that a quantitative appreciation became available to describe short-lived and long-lived events with clear definitions of short and long.

Essentially then, the start of modern geological descriptions of Earth processes owes its genesis to this absolute age dating procedure. Because different radionuclides have different half-lives they are useful in describing different age groupings. Short-lived nuclides are best for modern processes, while the further back in time one wishes to explore use is made of nuclides that decay on the time scale of interest. Thus, one develops an overlapping sequence of time markers that are grounded in the age determinations from the appropriate nuclides. Of great help in this regard is the ability to use more than one nuclide to determine ages so that better control is achieved than might otherwise have been the case.

However, while the principles of nuclide age dating are sound, there are numerous complications that can arise. For instance, if there is an intrinsic variation with time in the flux deposited of a radionuclide then one has somehow to unscramble the flux and decay effects. Or again, if there is lateral transport of sediment with associated radionuclide tracers, then once more one has to unscramble such effects from the radioactive decay effects. Later diffusion and water solubility of nuclides, originally attached to deposited sedimentary particles, also cloud the picture and coarsen the ability to resolve ages to the most accurate levels possible. A host of such ancillary problems provide confounding and masking effects in the pursuit of ages. All such effects must be addressed for their contributions. One of the purposes of this volume is to show how one goes about unscrambling these problems.

Of greater importance is the use to which the flux variations and age dating are put, particularly for modern sediments, which are the main focus throughout the volume. In essence one is concerned with contaminant transport, burial and potential later mobilization. The environmental impact of such chemical, nuclear, or other waste material is of concern in a variety of areas around the world today and influences the direction of involvement not only of scientists but also governments and industry alike.

As we will see in later chapters of the volume, it is the desire to understand such contaminant behaviors with time and space that forces a detailed understanding of the radionuclide-determined age and flux variation questions. Without an appreciation of the assumptions and consequences of models and their behaviors, and also the attendant uncertainty so brought to an interpretation of nuclide data, there is little that one can be precise and accurate about. Rather one would then have an unknown uncertainty that one could not quantify or justify, which could lead to a serious mis-statement of the depositional rates of sediments and also to the flux variations of contaminants. In turn, proposed remediation schemes would then be in jeopardy of being incorrect for the problem to hand.

It is this latter aspect that triggered a goodly amount of the work reported here. The book is, in our opinion, set at a level where an able student should have no difficulty in following the procedures and methods developed and that are also illustrated in each chapter by copious case histories showing the strengths and weaknesses of a given procedure with respect to a data example. It is also the case that professionals in the field should find much that they can use here and also much that warns them about over-applying methods far outside their domains of validity and assumptions. In that sense the volume serves to show the procedures to use with different types of situations.

We are painfully aware that there is a long way to go before we have a clear understanding of all processes that are relevant to the goal of disentangling flux and radioactive decay variations. Indeed this point is clearly made in many of the case histories exemplified throughout the volume. Nevertheless, there are a sufficient number of situations given where it is clear that one can disentangle flux and age effects cleanly, and that particular development is certainly indicating the directions needed for the future.

An introductory chapter provides both an historical background and also the basic nuclear decay scheme for the use of ^{210}Pb, and man-made ^{137}Cs is also invoked as an age marker for modern sediments. These two nuclides are powerful workhorses for attacking the problem of modern sediment deposition rates and flux variations with time. Some simple illustrations are also given of what one can learn by following different approaches and assumptions.

After this background chapter, it becomes clear that integration is needed of the various end-member assumptions of Chapter 1. In this way a systematic determination can be made of whether sediment deposition rates or surficial flux variations are dominant in particular cases, based on the resolution provided by the available data in each such case. This integrated procedure is developed in Chapter 2 with some synthetic case histories provided so that one can see how the various aspects of the integrated method can be used to analyze and disentangle the competing effects of the combined contributions of source variation and sediment supply rates to observed radionuclide profiles with sediment depth. This chapter is then followed (Chapter 3) by analyses of real case histories done at single sites. These applications have at least two advantages: they show how well the integrated procedure works in a variety of cases that would otherwise be difficult to understand; they show how well the procedure can determine sediment ages because there is often anthropogenic knowledge of true ages from direct sampling of sediments and their deposition over the course of decades.

In general, the integrated procedure is found to pass these stringent tests with flying colors, and also to resolve aspects of individual studies that would otherwise be almost impossible to handle correctly.

In the case of the Galapagos Islands example, the two nuclides available were not those of lead and cesium but rather ^{230}Th and ^{231}Pa. While each may have a different and time variable concentration at deposition, nevertheless both must provide the same sedimentation rate with time. So in a sense one can use each radionuclide as a "known" age marker for the other. In this way a consistent age to depth determination can be made for both nuclides. The point of this example is to show that radionuclides can be used to constrain the system instead of anthropogenically known dates of sediments, thereby extending the method to a much greater class of cases than if one were forced to rely solely on known age horizons as ground-truth values.

Chapter 4 turns attention to study areas where there are several profiles of the lead and cesium nuclides with sediment depth. The basic idea here is to determine the variation across a region of sediment deposition rates and nuclide flux variations with time. An example from the Amazon Delta and one from Lake Baikal demonstrate the efficacy of the disentanglement procedure under extremely different conditions, thereby proving clues concerning temporal sediment transport across the regions-something that is otherwise extremely difficult to obtain. A third multiple site study uses strontium nuclide ratios to indicate that the general procedure is not limited to just the lead and cesium nuclides. In each case extraneous information (when available) is used to tie down the resolution of the nuclide results. In general, such extraneous information provides corroborative confirmation of the accuracy of the nuclide results for the cases studied; thus one can use the radionuclide method with confidence in places where no additional information is available.

Attention is turned in Chapter 5 to the effects of biological or physical mixing of sediments after they have been laid down. These effects transport both the sediments and the contained nuclides in vertical and lateral directions. The vertical mixing acts to obliterate the intrinsic nuclide variations laid down with the sediments, making it difficult to reconstruct not only the ages of such sediments but also difficult to determine if there was any intrinsic source variation of the nuclide concentrations. In turn, such effects make it hard to evaluate the likely sources of radionuclide flux attached to deposited sediments, of great concern in environmental studies of contamination. The lateral transport by turbidites, or other erosional processes, can lead to nuclide variations with depth at particular locations that have little to do with intrinsic source or intrinsic age effects, but that are more controlled by subsequent dynamical instability of the deposited sediments. Biological applications of vertical mixing are well represented by two case histories, one from North Inlet, South Carolina, with the other being a multi-site study from Lake Baikal, Russia.

The rather nasty problems of post-depositional diffusion of nuclide concentrations relative to deposited sediments, and of partitioning of nuclide concentrations between sedimentary particles and surrounding water as the sediments are deposited through the water column, are both examined in Chapter 6. These effects limit the degree with which one can be at all accurate concerning the origin of the intrinsic radionuclide fluxes from measurements in sediments, and they also confound stratigraphic age determinations as well. Case histories are presented to show when such effects are important to understand and how one goes about constructing model behaviors for such effects.

One of the major modern reasons for studying lead and cesium radionuclide variations in sediment core is directly tied to present-day contaminant studies in the world's oceans, rivers and lakes. Here one is interested in determining the sources of contaminants and their influence on the water bodies. Chapter 7 presents four case studies, from the Kara Sea, the Devil's Lake region of North Dakota, North Pond in Colorado – the site of uranium tailings deposition about fifty years ago, and the Eastern Arabian Sea respectively, to illustrate how the radionuclide disentanglement method can help unravel the various contributions and also address the timing and rate of contamination supply.

Finally, in Chapter 8 concern changes to methods for assessing ages of horizons that cannot be dated by radionuclide methods but which lie between stratigraphic horizons that have been dated with known degrees of precision. The aim here is to show that different age dates for such non-nuclide horizons are available depending on the assumptions one makes. A procedure is given for determining the age date with the smallest uncertainty based on the underlying and overlying known dated horizons (with error assignments), and the procedure can be generalized to incorporate multiple horizons. But one is still at the mercy of the assumptions made. Without a true age-dating procedure there is no certainty that the assumptions made, or indeed any assumptions made, reflect the true unknown age. And that cautionary point is the essence of the chapter.

By bringing together in one place the processes currently used in nuclide investigations of modern sediments, it is our hope that sufficient purpose will be shown of the need for urgent research on procedures and methods of greater power than we currently enjoy. We hope that such a development will not be long delayed and also hope that this monograph provides some of the impetus for such an undertaking.

The task of writing a book is seemingly unending and many people have contributed to the final written product. Perhaps most significant has been the support of family and friends over the years. In addition Donna Black is thanked for typing large parts of the first draft of the book. Akvaplan-niva AS is thanked for permitting JLC to carry on with the volume and for providing financial support. Also thanked are the Departments of Energy and Defense of the United States Government for various grants over the years. The IAEA is thanked for support to JLC during a period spent at their Marine Laboratory, Monaco. Thanks also go to Alexander von Humboldt Stiftung and the University of Halle-Wittenberg for financial support for IL during a sabbatical year at Halle, during which period the bulk of the book was written. The courtesies of Prof. Gerhard Bachmann during the stay in Halle are gratefully appreciated. Of the many individuals who have contributed comments, ideas, criticism, and have otherwise caused us many a sleepless night, we single out here for special thanks Jiuliang Liu, Jared Abraham and Darrin Cisar; they wrote both the first version of the SIT code and also the currently operating version. Without their help we would truly have had a much poorer understanding of how variable sediment accumulation rates and varying surficial flux with time interact and can be disentangled.

JoLynn Carroll
Tromsø, Norway

Ian Lerche
Columbia, SC
USA

Chapter 1

Introduction

The shift of geological science, from a qualitative description of the ordering of events to a quantitative basis dealing with processes and the times over which processes act, occurs only when some way exists for attaching a true age determination to the sedimentary and water records. Such a procedure is, apparently, only available through the nuclear decay of unstable nuclides of various elements that are part and parcel of the Earth system. To be sure, long before the era of radionuclide measurements there were available non-radiometric methods for ascertaining the paragenetic ordering of events and the relative scales over which such sequences of events took place. But without a radiometric method for providing ages there is no way a quantitative procedure can be devised to age-date events absolutely.

This monograph is concerned with modern sediment processes only and not with the total panoply of geological time over the total age of the Earth. The reason for this focus is that modern sediments are the most subject to a host of anthropogenic influences, including contamination by biological, chemical and radioactive sources. It becomes of some concern to determine where such sources might be, how the contaminants are transported and deposited, and how they influence the sediments and waters into which they are brought. Part of the difficulty in such determinations is that the sediments and waters do not remain in place after they are charged with contaminants nor do the contaminants stay attached to the sediments but can later sorb differently or diffuse or be advected by wind or water. Thus some procedure is required not only to date the sediments but also to unscramble from the observations one makes those components that are indeed in situ and those that are mobile. And the mobility time and distance scales must also be provided somehow.

Because of the serious nature and effects on humanity of contaminants, there are major efforts underway to categorize such effects. Of course, part of that problem is also to unscramble naturally occurring processes from those produced anthropogenically so that there is no confusion as to the modification of the system under investigation by natural effects that could masquerade as being anthropogenically produced – and vice versa of course.

This chapter considers some of the non-radioactive time-scale procedures for modern sediments and then shows how they can be tied to true age scales by using radioactive methods. On the way to achieving that goal, one must also keep in mind the applications one is interested in handling, using both non-radioactive and radioactive procedures, so that a clear underpinning of each method in relation to the other can be provided. Later chapters will examine technical procedures and quantitative methods; this chapter is more concerned with setting the stage. In this way later chapters can be developed within an already familiar framework.

1. Non-radioactive dating methods

In essence, non-radioactive methods of ordering geological events almost define "classical" geology, for without some form of order there is little to say or do. Note here the word "ordering", which implies some idea that events occurred one after the other but does not provide a procedure to determine quantitative ages. The basic underpinning argument used is that if one measures quantities in progressively deeper sedimentary formations, and if one can prove that the sedimentary layers have not been disturbed since deposition, then the ordering of measured quantities is an age ordering because older formations must then underlie the above younger formations. This essential argument has enabled a wide variety of quantities to be used over the years as markers of age ordering.

Perhaps the original use in the 1800's of fossil markers for particular groups of formations, and their extinction in underlying and overlying formations, provided one of the first procedures for relative ordering of formations in geology. Since that time, the use of biostratigraphic markers in general has seen a substantial expansion. There are available: pollens, which also have family extinctions depending on plant evolution; diatoms and foraminifera, both of which again show different extinctions of types with formation ordering; plants themselves, which are preserved as fossil remnants and also show extinction of types; and the relative preservation of skeletal structures of animals as fossil components also is used to trace the local ordering of formations. By combining one or more of these various indicators it is possible to set up equivalences of one sort of indicator to another so that a cross-correlation of, say, plant evolution and foraminifera types allows one to infill relative ordering of formations in localities where one or the other indicator of age is absent.

Apart from the biostratigraphic relative ordering, there are also ordering procedures based on magnetostratigraphy. Here the strength and direction of remnant magnetism is measured for rocks and so tied to the biostratigraphy, and vice versa. Thus, not only does one have a record of rock movement across the face of the Earth with time, but also a record of reversal of the Earth's magnetic field with time. By combining this sort of information from rock formations neighboring on each other, it is possible to put together a magnetic system of ordering alone, which then provides some indication of sequences of events. When locked into absolute age dating through the use of radioactive tracers one then can use magnetic reversals as an accurate clock. This sort of information is often tied to geological models of seafloor spreading so that one can track the reversal of magnetic field "striping" of sediments with position from the spreading center. When combined with a model of the evolution of spreading center behavior then one again has a means of providing relative, paragenetic, dating. Note here the introduction not only of data but also of a dynamical geological process in an attempt to provide relative ordering of events. Often such models and measurements can be combined with measured (or inferred) motion of continental masses and, if it is assumed that the speed of motion has not changed in the past, then one has a macroscopic rough timescale with a first measure of true age rather than absolute age. Unfortunately, there is no way to verify the intrinsic assumptions of the models or of the assumed constant speed of the continents in the past, so there is often considerable uncertainty in precisely what such a timescale means for the sea-floor spreading rates over long intervals of geological time.

Allied to this sort of model dependence is the continental reconstruction procedure, which uses not only the shape of present-day continents in attempts to reconstruct the combined motion from a single super-continent, but also uses the world-wide presence of flora and fauna for different formations and, indeed, even the presence of the same lithofacies on two continents for the same formation, to bring together the continental motion in the past. This motion can, in turn, then be tied again to magnetic striping and so the relative age scales are locked one to another. Included in this sort of reconstruction is also the climate question because, for instance, the presence of salt on a regional scale argues that the climate was then highly evaporitic so that one can tie plant types and fauna types to such climate variations including their expansion, radiation and final extinction. Again, then, one has a paragenetic correlation of one marker scale to another.

In addition, there is the use of Milankovich forcing to calibrate the stable nuclide profile of Quaternary sediments. This sort of astronomical "dating" of the stable nuclide profile can be construed as a second "absolute" dating method provided the astronomical orbital factors can be correlated one-to-one with their effects on the Earth and associated sediments. Some concerns about such secondary correlation procedures have been raised over the years (see Williams et al., 1988). While some degree of success seems recently to be in sight (Dicken, 1997) in tying radiogenic and Milankovich procedures one to another, there are still sufficient on-going concerns (Martinson et al., 1982; Williams et al., 1988; Schulz & Schaefer-Neth, 1997) that have yet to be resolved that we have not included such secondary "clocks" in this monograph for they are not yet anywhere near as sharp as radiogenic methods.

These various markers operate with varying degrees of success throughout the world and also with past time. Part of the reason is that the rocks may not be present for one or more reasons. Thus, non-deposition since Permian time in West Texas makes it impossible by direct observation to infer anything from later times; instead one must use peripheral information from neighboring areas to ascertain later evolution; or glacial scouring of previously deposited formations, such as occurs in the Barents Sea with erosion of just about all the Tertiary formations, makes it extremely difficult to reconstruct events during the Tertiary in that locale.

However, as a general set of relative ordering tools, such devices have proven inordinately helpful on their own in attempts to unscramble the relative history of the Earth. Because each such procedure can be successfully tied to the others, one has an age proxy scheme of universal worth even if not of absolute age.

2. Stable nuclides and other indicators of relative age

In addition to the macroscopic measures of relative events described above, there is also a host of stable nuclides that indicate some measure of age. For instance, one can measure the relative abundance of $^{87}Sr/^{86}Sr$ in different formations. Because one has the relative ordering of the formations, one can then draw a curve of the strontium nuclide ratio variation with the relative age order of formations. Hence any later sample of the nuclide ratio from an unknown relative age formation can be placed on the curve and its relative age so determined (Dicken, 1997).

Because there have been numerous periods of glaciation during the history of the earth, the polar ice caps have undergone waxing and waning with time, leading to the variation of sea-level and also to a variation in the ratio of $^{16}O/^{18}O$ in both the atmosphere and the oceans.

The absorption of oxygen by living marine creatures, with calcium carbonate skeletal remains, then means that the oxygen radionuclide ratio in such skeletal fossils indicates the climate type during the life of the fossil. Thus, by tying the formation fossil type to the oxygen ratio one obtains directly a measure of evolution of climate at the then location of the formation on the globe, at the very least in terms of ice versus non-ice age phenomena (Williams et al., 1988; Dicken, 1997).

Equally, the variation of the stable nuclide ^{13}C relative to a standard provides a measure of both shifts in carbonate content with depth of water and also with water temperature. Different species of foraminifera absorb carbon differently and also exhale CO_2 differently. Biological differentiation of carbon-bearing organic material also provides a source for differentiation of isotopic carbon. Perhaps the important point is that direct measurement of the ^{13}C anomalies in marine organisms provides a measure of variation relative to the known relative stages of the formations in which the carbon variations are found. In turn, these variations can be tied to geological events in a paragenetic sequence so that one not only can then use such variations on a world-wide basis as relative age markers in the absence of any other markers, but one can also identify major geological shifts in the carbon nuclide budget at particular marker times. More detail on such procedures is recorded in Williams et al. (1988).

Of direct concern to age dating, or at least event ordering, are other indicators of special events, such as the iridium anomaly at the Cretaceous-Tertiary boundary, or the presence of salt and anhydrites marking evaporitic dry conditions in the world at Permian age, or tree-ring dating, or direct observations by divers or underwater cameras and video equipment of modern processes. In short there is an enormous number of proxies for age that provide some arguments for event order and which, if tied one to another, provide a general paragenetic sequence of time order if not of time interval magnitude. What is needed to close the loop is a way of measuring absolute time and then of tying such measurements to the proxy scales and so of converting them to true age scales. This critical linkage is provided by the radioactive nuclides, which do provide a true absolute age-dating prescription (Dicken, 1997).

3. Radiometric dating

There are an enormous number of radioactive nuclides that are used for age dating sediments and seawater (see Dickin, 1997, for a comprehensive overview of the present-day state of knowledge). Half-lives range from seconds through minutes to hundreds of millions of years, so that an enormous range of true ages of sediments can be identified by using the appropriate radionuclides. In general, the radionuclides come in four forms: those that are extinct due to non-replenishment; those that are replenished by decay of naturally occurring parent nuclides in the sediments or waters of the earth or with parents brought up from deeper material in the course of geological time or by special events such as deep rooted volcanoes; those that are replenished in the atmosphere or the waters and sediments of the earth by cosmic ray bombardment; and those that are man-made both in the laboratory and, more widely spread, as a consequence of nuclear bomb testing since the Second World War, and by leakage from nuclear reactors, such as Windscale, Three Mile Island, and Chernobyl.

The sense of the general argument is easily given. Each unstable radionuclide produced from a parent decays with its own half-life. By measuring the amount of the parent and the

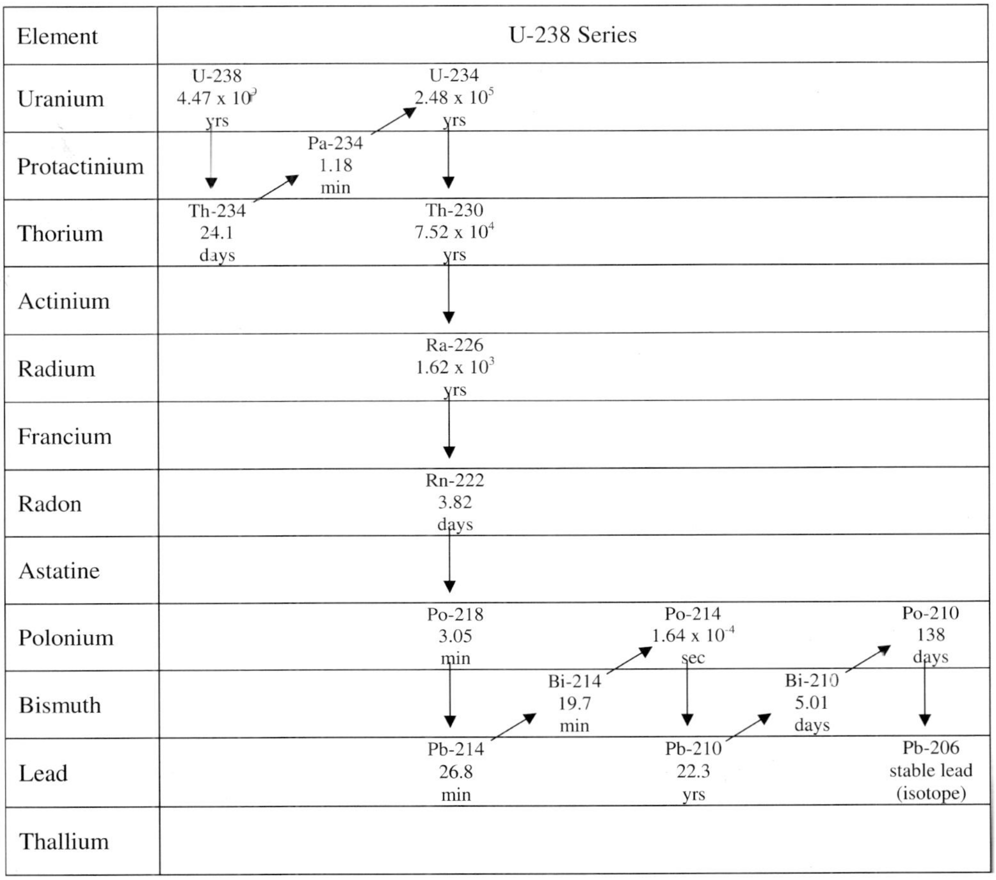

Element	U-238 Series						
Uranium	U-238 4.47 x 10^9 yrs		U-234 2.48 x 10^5 yrs				
Protactinium		Pa-234 1.18 min					
Thorium	Th-234 24.1 days		Th-230 7.52 x 10^4 yrs				
Actinium							
Radium			Ra-226 1.62 x 10^3 yrs				
Francium							
Radon			Rn-222 3.82 days				
Astatine							
Polonium			Po-218 3.05 min		Po-214 1.64 x 10^{-4} sec		Po-210 138 days
Bismuth				Bi-214 19.7 min		Bi-210 5.01 days	
Lead			Pb-214 26.8 min		Pb-210 22.3 yrs		Pb-206 stable lead (isotope)
Thallium							

Fig. 1.1. Decay series chain for arrival at ^{210}Pb from ^{238}U. Note that the half-lives from ^{226}Ra through to ^{210}Pb are so short in comparison to that for ^{210}Pb that one can effectively consider ^{226}Ra as the original parent of ^{210}Pb.

daughter for sedimentary formations one can then trace back the amount that was present at the time of deposition of the sediment. Essentially if one assumes too long an age for the sediment then the unstable daughter product will be less than observed while if one assumes too short an age for the sediment then the unstable daughter product will be greater than observed. Hence one can bracket the age of the sedimentary formation. With more than one radionuclide one can then do an exceedingly accurate job in determining formation ages. Thus true ages can be assigned to formations and so the proxies for event ordering can then be converted to true age scales as well.

For modern sediments (say to an age of around 100 years or less to fix ideas), the main thrust of this volume, the two dominant radionuclides of relevance are ^{210}Pb and ^{137}Cs because they have half-lives of about 22 and 30 yrs, respectively. Figure 1.1 presents the radioactive chain decay scheme from ^{238}U through to ^{210}Pb, showing that after ^{226}Ra the half-lives of intermediate radionuclides are extremely small in comparison to those for ^{226}Ra and ^{210}Pb, making

^{226}Ra the effective parent of ^{210}Pb. Other radionuclides can be used, as we will see in later chapters, but the worldwide prevalence of the lead and cesium radionuclides makes them the major workhorses for modern sediment age determinations. In addition, while the lead nuclide arises from natural decay of radium, cesium is the result of nuclear bomb testing which did not take place until 1945 at the earliest for atomic plutonium bombs, and no earlier than 1951 for hydrogen bombs, which are the source of the majority of the cesium radionuclide. Accordingly it is possible to use the cesium already as an age limiter even before one undertakes quantitative investigation of modern sediment events.

4. Applications of radiometric dating

It is not just the ability to obtain true radionuclide derived ages for modern sediments that is of relevance but also the applications with which one can utilize the radiometric ages. For instance, in a deltaic setting, where the sedimentation is highly variable spatially, measurement of the thicknesses of packages of sediments at various locations in the delta, and also of the ages of the sediments with depth, provide the ability to obtain the sediment accumulation rate across the delta and so tie river and estuarine inputs to the sediments observed. In this way one has the capability to estimate the directions of sediment transport by currents and also the patterns of change with the seasons of high and low river discharge including later erosion and redeposition of sediments.

Alternatively, the erosion by underwater turbidites of sedimentary packages can also be evaluated by measuring the sediment ages and thicknesses and so relating the lateral movement of a turbidite to the observed present locations of the sediments.

Further applications are legion: one can, for example, trace the motion of chemical elements in sediments that, while not themselves radioactive, do provide an indication of the scavenging of minerals by sediments and so of the direction of sources and sinks of such chemical elements – of great use when one is attempting to define likely causes for particular elements and their chemical products.

Of use in provenance studies are nuclides of Nd, Sr, and Pb to fingerprint sources of detritus to the sediment; of use in direct dating of sediment components or bulk sediment are ^{14}C, Ar/Ar, U/Th and U/Pb; while Sr nuclides are used to correlate sediments, and in this respect also of use are C, O and S nuclides; diffusion of elements and turbidite repetition in sediments can also be investigated using U-series nuclides (Dicken, 1997). Indeed, we present an example later using ^{231}Pa (half-life 3.25×10^4 yrs as shown in Fig. 1.2) to show that age determination can also be effected with longer-lived radionuclides in different settings, but that due diligence must be observed in handling origin problems of the nuclide (decay of U in the water column, for example). The different residence times of Th and Pa radionuclides in the water column allow such nuclides to provide some information about ocean circulation and past productivity, but also lead to difficulties with constant relative flux model assumptions to the sediments where one measures Th and Pa. This point will be brought out in the example in some detail.

However, as we shall show in later chapters, and has been investigated to some extent in simple fashion elsewhere (Williams et al., 1988; Dicken, 1997), the general arguments advanced, using mainly ^{210}Pb and ^{137}Cs as radionuclide tracers of events, can easily be taken

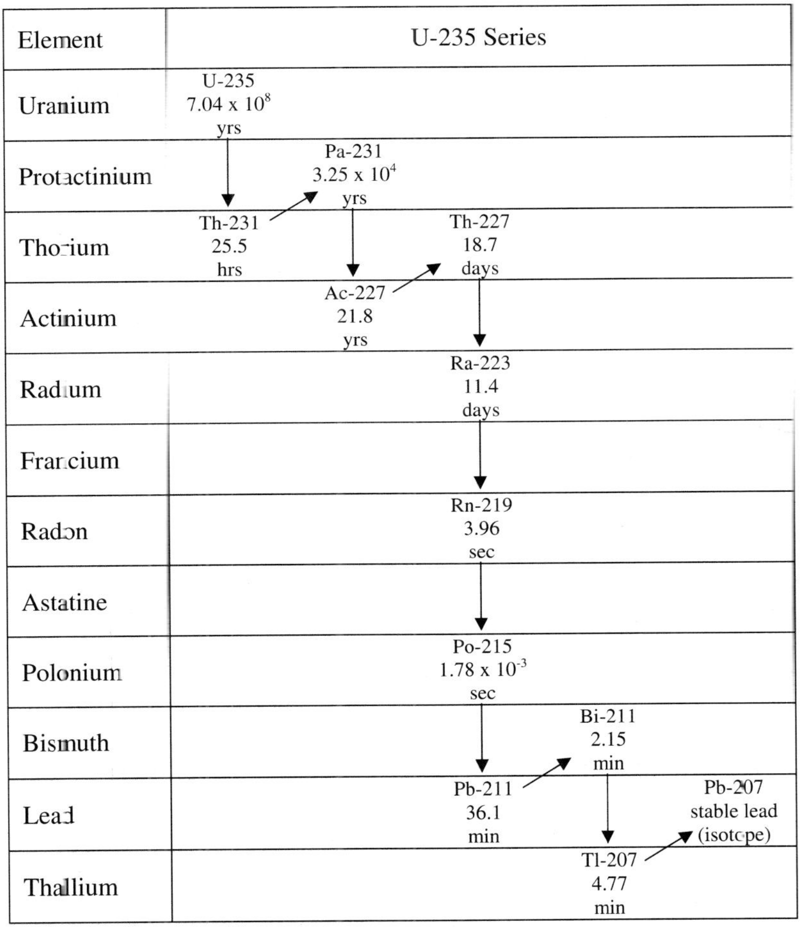

Element	U-235 Series			
Uranium	U-235 7.04×10^8 yrs			
Protactinium		Pa-231 3.25×10^4 yrs		
Thorium	Th-231 25.5 hrs		Th-227 18.7 days	
Actinium		Ac-227 21.8 yrs		
Radium			Ra-223 11.4 days	
Francium				
Radon			Rn-219 3.96 sec	
Astatine				
Polonium			Po-215 1.78×10^{-3} sec	
Bismuth				Bi-211 2.15 min
Lead			Pb-211 36.1 min	Pb-207 stable lead (isotope)
Thallium				Tl-207 4.77 min

Fig. 1.2. Decay series chain for arrival at ^{231}Pa from ^{235}U. Note that after ^{231}Pa the remaining chain radionuclide lifetimes are so short in comparison to that for ^{231}Pa that one can regard ^{235}U as the effective direct parent of ^{231}Pa, and one can also regard ^{231}Pa as the effective parent (on the timescale of hundreds of years) of ^{207}Pb.

over almost wholesale to any and every such nuclide indicator. Indeed, we present an example using Sr nuclides to show how the methods are extendible even to stable nuclides.

These studies, in both sediments and ocean waters, are of potentially great worth for evaluating the ability of the oceans to absorb elements from the sediments, and so one can follow ocean currents at significant water depth relatively easily. In addition, one can also determine the conditions under which the oceans deposit minerals and elements to the sediments and when they did so, so that the sinks of element deposition can be more easily traced.

Apart from these scientific pursuits, there are also serious studies underway for contaminant transport in modern oceans, rivers, and sediments. Such studies are necessary because the remediation procedures one would suggest depend to a large extent not only on identifying the sources of contaminant (and so remediating the sources) but also of remediating the already transported contaminant and of identifying whether the contaminant is dominantly in water or sediments. Clearly, in order to obtain an accurate idea of what is going on, one needs to age date the sediments with extreme precision. This problem is exacerbated when there can be variations of the very nuclide fluxes used for the age dating itself-leading to the possibility of wrong ages and so wrong estimates of contaminant flux. In addition, when there can be bioturbation effects in the sediments, which tend to homogenize the nuclide variations of the fluxes being deposited, and when there can be further diffusion and advection of prior deposited contaminants, then the unscrambling of all such effects is required before one can say with confidence what uncertainty one has for each separate component. This particular problem is one that started a goodly amount of the results reported here, because existing techniques make too many assumptions to be sure one has correctly unscrambled and understood signals in the nuclide records.

The chapters that follow analyze individual facets of these general problems in some detail, with applications given to individual situations where one can see the effects sharply. There are also case studies given where it is still not clear what is happening and the question arises as to whether current techniques for sorting out sediment ages and surficial flux variations are well enough developed to handle those cases. In addition, one of the prevalent concerns is the question of accuracy. It is all very well to develop techniques for sorting out the sedimentation accumulation rate effects from surficial variation effects. However the application of such techniques is limited by the sampling distribution and frequency of the data available, by the resolution of the data, and by the confounding effects of other processes not included in the analysis. Thus of prime concern is the extent to which one can provide quantitative measures of the resolution achievable with any given data set. This problem is also addressed so that, at the least, one has some idea of the worth of the data to hand as well as an indication of what one needs to do if one wishes, or needs, to improve matters.

While the presentation given here is not by any means complete, something extremely difficult to do in a rapidly developing field of study, it would seem that there is now sufficient understanding of the procedures and processes involved that an extraordinarily large number of diverse situations can be investigated in a single encompassing framework, due allowance being made for the differences between the situations. It is this fact that the developments given in the following chapters speak to most strongly.

Chapter 2

Radiometric methods

The evaluation of sediment ages using radioactive nuclides is the only way known of obtaining absolute ages; all other methods are either paragenetic or are eventually related back to some radiometric age scale. Thus it becomes a problem of some concern to understand precisely what is involved in assumptions underpinning the major methods used for sediment age determinations.

Profiles with depth of radionuclides in sediments are the result of unknown combinations of changes in sedimentation rate, nuclide flux, and losses or additions of the parent or daughter radionuclide after deposition. The basic premise of all sedimentation models for radionuclide tracers is that the shape of the radionuclide versus depth profile must reflect the history of deposition in some quantitative and extractable manner. The challenge, then, is to distinguish the signal of radioactive decay related to depositional events from all other effects. A number of mathematical procedures have been designed for this purpose all of which rely upon the basic assumptions of radioactive decay as described in Section 1 of this chapter. Yet, in practice, the basic assumptions of this procedure are often violated in nature.

However, a number of more sophisticated techniques have been developed to allow the basic utility of the approach to be preserved. These models can be grouped into two main categories, conceptual and signal theory methods. And these categories can be subdivided further into two types depending on whether sediments are mixed or unmixed (Fig. 2.1). Models used to describe tracer profiles influenced by mixing are presented in Chapter 5. In this chapter we describe the basic models used in sedimentation studies where mixing is not considered to be important. A summary of different model applications is given in Table 2.1. The explanation of the various approaches provides the foundation for later chapters where examples are given of the various methods, their application to real situations, and their successes and limitations.

1. Basic procedure

The basic method derives from the fundamental theory of radioactive decay of a nuclide. The decay with time of a radionuclide concentration is governed by the exponential law

$$P(t) = P_0 \exp(-\lambda t), \tag{2.1}$$

where P_0 is the initial concentration at time $t = 0$, and λ is the decay rate of the particular radionuclide. Note that equation (2.1) is independent of the units used for concentration

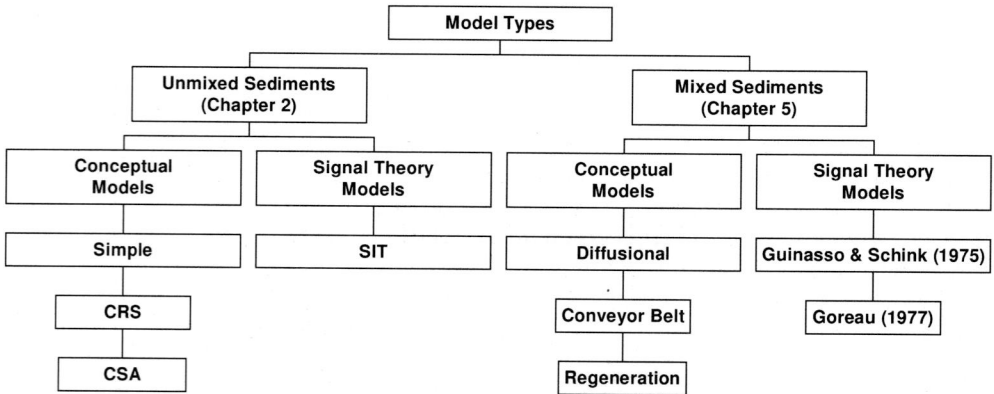

Fig. 2.1. Summary of model approaches.

so that different investigators have used a variety of units, depending in idiosyncratic preference and also ease of measurement in particular cases. Where possible we follow individual units originally used for each of the examples to be presented, ranging from dpm/g through to Bq/g. In this way a direct comparison with original research papers can most easily be made. If you wish to interconvert between these units, it is easy to remember that 1 Bq = 60 dpm.

In sedimentation studies equation (2.1) is translated from a scale based on time to one of depth, in this case, related to a column of sediment. The method is based on the assumption that there are no interruptions in sedimentation and resuspension throughout time and that there is no post-depositional mobility of the tracer nuclide. If the sedimentation rate is constant for all time and there is no mixing, then elapsed time t is connected to sediment burial depth x through $x = Vt$ (we avoid the use of the symbol d to measure sediment depth because of potential confusion with differential symbols that also use d). Accordingly we can then write

$$P(x) = P_0 \exp(-\lambda x / V),\tag{2.2}$$

where P_0 is the surficial concentration of a tracer nuclide at time $t = 0$ (corresponding to the sediment surface at location $x = 0$), V is the sedimentation velocity, λ is the decay constant, and the sedimentation velocity is obtained from an exponential fit to the measured tracer data profile, $P(x)$, with depth x. Hence, by measuring $P(x)$, the best velocity for an exponential fit with depth to the measured data can be obtained.

This simple and elegant method has been successfully applied in a variety of natural settings, and yet only satisfies a very limited class of depositional environments encountered in nature. Nonetheless, the approach sets the stage for the development of a number of more sophisticated analysis techniques and provides the basic premise upon which all such models are predicated.

2. Advanced procedures

2.1. *Conceptual models*

Conceptual models require specific knowledge of sediment accumulation history and of the processes that control changes in the activity concentration of a radionuclide with depth in a sediment profile. These models are based on mechanistic interpretations of sediment burial history (Robbins & Herche, 1993). We illustrate the procedures and assumptions here with three models, proposed and used extensively over the last two decades for investigating sedimentary ages in a variety of aquatic depositional systems. The different approaches were originally developed for lake environments but are often used in oceanographic settings as well.

In these environments, the approximately 22 yrs half-life of ^{210}Pb makes it an ideal indicator of modern sedimentation rates. Indeed, ^{210}Pb has been used extensively for this purpose since it was first introduced as a geochronometer by Goldberg (1963). The analysis of sedimentation rates by the ^{210}Pb method is based on the radioactive decay of unsupported ^{210}Pb with depth in a column of sediment. Unsupported ^{210}Pb is the ^{210}Pb activity that is not derived from the radioactive decay of its original parent radionuclide ^{226}Ra. This excess is determined as the difference between the ^{210}Pb and ^{226}Ra activities measured for each sediment interval. Because we deal mainly with only the excess ^{210}Pb we hereinafter refer to the excess solely as ^{210}Pb to avoid tedious repetition of "excess" throughout.

Conceptual models used in ^{210}Pb geochronology rely on assumptions concerning the relationships among the temporal rate of change of the specific activity, $A_s(t)$, of ^{210}Pb (dpm/g), the flux, $F(t)$, of unsupported ^{210}Pb from seawater to sediments (dpm cm^{-2} yr^{-1}) and the sediment accumulation rate, $R(t)$ (g cm^{-2} yr^{-1}), such that, at the time of sediment deposition,

$$A_s(t) = F(t)/R(t). \tag{2.3}$$

2.1.1. *Simple Model*
The Simple Model is applied to ^{210}Pb profiles when the flux of unsupported ^{210}Pb from seawater to sediment is constant and also the sedimentation rate is constant (Robbins, 1978). This model, also known as the Constant Flux/Constant Sedimentation Model (CFCS) (Appleby & Oldfield, 1978), is similar to equation (2.2); however ^{210}Pb activity, $P(m)$, is taken to vary with the cumulative dry-mass of sediment, m, instead of with sediment depth such that

$$P(m) = P_0 \exp(-\lambda m/U), \tag{2.4}$$

where the "velocity", U, is the rate of mass (in grams) added per unit time.

The Simple Model (equation (2.4)) eliminates the need to correct for changes in sediment porosity caused by the compaction of sediment after burial or from compressional disturbances due to coring. Differences in sediment porosity (fractional content of water by volume) and, correspondingly, in the percentage of solids with depth in a sediment profile as a result of compaction, lead to variations in sedimentation rate V (cm/yr) even when the sediment accumulation rate U (grams dry sediment cm^{-2} yr^{-1}) is constant. The sedimentation rate is related to the sediment accumulation rate through the weight of sediment (w) (grams dry cm^{-2}) above depth x (cm) (Robbins, 1975). Sediment weight is more frequently expressed including porosity (ϕ) whereby,

$$dw/dx = \rho_s(x)\big[1 - \phi(x)\big] \approx \langle\rho_s\rangle\big[1 - \phi(x)\big] \tag{2.5}$$

and where $\rho_s(x)$ is the density of sediment solids. Because the variation in ρ_s is typically less than the variation in ϕ, $\langle\rho_s\rangle$ (mean density of sediment solids) is substituted for ρ_s. Equation (2.3) is used when the values of cumulative weight are unknown. When the sediment accumulation rate is constant, the relationship between V at the sediment surface and V at any other depth is

$$V(x) = V(x = 0)\big[1 - \phi(0)\big]\big/\big[1 - \phi(x)\big]. \tag{2.6}$$

2.1.2. *Constant Flux Model*

The Constant Flux Model was first developed by Goldberg (1963). In later application Robbins (1978) coined the term Constant Flux Model (CF), and Appleby & Oldfield (1978) labeled the model as the Constant Rate of Supply (CRS) Model. The Constant Flux Model is applied when sedimentation rates are variable in time but the flux of ^{210}Pb to sediments remains *constant* (Robbins, 1978). The excess ^{210}Pb profile vertically integrated to a depth x (or alternatively, cumulative dry-mass m) will equal the flux (constant) integrated over the corresponding time interval. Integrating with depth, x, the governing equation is

$$A(x) = A_0 \exp(-\lambda t), \tag{2.7}$$

where $A(x)$ is the cumulative residual unsupported ^{210}Pb activity beneath sediments of depth x and A_0 is the total unsupported ^{210}Pb activity in the sediment column. The age of sediments at depth x is then

$$t = (1/\lambda)\ln\big[A_0/A(x)\big] \tag{2.8}$$

and the sedimentation velocity V (or alternatively, sediment accumulation velocity U) at each time is,

$$V = \lambda A(x)/P(x), \tag{2.9}$$

where $P(x)$ is the unsupported Pb activity at depth x.

2.1.3. *Constant Specific Activity Model*

As an alternative, a model assumption is that sediments have a constant, initial, unsupported ^{210}Pb activity. This model is applicable when the surface sediment ^{210}Pb activity is constant with time and sedimentation rate is variable with time. The unsupported ^{210}Pb activity, $P(x)$, at any depth x is then related to the surface sediment activity P_0 by the relationship

$$P(x) = P_0 \exp(-\lambda t). \tag{2.10}$$

This model is known as the Constant Specific Activity Model (CSA) (Robbins, 1978) or the Constant Initial Activity (CIA) Model (Appleby & Oldfield, 1978).

2.2. *Signal theory methods*

Signal theory methods, alternatively known as inductive modeling approaches, are predicated on the idea that the recovery of some initial information from sediments can be accomplished by comparing model profiles based on realistic values of the relevant parameters with actual radionuclide profiles. Inductive models allow calculation of the effect on sedimentation history of any series of time-varying input events to the sediment. These models are less restrictive in their assumptions and thus have been used to extend the current applications of radionuclides as geochronometers.

Liu et al. (1995) developed a model of sedimentation processes based on the signal theory approach. As shown later, inductive models were considered earlier for the quantitative prediction of sediment mixing effects (Goreau, 1977; Guinasso & Schink, 1975). The development of an inductive model that combines sedimentation and mixing to determine sedimentation histories is currently at an early developmental stage. Yet signal theory approaches represent a promising alternative to the current widespread use of conceptual models. Therefore considerable space is allocated in this book to presenting and examining this emerging field.

The basic strategy used in signal theory is to deduce the form of a model from the data at hand rather than relying entirely on a set of hypotheses to design a model taking all plausible interactions into account. Given a particular model structure, with the inputs and all their interactions defined, a search is performed for the set of model parameters that minimizes a modeling criterion for the known data. A modeling criterion is an equation to be minimized (e.g., equation (2.11)) that provides a quantitative estimate of the performance of a model relative to a given set of data.

2.3. *Sediment Isotope Tomography*

The inductive model, known as the Sediment Isotope Tomography (SIT) method was developed to allow the interpretation of ^{210}Pb profiles that result when both sediment accumulation rates and ^{210}Pb fluxes vary with time (Liu et al., 1995). In this method changes in radionuclide activity $P(x)$ with depth x in a sediment column are represented as:

$$P(x) = P_0 \exp\left[\left(-Bx + \overbrace{\sum_{n=1}^{N} \frac{a_n}{n\pi} \sin\left(\frac{n\pi x}{x_{\max}}\right)}^{\text{sedimentation term}}\right) + \left(\overbrace{\sum_{n=1}^{N} \frac{b_n}{n\pi}\left(1 - \cos\left(\frac{n\pi x}{x_{\max}}\right)\right)}^{\text{source term}}\right)\right],$$

(2.11)

where summations are for $n = 1, 2, 3, \ldots, N$, the terms involving a_n (b_n) refer to sedimentation (source), $x_{\max}$ is the greatest depth of measurements, and B is a trend coefficient related to the average sedimentation speed, V, by $B = \lambda/V$.

Over a given depth section there is some average sedimentation speed, V, plus variations around the average. The variations in velocity are represented by a Fourier cosine series because the variations around the average sediment accumulation rate must themselves have a zero average. A cosine series automatically accomplishes this task. Note that the time to depth variation is then described by a term linear in depth plus sine oscillations as given in equation (2.11).

The model represents intrinsic variations in specific activity (that are unrelated to variations in the rate of sediment deposition) by a Fourier cosine series (normalized to unit value at the sediment surface), as shown in equation (2.11). The value of the surface radionuclide activity, P_0, is also taken to be a variable to be determined – often within prescribed limits.

The combination of all three factors yields a predicted behavior for radionuclide variation with depth that depends linearly on P_0, and non-linearly on B and N. The factors P_0, B, and N define a linear set of $2N$ equations with $2N$ unknowns. These equations are solved by standard matrix inversion to determine the Fourier coefficients a_n and b_n. For each set of P_0, B, and N, there is a unique choice of values for the coefficients a_n and b_n, and hence a unique solution to equation (2.11).

A non-linear iteration scheme is then introduced that automatically produces values of P_0, B, and N at each iteration that are guaranteed to give a predicted radionuclide variation with depth that is closer and closer to the observed values as iterations proceed. The iteration scheme also guarantees to keep P_0, B, and N inside of any pre-set bounds at each iteration.

Iterations are stopped when a criterion of convergence is reached (e.g., the difference between predicted [Pred(x)] and observed values [Obs(x)] is less than the measurement resolution). Thus, introduction of X^2 with:

$$X^2 = \sum_{i=1}^{M} \frac{[\text{Pred}(x_i) - \text{Obs}(x_i)]^2}{M}, \tag{2.12}$$

where M is the number of data points, is a useful measure of fit.

Once a satisfactory fit to the data is achieved, then successful separation of sedimentation and source terms has been achieved. Sedimentation and source variations can then be plotted as functions of time or depth. Many combinations of model coefficients (P_0, V, a_n, b_n) will fit a given data profile to within the same level of uncertainty as is demonstrated in Fig. 2.2 with a contour plot of constant X^2 values in the parameter space of: average sediment accumulation rate, V, and surface ^{210}Pb concentration, P_0, for the synthetic data profile given in Fig. 2.6a. While the range of possible values for P_0 is relatively well constrained, the range for sedimentation rate is less well constrained, as reflected through the uncertainty in the sedimentation rate (0.7 to 2.4 cm/yr).

In order to reduce the number of model-determined answers, an independent time marker, such as ^{137}Cs, is input to the model. ^{137}Cs, a radionuclide injected into the atmosphere during the era of nuclear weapons production and aboveground testing, often reaches a maximum concentration in a sediment profile corresponding to the time period of 1963–1965, the peak of nuclear weapons testing. By providing the age of a particular depth horizon, many combinations of model parameters are eliminated that fit equally well to the ^{210}Pb profile alone. For example, if the depth of the ^{137}Cs maximum (50 cm for the data profile given in Fig. 2.6a) corresponds to 1963 (± 2) years, then the range of possible average sedimentation rates is limited to 1.6–2.3 cm/yr, as shown in Fig. 2.2 by the rectangular region. Any natural time markers can be used as independent calibration points. Other possibilities include known major drought or

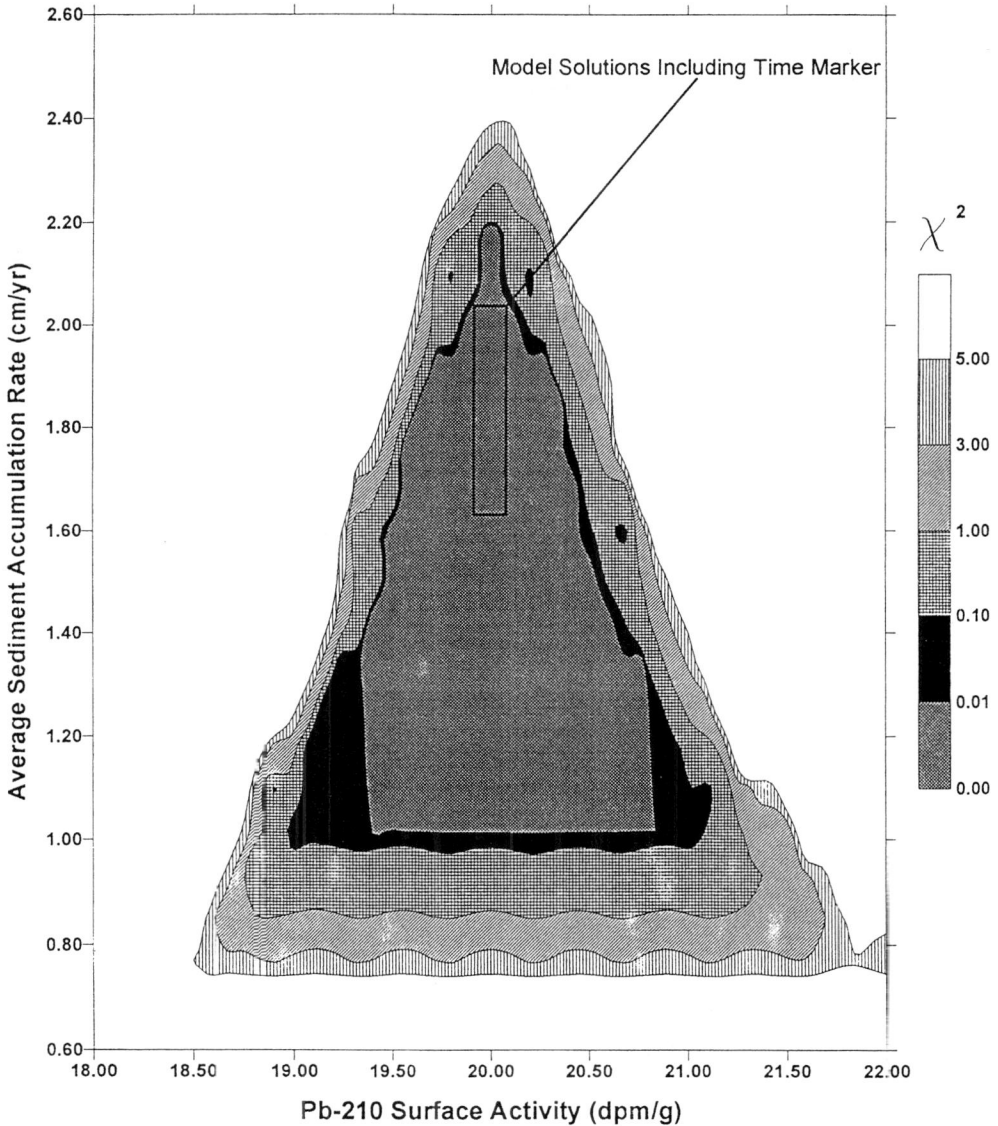

Fig. 2.2. Surface contour plot of the variation in X^2 (equation (2.13c)) with sediment accumulation rate and surface ^{210}Pb concentration. All acceptable model solutions are contained within the black box.

flooding events, or a contaminant discharge event such as the release of mining waters into a watershed (Varvus & Punning, 1993).

Available multiple time marker information in the form of (depth, time) pairs as well as the standard deviations of the ^{210}Pb data are combined into a new criterion of least-squares average mismatch between model and data as follows.

Consider that the ith observed ^{210}Pb measurement, Obs(x_i), at depth x_i has a standard variance $\sigma(x_i)^2$. Let there be n_2 $(i = 1, 2, \ldots, n_2)$ such measurements arranged sequentially with increasing depth. In addition let there be n_1 $(j = 1, 2, \ldots, n_1)$ measured ages with the jth age, t_j, corresponding to depth x_j, and let the jth known age have a variance $\sigma(t_j)^2$.

Then, suppose an initial set of values has been chosen for P_0, V and N. The corresponding initial predictions for ^{210}Pb values with depth at location x_i are then labelled IP(x_i) $(i = 1, \ldots, n_2)$, while the corresponding initial predicted ages at depths x_j $(j = 1, \ldots, n_1)$ are labelled IA(x_j). If the predictions of ^{210}Pb and of the ages were accurate then one would have IP(x_i) = Obs(x_i) and IA(x_j) = t_j. However, because the values for P_0, V, and N are arbitrary, there is no guarantee that the predictions will match the observations.

Measures of the average mismatch for all measurements for both predicted radionuclide values and ages are provided through X_v^2 and X_a^2, with

$$X_v^2 = \sum_{i=1}^{n_2} \left(\text{IP}(x_i) - \text{Obs}(x_i)\right)^2 / \sigma(x_i)^2 \tag{2.13a}$$

and

$$X_a^2 = \sum_{j=1}^{n_1} \left(\text{IA}(x_j) - t_j\right)^2 / \sigma(t_i)^2. \tag{2.13b}$$

With choices for P_0, V and N, different than the initial values, a measure of relative improvement in degree of fit to both the radionuclide measurements with depth, and to the known ages of particular horizons, is then provided by

$$X^2 = \left[n_1 \left(\sum_{i=1}^{n_2} (\text{Pred}(x_i) - \text{Obs}(x_i))^2 / \sigma(x_i)^2 \right) \middle/ X_v^2 \right.$$
$$\left. + n_2 \left(\sum_{j=1}^{n_1} (\text{Age}(x_j) - t_j)^2 / \sigma(t_j)^2 \right) \middle/ X_a^2 \right] \middle/ (n_1 + n_2), \tag{2.13c}$$

where Pred(x_i) is the predicted radionuclide value at depth x_i, and Age(x_j) is the predicted age at depth x_j.

This form of mismatch constraint "balances" the degrees of improvement relative to the number of values (n_1, n_2) available. Notice that if the values for P_0, V and N are chosen to be the initial values then $X^2 = 1$. Thus, the determination of "best" values for P_0, V and N is then controlled by the requirement that X^2 should be a minimum. Either a linear search procedure through P_0, V and N, or a non-linear tomographic iteration procedure in the manner of Liu et al. (1991), can be used to obtain the minimum value of X^2.

Specified maximum and/or minimum ages for a sediment core can also be applied to eliminate models that otherwise satisfy equation (2.13c) but which would contain unrealistic sediment ages for the top or bottom of the profile. The default maximum age is six times the ^{210}Pb half-life. Thus the final model-determined best answers satisfy not only the criteria of reproducing the shape of the ^{210}Pb versus depth curve, but also the criteria that the model-determined ages of the sediment horizons satisfy all available time information.

3. Synthetic tests of the SIT method

To show how signal theory approaches perform, synthetic data profiles of ^{210}Pb in sediments were created for different types of sedimentary regimes encountered in nature. These profiles do not exhaust all possibilities but provide examples of what can be accomplished using the SIT method.

Additional tests were conducted by adding random noise to each data profile, up to $\pm 20\%$ of the nominal data values. The random noise represents the uncertainty in radionuclide activities that result from imperfect data collection and/or measurement. For the synthetic data profiles presented here, the model results were constrained by specifying the depth of the ^{137}Cs maximum and taking the corresponding date to be 1963 ± 2 yrs. A minimum of three different random number distributions were applied to each data profile. For clarity, only one result is described here.

3.1. *Type I: Constant sediment accumulation rate*

In environments where available information suggests that the sediment accumulation rate is nearly constant with time and measured ^{210}Pb profiles show an exponential decrease in activity with sediment depth, the conditions satisfying the simple model (equation (2.6)) have been met. While there is no need to apply a model as complex as equation (2.11) in such circumstances, in practice uncertainties in the data associated with sample collection and analysis techniques often make it difficult to determine whether or not the above conditions have indeed been fully satisfied.

A synthetic ^{210}Pb profile was created representing a constant sediment accumulation rate of 0.5 cm/yr and a constant specific activity of 15 dpm/g (Fig. 2.3a). For simplicity, a constant sediment porosity of 0.65 and rock matrix density of 2.54 g cm^{-3} were assumed. Figure 2.3b shows the best fit model to the data together with the age ranges encompassed by all models satisfying the time marker limitation and the precision of the data (0.1 dpm). The results necessarily include non-zero a_n (variable sediment accumulation) and b_n (variable source) coefficients because the coefficients are determined by inverting a least squares matrix. However, the values determined by the model for a_n and b_n are appropriately small, as is the mismatch in the fluxes of ^{210}Pb (Fig. 2.3c). The criterion of acceptance for the best model is given by equation (2.13c). Note, however, for comparison purposes, the X^2 values given in Figs 2.3–2.7 correspond to equation (2.13c) in units of (dpm/g)^2.

Even with the addition of random noise, the model was able to reconstruct the data profile and recover the correct time to depth relationship (Fig. 2.3 (d–e)). The slight mismatch observed in the time to depth profile is reflected in a small model-determined time variation in the flux of ^{210}Pb (Fig. 2.3f).

For comparison, Figs 2.3 (g–i) demonstrate the results of the alternative models when applied to the synthetic data profile in Fig. 2.3a (without random noise). With the exception of the Constant Flux (CF) model (Fig. 2.3i), both the Constant Flux/Constant Sedimentation

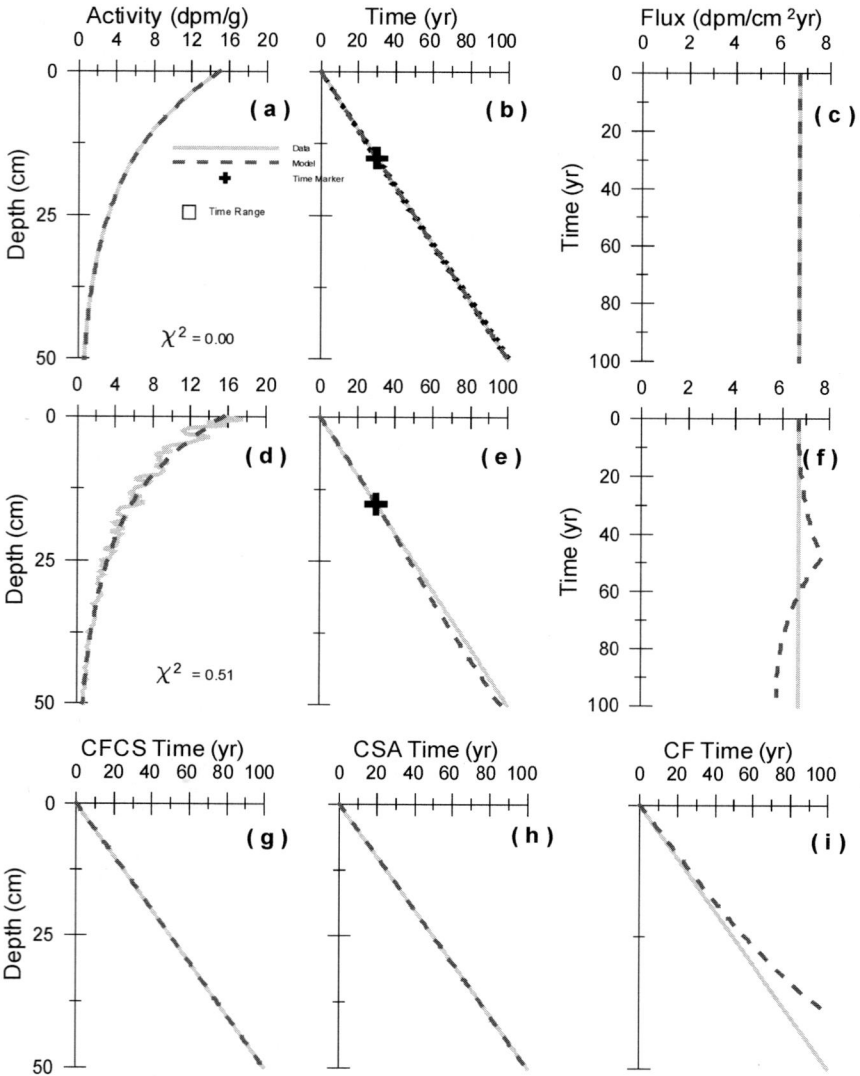

Fig. 2.3. A synthetic ^{210}Pb profile was created with a constant sediment accumulation rate of 0.5 cm/yr. Sub-figures (a–c) are the results for a synthetic profile with no random noise added to the data, while sub-figures (d–f) are the results for the synthetic profile shown in (a) with 20% random noise added to the data. Sub-figures (g–h) compare the time to depth distributions for the CFCS, CSA and CF models. With the exception of the CF model, which overestimates sediment ages at the depth base of the profile, the other models reproduce the age to depth relationship acceptably.

(CFCS) model and the Constant Specific Activity (CSA) model recover the time to depth relationship for the data profile, as expected. The CF model works reasonably well at younger sediment ages but predicts ages that are too old for the bottom half of the core. Unfortunately, none of the alternative models works well if ^{210}Pb activities increase with depth within any portion of the data profile, because such an increase is then interpreted to correspond either to negative sediment accumulation rates or to a decrease in sediment age with depth. For this reason, these models were not applied to the data profile given in Fig. 2.3d.

3.2. *Type II: Variable sediment accumulation rate*

In environments where changes in the ^{210}Pb content of the sediments depend upon changing sediment accumulation rates alone, the flux of ^{210}Pb to bottom sediments is constant. This situation is encountered when water residence times are long compared to particle residence times and sediment/water distribution coefficients (K_d values) are constant. A ^{210}Pb profile was created by inputting model coefficients representing variable sedimentation ($a_n \neq 0$, $N = 2$). The ^{210}Pb activities at each sediment depth were calculated assuming a constant flux of 21 dpm/cm^2 yr. Activities were then corrected for radioactive decay due to the elapsed time since deposition.

The model-determined best-fit and time-to-depth profiles show only slight differences from the synthetic answers (Fig. 2.4 (a, b)), with a slight variation in the flux (Fig. 2.4c). Even with the addition of random noise to the ^{210}Pb profile, the model-determined best answer (Fig. 2.4 (d–f)) is nearly identical to that achieved without the addition of random noise.

Model-fit results obtained by applying other models (CSCF, CSA, and CF) are also good (Fig. 2.4 (g–i)). However, as previously observed, the CF model significantly overestimates the known input sediment ages near the bottom of the core.

3.3. *Type III: Variable flux*

Oftentimes there are changes in the specific activity of deposited sediments due to changes in the flux to the seabed of ^{210}Pb in response to quasi-regular fluctuations in sediment type or water mass transport. Accordingly, a synthetic profile was constructed where the specific activity changed but sedimentation rate was held constant.

Appropriate model coefficients representing constant sedimentation velocity (V) with $a_n = 0$ and variable source ($b_n \neq 0$) were chosen. The input "forward model" data profile was created by the equation

$$P(x) = P_0 \exp\left[-Bx + \left(\sum_{n=1}^{N} \frac{b_n}{n\pi}\left(1 - \cos\left(\frac{n\pi x}{x_{\max}}\right)\right)\right)\right]. \tag{2.14}$$

The resulting ^{210}Pb versus depth profiles were then analyzed using the full model. The model input and output ^{210}Pb versus age profiles are indistinguishable as shown on Fig. 2.5a. The model-determined *best* answer reproduced the time to depth relationship perfectly. However, the *range* of variation in age at the bottom of the profile is 15 years (Fig. 2.5b) The variation in flux is identical to the input profile (Fig. 2.5c). Even with the addition of random noise to the data profile, the model successfully reproduced the correct fits to data, time to depth, and flux variations for this synthetic data profile (Fig. 2.5 (d–f)).

The assumptions of the commonly used models (CSCF, CSA, and CF) do not apply in this situation. If one were to assume that one of these models were appropriate, the resulting sediment accumulation rates would be poorly determined, as shown in Fig. 2.5 (g–i).

20 J. Carroll, I. Lerche

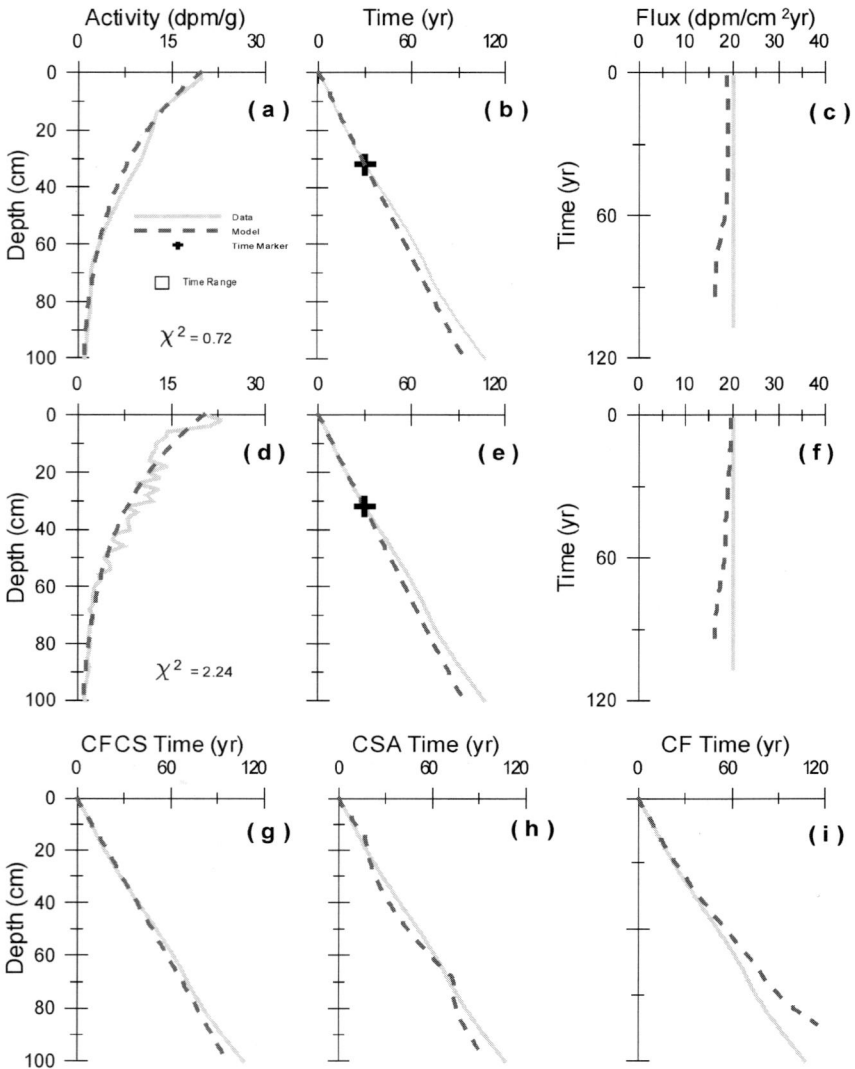

Fig. 2.4. A synthetic ^{210}Pb profile was created with a variable sediment accumulation rate and a constant flux with time. Sub-figures (a–c) are the results for a synthetic profile with no random noise added to the data, while sub-figures (d–f) are the results for the synthetic profile shown in (a) with 20% random noise added to the data. Sub-figures (g–h) show the time to depth distributions for the CFCS, CSA and CF models.

3.4. Type IV: Variable sediment accumulation rate and flux

The most common (and most difficult) situation encountered in geochronology studies is one in which *both* sediment accumulation rate and flux vary with time. To simulate this situation, ^{210}Pb versus depth profiles were created assuming a constant sedimentation rate (V)

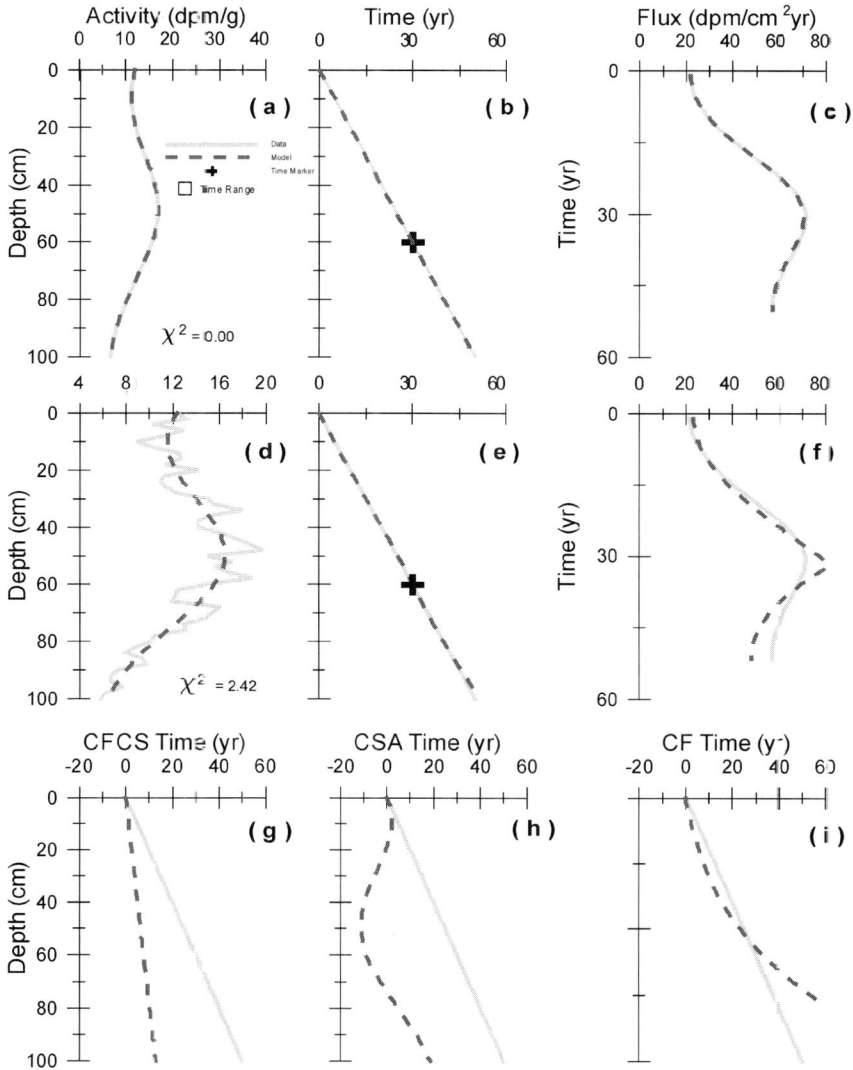

Fig. 2.5. A synthetic ^{210}Pb profile was created with a constant sediment accumulation rate and a variable flux with time. Sub-figures (a–c) are the results for a synthetic profile with no random noise added to the data, while sub-figures (d–f) are the results for the synthetic profile shown in (a) with 20% random noise added to the data. Sub-figures (g–h) show the time to depth distributions for the CFCS, CSA and CF models; none of these models provides reliable sediment ages.

together with $a_n \neq 0$ and $b_n \neq 0$. The resulting ^{210}Pb profiles were then analyzed. Two simulations were performed: one for a data profile exhibiting high average sedimentation rates (1–5 cm/yr) (Fig. 2.6 (a–i)); and one for a data profile exhibiting low average sedimentation rates (0.1–0.5 cm/yr) (Fig. 2.7 (a–i)).

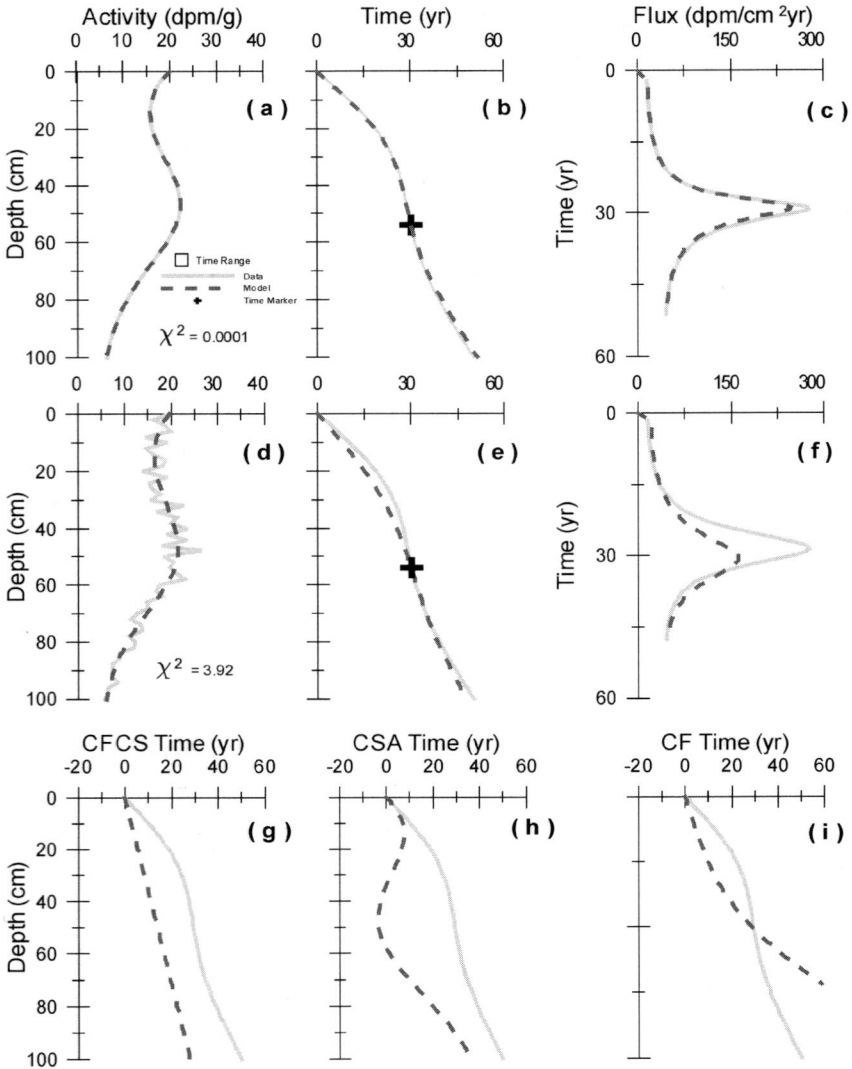

Fig. 2.6. A synthetic ^{210}Pb profile was created with variable sediment accumulation rate and variable flux with time. Sediment accumulation rates varied from 1 to 5 cm/yr. Sub-figures (a–c) are the results for a synthetic profile with no random noise added to the data, while sub-figures (d–f) are the results for the synthetic profile shown in (a) with 20% random noise added to the data. Sub-figures (g–h) show the time to depth distributions for the CFCS, CSA and CF models; none of these models provides reliable sediment ages.

From Fig. 2.6, the model-determined best fit to the data is nearly identical to the input data profile, and the time to depth, and flux variations with time are likewise well determined (Fig. 2.6 (a–c)). Even with the addition of random noise to the data profile, the model-determined best answer is nearly identical to the input profile, and the range of acceptable answers is limited, making the time to depth profile well-constrained (Fig. 2.6 (d–f)). When

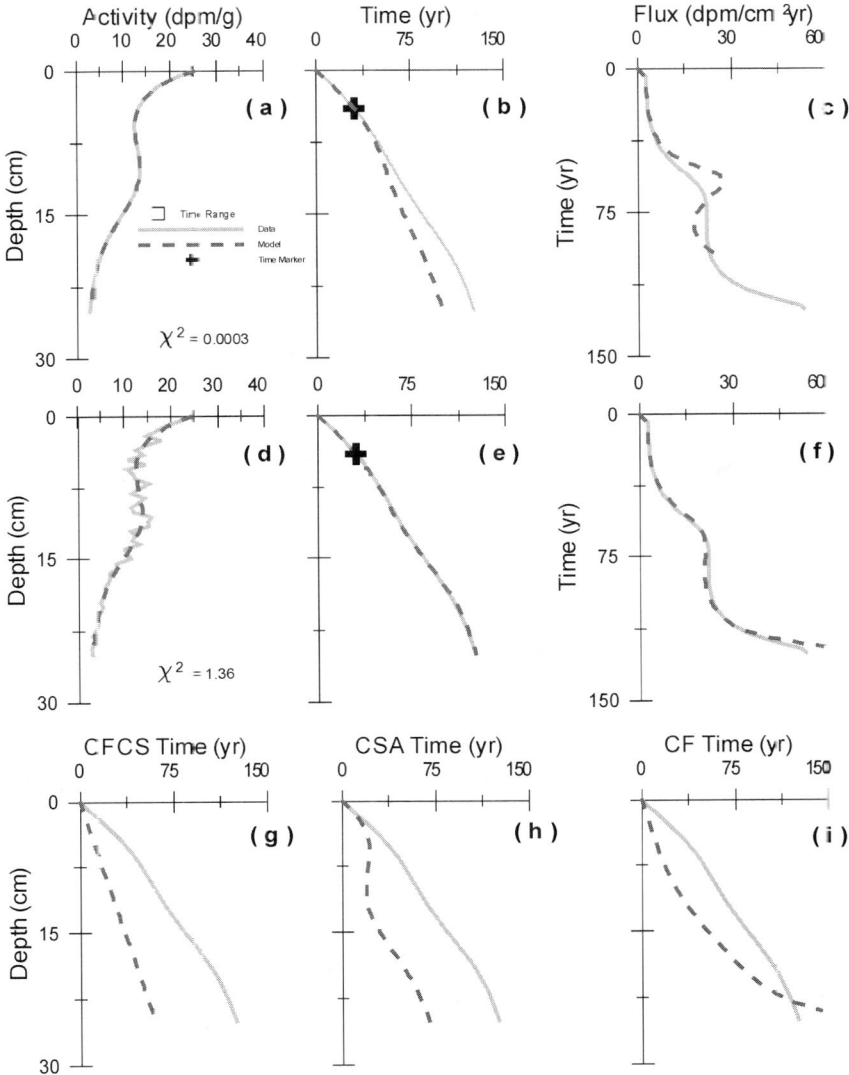

Fig. 2.7. A synthetic ^{210}Pb profile was created with variable sediment accumulation rate and variable flux with time. Sediment accumulation rates varied from 0.1 to 0.5 cm/yr. Sub-figures (a–c) are the results for a synthetic profile with no random noise added to the data, while sub-figures (d–f) are the results for the synthetic profile shown in (a) with 20% random noise added to the data. The location of the time marker at a shallow depth results in greater uncertainty in sediment ages at the bottom of the profile. Sub-figures (g–h) show the time to depth distributions for the CFCS, CSA and CF models; none of these models provides reliable sediment ages.

the alternative models were applied to this profile, the resulting time to depth profiles are incorrect, as shown in Fig. 2.6 (g–i).

If, on the other hand, a data profile exhibited much slower sediment accumulation rates, the model-determined best-fit profile is still well determined. However, the time to depth profile

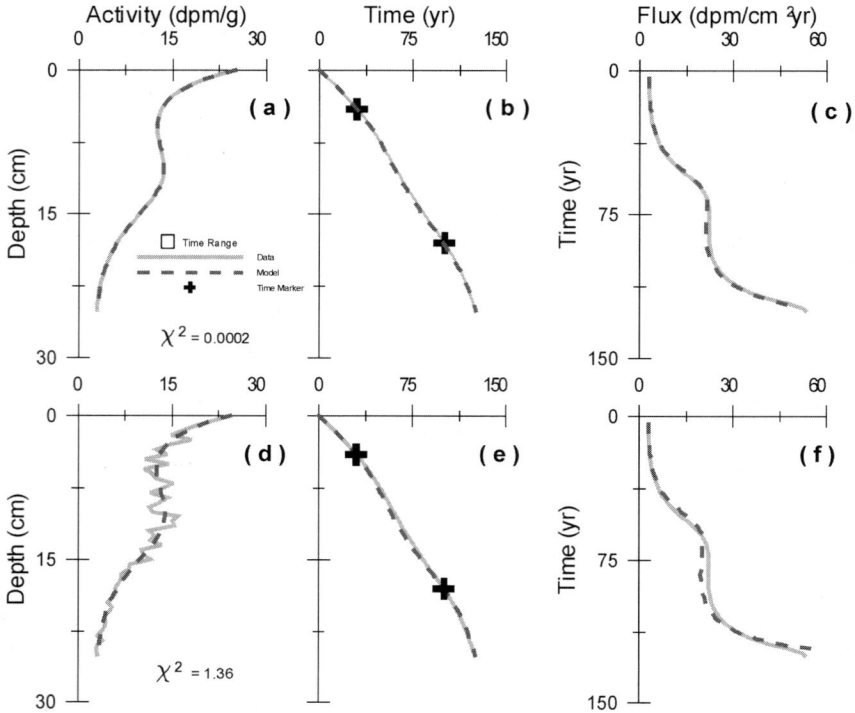

Fig. 2.8. Two time markers were used to analyze the synthetic ^{210}Pb profile given in Fig. 2.7 (a–c). The addition of a second time marker near the bottom of the profile resulted in a near perfect model solution for the sediment ages with depth.

may not be well constrained (Fig. 2.7 (a–c)). For a ^{137}Cs time marker located at a shallow depth in the data profile, the bottom of the profile is not well constrained and the range of ages at the bottom can be large, in this case 50 years. With the addition of random noise to the data profile, the model-determined answers are similarly poorly constrained (Fig. 2.7 (d–f)). As expected, none of the alternative models produces the correct time to depth relationships (Fig. 2.7 (g–i)). When a second time marker is made available for this profile at depth, the model-determined answers are almost identical to the input profile and are well constrained even when random noise is added to the profile (Fig. 2.8 (a–f)).

4. Method comparison

The use of any computerized literature search engine today reveals countless examples of published articles dealing with ^{210}Pb and mostly in terms of its use as a geochronometer. In 1993 Robbins and Herche estimated the number of articles at over 2000. This widespread application of the ^{210}Pb technique demonstrates both the need for and general applicability of methods in modern radionuclide geochronology. Applications in different aquatic systems include studies conducted in lakes and ponds (Carroll et al., 1999; Waugh et al., 1998), coastal fjords (Savinov et al., 2000), bays and estuaries (Hancock & Hunter, 1999), submarine canyons (Buscail et al., 1997) and continental margins (Radakovitch & Heussner, 1999).

Due to their relative simplicity, conceptual models are most often used to establish sediment chronologies. The method has been most successful in dating minimally disturbed sediment cores because the models require one to know a priori the causes and effects of any confounding influences on a radionuclide profile. In cases where model restrictions are not fully adhered to, significant errors may result in the age of sediment layers. A common practice has been to use the CFCS model to determine a constant sediment accumulation rate even when a data profile shows some variations in activity with depth. Yet even small changes in the accumulation rate translate into large errors in the age of the sediments at the bottom of a core. For example, a 1–2 cm/yr variation in accumulation rates doubles the age of sediment at the bottom of a 50 cm sediment core from 50 to 100 years. It is clear from these studies that conceptual models work well when specific knowledge is available of the sediment accumulation processes and processes affecting non-exponential changes in radionuclide profiles. However, when such supporting information is not available, data may be uninterpretable or assumptions must be made that may result in unreliable conclusions.

To extend the use of conceptual models to more complex profiles, a number of adaptations to the basic procedures have also been explored over the years. One simple approach is to apply several models to different portions of a single sediment profile. The choice of where to use a particular model is based on supportive evidence, such as sediment structures, porosity changes, grain size changes, mineralogy changes, additional radionuclide profiles (e.g., ^{137}Cs, ^{228}Th, ^{7}Be), and correlation with sediment layers laid down during known depositional events. An example of this approach is a study conducted at a back-barrier site in Cape Lookout Bight, N.C. (Chanton et al., 1989). The accumulation history in Cape Lookout Bight is complicated by storm deposits of sand and by the supply of silts and clays from multiple sources. Several ^{210}Pb profiles showed exponential decreases in ^{210}Pb in the upper one meter of sediment but showed variable ^{210}Pb activities below one meter. For the upper one meter of sediment, the Simple Model was employed. Below one meter, the authors employed the Constant Specific Activity Model.

Other techniques include monitoring of ^{210}Pb and particle fluxes to the sediment bed and comparison with sediment core ^{210}Pb distributions (Weyhenmeyer et al., 1997; Paulsen et al., 1999). These, and earlier studies (e.g., Boudreau, 1986; Smith et al., 1986; Wheatcroft, 1992; Crusius & Anderson, 1995), have helped to clarify some of the confounding influences that hinder the determination of depositional histories using conceptual models. In the Bay of Biscay (northeastern Atlantic), for example, the majority of mass flux to the seabed (65–89%) resulted from lateral (non-local) inputs to sampling sites (Radakovitch & Heussner, 1999). Similarly, the redistribution of particles in lakes, or sediment focussing, results in higher sediment accumulation rates in deep portions rather than in shallow areas of lakes (Crusius & Anderson, 1995). These, and other site-specific processes, have been shown to impact the shapes of radionuclide profiles in sediment cores. All such processes must be explicitly accounted for in the design of a conceptual model for use in a particular system.

Inductive models can be applied to interpret a wider variety of profile types than has been previously possible using conceptual models (Table 2.1). But shape alone of a radionuclide profile with sediment depth may have several equally plausible explanations. The challenge in signal theory approaches is to gather sufficient information in order to constrain the number

Table 2.1
Summary of model assumptions

Model name	Specific activity	Accumulation rate	Flux of ^{210}Pb
Simple	constant	constant	constant
Constant Flux	variable	variable	constant
Constant Specific Activity	constant	variable	variable
SIT	variable	variable	variable

of explanations to only a few that are similar. Additionally, the performance of any inductive model is limited by the extent to which the relevant information is revealed in the data. That is, one needs to know the extent to which everything necessary to identify sediment ages is reflected somehow in the radionuclide profile.

Tests conducted on one model (SIT) using a variety of synthetic data profiles shows that the signal theory method reproduces satisfactorily the correct time to depth profiles when constrained with a time marker. The model tested was most accurate when analysing synthetic data profiles created with both sedimentation and flux variations ($a_n, b_n \neq 0$), and was less accurate when analysing the data profile exhibiting constant sedimentation with flux variations (Table 2.1), but, even then, was remarkably close to the synthetic data profile. With knowledge of at least one time horizon, such as the depth of the peak in ^{137}Cs activity, restriction of the range of possible solutions to a manageable number is achieved.

The example profiles analysed by the inductive model were created for the radionuclide ^{210}Pb. However, the method is applicable to other radionuclides in geochronology studies, and has been applied to ^{232}Th and ^{231}Pa data for deep-sea sediments (Chapter 3). The method has also been used to interpret the evolution of seawater ^{87}Sr/^{86}Sr ratios from measurements in deep-sea cores (Chapter 4).

Appendix A. Technical development of SIT method for unmixed sediments

Measurements of radioactive decay of a nuclide are taken with respect to sedimentary depth, x, measured from the surface ($x = 0$) downward.

A.1. *Decay only*

Suppose, initially, that the radionuclide concentration per unit mass of solid material deposited is fixed for all time (we shall relax this constraint later). Then the decay with time (not depth) of a radionuclide concentration is governed by the exponential law

$$P(t) = P_0 \exp(-\lambda t), \tag{A.1}$$

where P_0 is the surficial concentration at time $t = 0$, and λ is the intrinsic decay rate of the particular nuclide. If the sedimentation rate is constant for all time, then elapsed time t is connected to burial depth through $x = Vt$, where V is the sedimentation velocity. Accordingly we can then write

$$P(x) = P_0 \exp(-\lambda x/V). \tag{A.2}$$

Hence, by measuring $P(x)$, the best velocity for an exponential fit with depth to the measured data can be obtained.

Two theoretical factors, and one observational fact, indicate that equation (A.2) is not adequate to describe a large percentage of radionuclide data fields. Theoretically, there are the problems that, even if the sedimentation velocity is constant, later compaction of sediments after burial will ensure that the depth to age conversion is not constant, as assumed in equation (A.2); also, even if the compaction effect could be ignored, the rate of sedimentation can itself vary with time and is not necessarily constant, as assumed in equation (A.2). The observational fact is that in some radionuclide records the value of nuclide concentration at depth can show maxima; something that is not possible with a constant velocity model. To handle all of the above problems simultaneously, assume the sediment velocity to vary with depth (or time t) so that deposition is controlled by

$$\mathrm{d}x/\mathrm{d}t = V(t), \tag{A.3a}$$

or, alternatively, by

$$\mathrm{d}t = \mathrm{d}x\, S(x), \tag{A.3b}$$

where $S(x)$ is the slowness, i.e., $S = 1/V$.

Throughout this treatment we deal with the "slowness" formulation, equation (A.3b). Then, the age of sediments currently at depth x is given by

$$t = \int_0^x \mathrm{d}x'\, S(x'), \tag{A.4}$$

where age zero ($t = 0$) corresponds to the present-day at the sediment surface, $x = 0$.

Suppose then that we split the "slowness" into a constant average value, B, and a fluctuation around the average value that has zero average value, i.e.,

$$S(z) = B + \Delta S(z) \tag{A.5a}$$

with

$$\int_0^1 \mathrm{d}z'\, \Delta S(z') = C. \tag{A.5b}$$

where we have normalized the depths of the radionuclide measurements so that the physical depth, $x_{\max}$, of the deepest measurement corresponds to $z = 1$, i.e.,

$$z = x/x_{\max}. \tag{A.6}$$

The choice

$$\Delta S(z) = \sum_{n=1}^{N} a_n \cos(n\pi z), \tag{A.7}$$

always satisfies equation (A.5b) as well as using the fundamental Fourier series theorem that any bounded function of zero mean value can be represented by a cosine series.

Equation (A.4) then yields the age to depth equation

$$t = x_{\max}\left[Bz + \sum_{n=1}^{N}\frac{a_n}{\pi n}\sin(n\pi z)\right]. \tag{A.8}$$

Note that we have, as yet, made no statement on a procedure for determining B, a_n or the number of terms, N, needed to define a given data field. Equation (A.8) provides a general connection between depth and age that removes the restriction of constant velocity.

A.2. *Source only*

However, a further intrinsic assumption has now to be addressed. The implicit assumption of a constant radionuclide concentration per unit mass of sediment deposited is manifestly invalid in a large number of cases because nuclide concentrations do show localized maxima with depth. If the effects of radioactive decay and of sedimentation rate variations are put to one side for the moment, it then follows that a variation of radionuclide concentration with depth would be due to intrinsic source variations with time. In the case of such source variations there is no fundamental physical reason that the source should be larger or smaller than the present day surficial value, P_0. The only requirement is that any source variation should be positive.

To handle all such problems simultaneously introduce the function

$$F(z) = \sum_{n=1}^{N}\frac{b_n}{n\pi}[1 - \cos n\pi z], \tag{A.9}$$

and write the source variation alone as

$$P(z) = P_0 \exp[F(z)]. \tag{A.10}$$

This choice has several advantages. Note that on $z = 0$, the sediment surface, $F = 0$ so $P(z = 0) = P_0$, the surficial value. Note also, that $\exp(F(z))$ is positive no matter what choices are made for the coefficients b_n and the number of terms, N, to use in describing the source variation. Note further that the fundamental Fourier theorem guarantees that the cosine series will describe any bounded variation of a function with zero mean value.

A.3. *Decay and source combined*

Combined, the radioactive decay behavior and the source variation behavior enable us to write the general form

$$P(z) = P_0 \exp\{-\tau(z) + F(z)\}. \tag{A.11}$$

If the coefficients B, a_n, and b_n ($n = 1, 2, \ldots, N$) can be determined from a data set then the source variation with time is determined parametrically through

$$\tau(z) \equiv \lambda t = Bz + \sum_{n=1}^{N} \frac{a_n}{n\pi} \sin n\pi z \tag{A.12}$$

and

$$\text{source}(t) = \exp\left[\bar{F}(t)\right], \tag{A.13}$$

where z values are used in equation (A.12) to construct a conversion of z to time, t, and then used in equation (A.13) to express directly the source with time variation.

A.4. *Coefficient determination*

The problem before us then would appear to be to obtain a procedure which can determine the parameters P_0, B, N, a_n, b_n from a data set of radionuclide measurements with depth and to ascertain the best correspondence to the data. Once this is done the variation of sediment rate, dz/dt, with time or depth is easily determined from equation (A.12) as is the source variation (from equation (A.13)).

The development of a least-squares procedure to handle such problems is relatively straight-forward, albeit tedious.

Equation (A.11) is differentiated with respect to z:

$$\frac{dP}{dz} = P(z)\left[-\frac{d\tau(z)}{dz} + \frac{dF}{dz}\right]. \tag{A.14}$$

Equation (A.14) is now multiplied by $P(z)\exp(ik\pi z)$ and integrated over the domain $0 \leqslant z \leqslant 1$. Then, the real and imaginary parts for each value of k are separated

From the real part we obtain the equation

$$\sum_{n=1}^{N} a_n M_{nk} + \sum_{n=1}^{N} b_n Q_{nk} = R_k \quad (k = 0, 1, \ldots), \tag{A.15}$$

with

$$M_{nk} = 2\int_0^1 P(z)^2 \cos(n\pi z)\cos(k\pi z)\,dz, \tag{A.16a}$$

$$Q_{nk} = 2\int_0^1 P(z)^2 \cos(k\pi z)\sin(n\pi z)\,dz, \tag{A.16b}$$

and

$$R_k = P(z=1)^2(-1)^k - P_0^2 + k\pi \int_0^1 P(z)^2 \sin(k\pi z)\,dz$$

$$- 2B\int_0^1 P(z)^2 \cos(k\pi z)\,dz \tag{A.16c}$$

while from the imaginary part we obtain

$$\sum_{n=1}^{N} a_n S_{nk} + \sum_{n=1}^{N} b_n T_{nk} = U_k \quad (k = 0, 1, \ldots), \tag{A.17}$$

with

$$S_{nk} = 2 \int_0^1 P(z)^2 \cos(n\pi z) \sin(k\pi z) \, dz, \tag{A.18a}$$

$$T_{nk} = 2 \int_0^1 P(z)^2 \sin(n\pi z) \sin(k\pi z) \, dz, \tag{A.18b}$$

and

$$U_k = -k\pi \int_0^1 P(z)^2 \cos(k\pi z) \, dz - 2B \int_0^1 P(z)^2 \sin(k\pi z) \, dz. \tag{A.18c}$$

We recognize equations (A.15) and (A.17) as providing a linear set for determination of the a_n and b_n. Indeed for $k = 2N$ (and P_0 and B prescribed) equations (A.15) and (A.17) provide precisely $2N$ equations for the $2N$ unknowns $(a_1, a_2, \ldots, a_N; b_1, b_2, \ldots, b_N)$. Hence, standard matrix inversion methods (Menke, 1984) can be used to determine the coefficients a_n and b_n. This procedure is followed initially to make first estimates of the a_n and b_n for given, but variable, values of P_0, B and N.

With the sets $\{a_n\}$, $\{b_n\}$ $(n = 1, \ldots, N)$ determined as above, it might be thought that the separation of source and sediment variations is completely disentangled. However, three factors still need to be addressed to maximize resolution. They are (i) an implicit requirement that the observed sediment deposition rate, deduced from the observations and the signal theory matrix procedure, be intrinsically positive, i.e., dz/dt (from equation (A.12)) must be greater or equal to zero. But this requirement has nowhere been enforced in the least squares approach. It can, and does, happen that finite data sampling or uncertainty in data quality, together with different choices for P_0, B and N, can lead to a situation in which dt/dz (from equation (A.12)) becomes negative over select intervals of z. Such behavior is physically unacceptable; (ii) there has still been no procedure provided for selecting P_0 and B in such a way as to guarantee that a minimum mismatch of predictions and observations can be attained which will maximize the statistical sharpness of separating source and sediment variation effects; (iii) there has been no procedure given for constructing the number of terms, N, needed to best satisfy a given data set. We now consider in turn each of these remaining facets.

A.5. Positivity considerations

Introduce

$$\tau_0(z) = Bz + \sum_{n=1}^{N} \frac{a_n}{n\pi} \sin(n\pi z), \tag{A.19a}$$

$$\frac{d\tau_0}{dz} = B + \sum_{n=1}^{N} a_n \cos(n\pi z) \equiv B\left(\frac{d\tau_0'}{dz} + 1\right). \tag{A.19b}$$

Then in place of $\tau(z)$ in equation (A.14) write

$$\tau(z) = B \int_0^z \left[1 + 2\left|\frac{d\tau_0'}{dz}\right|\right]^{1/2} dz' \equiv B\tau'(z) \tag{A.20}$$

and

$$\delta\tau(z) = \tau(z) - \tau_0(z), \tag{A.21}$$

so that the age to depth conversion is provided by equation (A.20) with $t(z) = \tau(z)/\lambda$.

Notice that the non-linear function (A.20) guarantees that $\tau(z)$ will always be a positive increasing function of depth, i.e., the sediment rate will be positive.

But the introduction of this non-linear variation then changes the determination of the co-efficients a_n and b_n from the procedure outlined in equations (A.14) through (A.18). Tracing through the derivation of the least squares difference procedure above, with form (A.20) for $\tau(z)$, yields the left-hand sides of equations (A.15) and (A.17) for determining a_n and b_n; the right-hand sides have additional terms added from the non-linear behavior of $\tau(z)$. Thus, we add to R_k the term

$$\delta R_k = 2B \int_0^1 P(z)^2 \left[\frac{d\tau'(z)}{dz} - 1 - \frac{d\tau_0'(z)}{dz}\right] \cos(k\pi z)\, dz$$

$$\equiv 2B \int_0^1 P(z)^2 \left[\left[1 + 2\left|\frac{d\tau_0'}{dz}\right|\right]^{1/2} - 1 - \frac{d\tau_0'}{dz}\right] \cos(k\pi z)\, dz, \tag{A.22}$$

and we add to U_k the term

$$\delta U_k = 2B \int_0^1 P(z)^2 \left[\frac{d\tau'(z)}{dz} - \frac{d\tau_0'(z)}{dz} - 1\right] \sin(k\pi z)\, dz. \tag{A.23}$$

For given values of $\bar{F}_0$, B and N the procedure for solving equations (A.15) and (A.17) (modified by the non-linear terms δR_k and δU_k) is as follows. First ignore the terms δR_k and δU_k, solve equations (A.15) and (A.17) to obtain first approximations to a_n and b_n. Use these approximate values in equations (A.22) and (A.23) to obtain first approximation to δR_k and δU_k. Add these corrections to R_k and U_k, respectively, and then resolve equations (A.15) and (A.17) with the modified R_k and U_k terms, obtaining the next approximations to a_n and b_n. Use these to update δR_k and δU_k. Repeat the procedure until no further significant change takes place in a_n and b_n, at the level of numerical precision of the least squares deter-mination of $\tau(z)$, i.e., if, after q iterations we have

$$\left|\tau_q(z) - \tau_{q-1}(z)\right| < \varepsilon,$$

where ε is a pre-set level of significance, then the determination of a_n, b_n is deemed satisfactory.

Thus a $\tau(z)$ has now been obtained, for any values of P_0, B and N, which is manifestly positive and increasing with increasing z, as required.

In practice, about 3–8 non-linear iterations is sufficient to determine accurate values for a_n and b_n.

A.6. *Determination of the best P_0 and B values*

While the determination of a time to depth conversion and a positive source are now guaranteed, we have yet to provide a method of estimating the values of P_0 and B most consistent with the observed radionuclide variations with depth for a specified number, N, of terms. A procedure for guaranteeing that progressively better values of P_0 and B are determined operates as follows.

Specify a search range for each of P_0 and B say

$$P_{min} \leqslant P_0 \leqslant P_{max}; \qquad B_{min} \leqslant B \leqslant B_{max}.$$

Introduce the normalized variables A and b through

$$A = \frac{P_{max} - P_0}{P_{max} - P_{min}}, \qquad b = \frac{B_{max} - B}{B_{max} - B_{min}} \tag{A.24}$$

so that $1 \geqslant A \geqslant 0$, $1 > b \geqslant 0$. For later simplicity set $a_1 = A$, $a_2 = b$.

Then construct

$$X^2(A, b) = \frac{1}{Q} \sum_{r=1}^{Q} [P(z_r) - p_r]^2, \tag{A.25}$$

where Q is the number of measurement points, p_r is the measured value at depth z_r, and $P(z_r)$ is the predicted value of $P(z)$ from equation (A.11) for a specified A, $b(P_0, B)$ pair for a given N.

Then, to determine the best parameter values that minimize the global control function, X^2, is relatively simple. Use the fact that $0 \leqslant a_j \leqslant 1$ ($j = 1, 2$) to write

$$a_j = \sin^2 \theta_j, \quad \text{with } \theta_j^{(0)} = \sin^{-1}[(a_j^{(0)})^{1/2}]. \tag{A.26}$$

Then construct the following iteration loop to be passed through L times, with the nth iteration loop as follows

$$a_j^{(n)} = \sin^2 \theta_j^{(n)}, \tag{A.27a}$$

$$\theta_j^{(n+1)} = \theta_j^{(n)} \exp\left[-\tanh\left\{\Gamma_j \delta_j^{(n)} \frac{\partial X^2(a^{(n)})}{\partial a_j^{(n)}}\right\}\right], \tag{A.27b}$$

$$a_j^{(n+1)} = \sin^2 \theta_j^{(n+1)}, \tag{A.27c}$$

with

$$\Gamma_j = \left| \frac{\partial X^2(a_j^{(0)})}{\partial a_j^{(0)}} \right|^{-1} \ln\left[1 + (La_j^{(0)})^{-1}\right], \tag{A.28}$$

$$\delta_j^{(n)} = q_j^{(n)} / \left\{ P^{-1} \sum_{l=1}^{P} q_l^{(n)} \right\}, \tag{A.29a}$$

$$q_j^{(n)} = \left\{ |a_j^{(n)} - a_j^{(n-1)}| / |a_j^{(n-1)}| \right\} + \beta^2, \tag{A.29b}$$

where the dependence of X^2 on $a_j^{(0)}$ has been suppressed, and where β^2 measures the degree to which one numerically calculates the partial derivatives, i.e.,

$$\frac{\partial X^2(a^{(n)})}{\partial a_j^{(n)}} \cong \left[X^2[a_1^{(n)}, a_2^{(n)}, \ldots, a_j^{(n)} + \beta a_j^{(n)}, a_{j+1}^{(n)}, \ldots] - X^2(a_j^{(n)}) \right] (\beta a_j^{(n)})^{-1}. \tag{A.30}$$

This procedure guarantees to find the minimum value of X^2 in the range chosen for p; guarantees that X^2 will be smaller at each iteration; and guarantees (through the factor $\delta_j^{(n)}$) to treat first with those components of a which need to be changed the most and, as the minimum is approached, treats all components equally strongly within the domain. The procedure also guarantees to keep each component of a in the range $\{0, 1\}$ and, as a bonus, if any component of (P_0, B) needs to have its range adjusted this fact will show up with the corresponding component of a tending to zero ($\sin^2 \theta_j^{(n)} \to 0$) for (P_0, B) needing to be lowered, or to unity ($\sin^2 \theta_j^{(n)} \to 1$) for (P_0, B) needing to be increased. Thus one also determines in which direction to shift a pre-chosen range of each component. The initial choice of (P_0, B) is still arbitrary, but its range is rapidly determined by the shift requirements, so that a precise choice for (P_0, B) is not critical as long as it is in a region consistent with a broad, but murky, idea of where parameter values should lie.

It is also clear that because the signal theory procedure uses the initial values as a spring-board to minimize the control function, then a fast path to a minimum is to perform the iteration procedure L times, update the initial values by the Lth values, and iterate again L times; this procedure is far superior to doing $2L$ iterations around a given initial value set. Practically, a loop pass through 10 iterations ($L = 10$) is a useful value, so for multiple passes one uses loops of 10, updating after each loop of 10.

Thus this procedure guarantees that if A and b are chosen to be initially positive then they stay positive, and at each iteration the new values of A and b will come closer to providing a minimum in X^2, i.e., in the mismatch between observations and predictions. A limiting behavior is obtained when the numerical accuracy is reached with which one can calculate derivatives of X^2 with respect to A and b.

A.7. *Determination of L, the number of terms*

For each L value used, application of the above procedures provides a least squares match to the data which is minimal. The mismatch, at each L, is recorded by X^2. Determination of an optimal number of terms to use is then achieved by sequentially increasing L, running through the above procedures and recording $X^2(L)$. As L increases $X^2(L)$ first goes through a minimum and then either levels off or rises again. If X^2 rises as L increases then it is fairly obvious which values of L gives the lowest value of X^2, providing a minimum mismatch. If X^2 levels off so that no further significant improvement in mismatch to the data can be obtained, then the proper inference is that there is no point in continuing to increase the number of terms beyond that value of L which first reaches the leveling position, because the coefficients are then under-determined.

The application of all of these procedures guarantees that source and sediment variations are maximally disentangled; guarantees that source variations with time and sediment deposition rates with time are intrinsically positive; guarantees that the present-day surface value and the trend of the mean sedimentation rate are maximally consistent with the data; and guarantees that the optimal number of terms have been used consistent with data resolution.

A.8. *Determination of weighting factors*

Two factors dictate that, while the procedures above guarantee a minimum least square fit, yet the fit to the data is not necessarily optimal. First, if the data ever show an increase over restricted ranges of depths then we know for certain that there was a source variation, yet nowhere have we used this information. Second, it will be noted that equation (A.14) was multiplied by $P \exp(ik\pi z)$ and integrated to obtain a least squares set of equations for the unknown coefficients. But we could have multiplied by any function, not just by P. Thus, we could have directly integrated equation (A.14), for example. Note also that if P is increasing then equation (A.14) requires $dF/dz > dt/dz$ in that domain of increase so that equation (A.14) contains the information on source variation. Yet this information is distorted in some manner when the least squares approach is taken (the information is distorted in some manner no matter what approach is taken).

To allow for these problems simultaneously in a manner that improves the data fit, we proceed as follows. Introduce the weight factor w so that

$$P(z) = P_0 \exp\left[-\tau(z)(1-w) + F(z)(1+w)\right]. \tag{A.31}$$

If $w = 0$, source variation and radioactive decay factors are equally balanced. If $w = 1$ then only the source variation controls the behavior, while for $w = -1$ only the radioactive decay plays a role.

The source variation (relative to a surface value of P_0) is then $\exp[F(z)(1+w)]$ while the time to depth conversion is provided through $t(z) = \lambda^{-1}(1-w)\tau(z)$.

The weight factor w is to be constrained to lie in the domain $-1 \leqslant w \leqslant 1$.

Pragmatically what is done now is to allow w to be a parameter to be determined as in Section A.6 at the same time as the parameters P_0 and B are determined. This procedure then produces a weight factor and a best fit to the data that maximally resolves the sediment and source variations with time. The procedure also allows for uncertainties produced both by errors and finite sampling of the data, as well as from upturn of the radionuclide variations with depth and the possible departure from the demands of a least squares mismatch to the data.

A.9. *Sectioning the data field*

Two problems suggest that the use of all of the radionuclide data with all of the depths of measurements at once may be highly inappropriate.

First, while it is true that sine and cosine series will, in principle, accurately describe any shape with an infinite number of terms, it can, and does, happen that with an extremely variable profile with depth it may take an extraordinarily large amount of computer time to find the best number of terms to use in conjunction with simultaneous determination of the best surface value, overall decay constant with depth, weight factor and individual coefficient $\{a_n\}$ and $\{b_n\}$. It is preferred to minimize the computer time.

Second, it happens that the data set available to a depth z_*, is run, best parameters computed, and then additions to the data set at depths in excess of z_* later become available. These later additions should surely not change the determination of present day surface value, of the age to depth conversion and/or the source variation from the surface ($z = 0$) to the previously computed depth at $z = z_*$. Yet, because the non-linear procedure is a variant of a least squares technique, complete restart of the system is required to determine de novo all parameter values as well as the new number of terms required to fit the new total data set. This requirement also causes major increases in computer time.

Suppose then that the data field is initially and arbitrarily considered in two pieces: D1 occurring in $0 \leqslant x \leqslant x_*$. D2 in $x_{max} \geqslant x \geqslant x_*$, where x_{max} is the greatest measurement depth.

Consider the data field D1 first. Assume that the inverse procedure has been carried through to completion yielding best parameter values $P_0^{(1)}$, $B^{(1)}$, $N^{(1)}$, $\{a_n^{(1)}\}$, $\{b_n^{(1)}\}$ and $w^{(1)}$. Then the prediction for the data behavior in $0 \leqslant x \leqslant x_*$ (i.e., $0 \leqslant z \leqslant 1$, where $z = x/x_*$) is

$$P^{(1)}(z) = P_0^{(1)} \exp\left[-\tau^{(1)}(z)\left(1 - w^{(1)}\right) + F^{(1)}(z)\left(1 + w^{(1)}\right)\right], \tag{A.32}$$

where the superscript parameter (1) refers to values determined from D1. Now consider the data field D2 in $x > x_*$. Write $y = x - x_*$ so that D2 extends from $y = 0$ to $y = y_{max} = x_{max} - x_*$. Again fit the field D2 by the inverse procedure generating best parameter values $P_0^{(2)}$, $B^{(2)}$, $N^{(2)}$, $\{a_n^{(2)}\}$, $\{b_n^{(2)}\}$ and $w^{(2)}$ to obtain

$$P^{(2)}(y) = P_0^{(2)} \exp\left[-\tau^{(2)}(y)\left(1 - w^{(2)}\right) + F^{(2)}(y)\left(1 + w^{(2)}\right)\right], \tag{A.33}$$

with $0 \leqslant y \leqslant 1$ where $y = Y/Y_{max}$.

Thus the "surface" value $P_0^{(2)}$ actually refers to the physical depth point $x = x_*$ while $\tau^{(2)}(y)$ and $F^{(2)}(y)$ both are zero on $y = 0$ ($x = x_*$).

However, from D1, we already have a value for $P_0^{(2)}$, viz.

$$P_0^{(2)} \equiv F^{(1)}(1) = P_0^{(1)} \exp\left[-\tau^{(1)}(1)\left(1 - w^{(1)}\right) + F^{(1)}(1)\left(1 + w^{(1)}\right)\right]. \tag{A.34}$$

The procedure is then clear. Use the data field D1 to construct a separation into sedimentation rate and source variations. Take the value $P^{(1)}(1)$ and use it to be the constrained "surface" value for the data field D2. Obtain the best parameters for the sedimentation rate and source variations of D2.

Then the time to depth variation for the total data field is

$$t^{(1)}(x) = \left(1 - w^{(1)}\right)\tau^{(1)}(z)/\lambda, \qquad\qquad \text{in } 0 \leqslant x \leqslant x_* \ (0 \leqslant z \leqslant 1), \qquad (A.35a)$$

$$t^{(2)}(x) = \left(1 - w^{(2)}\right)\tau^{(2)}(y)/\lambda + t^{(1)}(x_*), \quad \text{in } x_{\max} \geqslant x \geqslant x_* \ (0 \leqslant y \leqslant 1), \qquad (A.35b)$$

with the source component variation

$$S^{(1)}(x) = \exp\left[F^{(1)}(z)\left(1 + w^{(1)}\right)\right], \qquad\qquad \text{in } 0 \leqslant x \leqslant x_*, \qquad (A.36a)$$

$$S^{(2)}(x) = \exp\left[F^{(2)}(y)\left(1 + w^{(2)}\right)\right]S^{(1)}(x_*), \quad \text{in } x_{\max} \geqslant x \geqslant x_*. \qquad (A.36b)$$

This sectioning of the data not only cuts down computer time, but also ensures that the addition of data from a greater depth has no effect on the time to depth conversion or source variations from shallower depths (and so younger times). Now, the sectioning into two separate partitioned data fields at $x = x_*$ is still arbitrary, both in terms of the number of partitioned data fields and the positioning of the partition depths. Indeed one could partition the total data field into m sequential data sets D1, ..., Dm occupying $0 \leqslant x \leqslant x_1, x_1 \leqslant x \leqslant x_2, \ldots, x_{m-1} \leqslant x \leqslant x_{\max}$. But there is a practical limit. The resolution of individual data measurements suggests that it is pointless to subdivide further the data fields when the least mean squares residual mismatch per average point in a given partition is less than the measurement resolution. In addition, when adding one more partition no longer improves the mismatch to the data then again there is little point in continuing this procedure.

Pragmatically we have found that arbitrarily partitioning the data field at the median point, or into thirds, normally provides more than enough partitions to handle all except the most exceptional variations of data. The most partitions we have ever needed is four (quarters).

Chapter 3

Single site studies

The previous chapter provides the theoretical background for different modelling approaches and a discussion of both the capabilities and limitations of individual methods for assessing sediment ages and depositional flux variations. In this chapter, practical examples are given demonstrating applications of inductive models to 'real world' situations. These examples serve to illustrate further the different approaches of conceptual and inductive modelling procedures.

Four different lakes with excess ^{210}Pb radionuclide profiles, one of which also has a ^{137}Cs profile, and constraining historical age markers, are investigated; in addition, a marine back-barrier site in North Carolina, also with ^{210}Pb, is examined. Finally, the utility of inductive modelling techniques to other radionuclides (^{231}Pa and ^{230}Th) is demonstrated for a deep water marine core from the Galapagos micro-plate region. While no historical age markers are available to act as constraining information in this deep marine situation, nevertheless both of the radionuclides must provide a common depth-to-age conversion, even through their surficial deposition fluxes may differ and also be time-dependent. In this way, each radionuclide acts as a control on results from the other. As we shall see, this information is the equivalent of having available multiple "historic" age-markers to use as constraints.

Measures of resolution, uniqueness, precision and sensitivity are addressed in Appendix A to this chapter with quantitative descriptions provided there of cumulative probabilistic assessment of results.

1. Lake sediments and ^{210}Pb geochronology

1.1. *Lake Kurtna District*

The Kurtna Lake District in northeast Estonia is part of a kame field produced on the marginal formation of the last continental ice shield, about 12,200–12,300 years ago. Prior to the Second World War, human impact in the region was minimal, consisting mainly of lake water-level regulation in order to drain the surrounding bogs and pastures. At the beginning of the 1950s anthropogenic influences increased sharply. In 1951, an oil-shale operating Ahtme power plant was erected only a few kilometers from the District. The chimneys of the Ahtme plant were not equipped with electric filters until 1976. During the 1960s and 1970s several more powerful plants, emitting a large amount of fly ash to the atmosphere, were built in northeast Estonia. Fly ash deposition caused an increase in the accumulation rates of several chemical elements in lake sediments (Varvus & Punning, 1993). The water bodies became useful as reservoirs for industrial water supplies, as well as for the construction of discharge systems for water-borne industrial wastes. In the 1970s, lake levels decreased as a result of groundwater pumping, resulting in shore and slope erosion. Two dystrophic lakes located in the Kurtna Lake District, Lake Nômmejärv and Lake Kurtna, are examined here.

1.1.1. *Lake Nômmejärv*

Lake Nômmejärv lies on the western edge of the kame field and is surrounded by kames and morainic hills. The area of the lake is about 15×10^4 m^2 (0.15 km^2) and its maximum depth is 7 m. Since 1970 the lake has had a strong inflow due to the water pumped out of oil-shale mines and discharged to the lake by an artificial channel. Inflowing mine waters have strongly influenced the mineral content of Lake Nômmejärv (HCO$_3$ = 4.9 mg/l, SO$_4$ = up to 150 mg/l, Na = 21 mg/l). The input of suspended matter also increased to about 40 T/yr, of which 30 T accumulates in the lake (Sagris, 1987; Punning et al., 1989).

The distribution of ^{210}Pb with sediment depth and associated 1 sigma errors for a sediment core collected from the lake are given in Fig. 3.1a. The sediment profile of ^{210}Pb does not exhibit an exponential decrease in ^{210}Pb activity with increasing depth; there is, however, a major increase in ^{210}Pb activity deeper than a depth of 22 cm. The Constant Flux (CF) Model (Appleby & Oldfield, 1978) was applied to quantify sedimentation rates for this non-exponential ^{210}Pb profile (Varvus & Punning, 1993). Based on the CF model, the age of the sediments at 22 cm is 43 years and the age of the sediments at 45 cm is 111 years. However, based on the onset of industrial activities in the region, the age of the sediment layer at 22 cm is 16 years, and the layer at 45 cm is 35 years old. Thus sediment ages are overestimated by the CF model. Apparently, this problem is commonly observed with the CF model (McCall et al., 1984). The problem occurs because small changes in radionuclide activities near the bottom of a sediment core are translated into large (and uncertain) values in the CF model equation for time, when they are "corrected" by an exponential function to allow for the time elapsed since deposition.

An alternate explanation of the sedimentation history of the lake is now given based on the inductive modelling procedure. Based on the timing of the onset of known anthropogenic events, two time markers were used to constrain the interpretation of the data by the model (Fig. 3.1b and Table 3.1). The model fit to the data given in Fig. 3.1a shows that the model was able to reproduce the broad maximum in ^{210}Pb activity centered around a sediment depth of 29 cm. The associated probability distribution for the age-to-depth profile is given in Fig. 3.1b together with the results determined from the CF model by Varvus & Punning (1993). Upper and lower error bars represent the $P(90)$ and $P(10)$ values from the probability distribution method (see Appendix A). The inductive model produced answers that reproduced the shape of the data profile and, most importantly, satisfied the known time markers.

The ^{210}Pb flux versus the most probable age (Fig. 3.1c) shows a slight increase in the ^{210}Pb flux to the lake bottom in and after the early 1950s, corresponding to the onset of operations at the Ahtme plant. A much larger flux increase occurs around 1970, in agreement with the known timing of the inflow of waters from the oil-shale mines. The model-determined flux of ^{210}Pb to the sediments has systematically decreased since the 1970s, consistent with the installation of electric filters on the oil shale power plants in 1976.

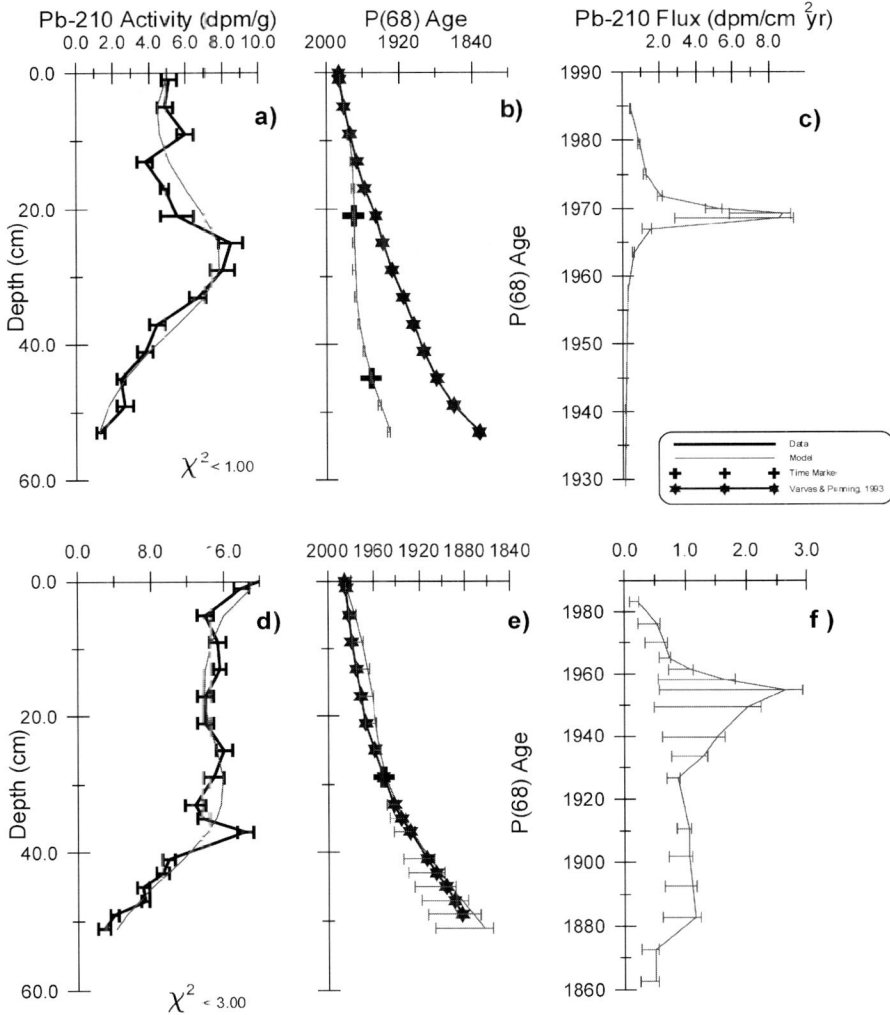

Fig. 3.1. (a–c) Lake Nômmejärv: (a) Measured ^{210}Pb (dpm/g) versus depth (m) and associated errors (from Varvus & Punning, 1993). The smooth curve represents SIT model $P(68)$ ^{210}Pb activities versus depth ($\chi^2 < 1.0$). (b) Probability distribution of age-to-depth variations consistent with the smooth curve in Fig. 3.1a. The two crosses represent age markers of oil shale fly ash in 1951 and oil shale plant waste water input onset in 1970, respectively. The starred curve is the age-to-depth determination of Varvus & Punning (1993) based on the Constant Flux model. (c) Probability distribution of surficial flux (dpm cm^{-2} yr^{-1}) variation with time. A minor flux increase related to the input of oil shale fly ash and a major flux increase caused by the addition of mining wastes to the lake are observed. (d–f) Lake Kurtna: (d) Measured ^{210}Pb (dpm/g) versus depth (m) and associated errors (from Varvus & Punning, 1993). The smooth curve represents SIT model $P(68)$ ^{210}Pb activities versus depth ($\chi^2 < 3.0$). (e) Probability distribution of age-to-depth variations consistent with the smooth curve in Fig. 3.1d. The cross represents an oil shale fly ash layer deposited in 1951. The starred curve is the age-to-depth determination of Varvus & Punning (1993) based on the Constant Flux model. (f) Probability distribution of surficial flux (dpm cm^{-2} yr^{-1}) variation with time. Note the change-over of flux variation (from steadily increasing to steadily decreasing with time) at about 1953 – representing the timing of man-made channels connecting Lake Kurtna to other lakes and so altering water and sediment input.

Table 3.1
Characteristics of SIT models for ^{210}Pb ($t_{1/2} = 22.3$ years) and ^{137}Cs ($t_{1/2} = 30.2$ years) distributions in freshwater and marine sediment cores

Location	Time marker (age ± yr)	Time marker description	Surface activity (dpm/g)	Velocity (cm/yr)	Mismatch (dpm/g)^2	Reference
Lake Nômmejârv	22 cm = 1970 ± 3 45 cm = 1951 ± 3	fly ash deposit oil-shale mining discharge	^{210}Pb$_{max}$ = 4.9–5.2	0.9–1.0	< 1.0	Varvus & Punning, 1993
Lake Kurtna	29 cm = 1951 ± 3	fly ash deposit	^{210}Pb$_{max}$ = 17.8–20.9	0.4–0.6	< 3.0	Varvus & Punning, 1993
Lake Rockwell	18.5 cm = 1963 ± 3 29 cm = 1951 ± 5	^{137}Cs peak ^{137}Cs initial	^{210}Pb$_{max}$ = 49.5–51.2	0.8–1.0	< 9.0	McCall et al., 1984
Cape Lookout Bight	90 cm = 1971 ± 3 150 cm = 1963 ± 3 268 cm = 1951 ± 3	Hurricane Ginger ^{137}Cs peak ^{137}Cs initial	^{210}Pb$_{max}$ = 10.6–11.0	9.2–11.9	< 1.0	Chanton et al., 1989

Table 3.2
Characteristics of SIT models for ^{231}Pa ($t_{1/2} = 75,200$ years) and ^{230}Th ($t_{1/2} = 32,500$ years) distributions in a deep-sea sediment core, KLH093 from Frank et al. (1993)

^{231}Pa$_{max}$ (dpm/g)	Velocity (cm/1000 yr)	Coefficients	Mismatch (dpm/g)^2	^{230}Th$_{max}$ (dpm/g)	Velocity (cm/1000 yr)	Coefficients	Mismatch (dpm/g)^2
1.9	1.5	$a_1 = -2.10; b_1 = 2.90$ $a_2 = 1.70; b_2 = -4.26$	0.03	10.1	1.5	$a_1 = -0.66; b_1 = 2.04$ $a_2 = 0.35; b_2 = 1.95$	1.1
1.9	2.0	$a_1 = -1.86; b_1 = 1.61$ $a_2 = 2.22; b_2 = -4.02$	0.03	10.3	2.0	$a_1 = -0.43; b_1 = 1.35$ $a_2 = 0.71; b_2 = 1.75$	1.2
2.0	2.4	$a_1 = -0.51; b_1 = 0.75$ $a_2 = 2.54; b_2 = -2.67$	0.03	10.5	2.6	$a_1 = -0.20; b_1 = 0.87$ $a_2 = 0.98; b_2 = -1.55$	1.3

1.1.2. *Lake Kurtna*

Lake Kurtna is situated on the western edge of the kame field. The area of the lake is 34×10^4 m^2 (0.34 km^2) and the maximum depth is 6.9 m. The distribution with depth of ^{210}Pb and associated 1 sigma errors are given in Fig. 3.1d (Varvus & Punning, 1993). Once again, the sediment profile of ^{210}Pb does not exhibit an exponential decrease in ^{210}Pb activity with depth.

Varvus & Punning (1993) used the Constant Flux (CF) model to determine sediment accumulation rates from the distribution of ^{210}Pb in the sediment core. They found good agreement between the ^{210}Pb data and the known dates of operation of the Ahtme power plant (Fig. 3.1e).

The age-to-depth distribution (Fig. 3.1e) derived from the inductive model also shows good agreement with the history of the anthropogenic influence on Lake Kurtna. Model answers were calibrated using one time marker of 1951 (±3 years) at a depth of 29 cm, corresponding to discharge of fly ash from the Ahtme power plant. While the cumulative probability distributions for sediment ages ($P(10)$ to $P(90)$)[1] at each depth are fairly wide toward the base of

---

[1] See Appendix A for details.

the profile (as shown on Fig. 3.1e), the most probable age $P(68)$ is near to the $P(90)$ limit, implying little age uncertainty at the 90% confidence value.

Based on these modelling results, the flux history of ^{210}Pb for Lake Kurtna (Fig. 3.1f) shows a maximum in the early 1950s, corresponding to the development of artificial channels in 1953 to connect the lake to several others; these channels significantly altered the flow regime of Lake Kurtna (Varvus & Punning, 1993).

According to Varvus & Punning (1993), the use of ^{210}Pb geochronology in the Kurtna Lake District was problematic. In the case of Lake Nômmejärv, problems they encountered with the CF model were in the dating of lake sediments due to changes in the hydrologic regime through time. However, the results of the inductive model show that the distribution of ^{210}Pb in the sediments is consistent with the timing of historical events, and the corresponding ^{210}Pb flux variations with time consistently reflect the various stages of anthropogenic influence in this region.

1.2. *Rockwell reservoir, Northeastern Ohio*

Lake Rockwell, a man-made freshwater reservoir in Northern Ohio, was excavated in 1914 to store and supply water for the City of Akron, Ohio. The surface area of the lake is 2.78 km^2 and has a drainage basin of 531 km^2 (Hahn, 1955). A 75 cm sediment core, collected from the bottom of Lake Rockwell, was extruded and sectioned into 1 cm sections to a depth of 24 cm, 2-cm sections to a depth of 40 cm, and 5-cm sections to the bottom of the core (McCall et al., 1984). The ^{210}Pb activity in the 45–50 cm interval decreases to below the ^{210}Pb supported activity (^{226}Ra $= 67 \pm 3.3$ dpm/g), suggesting that the transition from freshwater lake sediments to soils occurs within this depth interval (G. Matisoff, pers. comm.), which must then correspond to the time of creation of the reservoir, i.e., 1914.

Two age-markers are available for the core. The ^{137}Cs peak, occurring at 18.5 cm depth, is taken to correspond to the age 1963 ($\pm$3 yrs); while direct sediment thickness measures in 1938–1956 provide an age of 1949 at about 30 cm depth.

The data profile of ^{210}Pb (Fig. 3.3a) exhibits a roughly linearly decreasing activity of ^{210}Pb with increasing sediment depth. The inductive model reproduces the shape of the data profile well. The best surface value for ^{210}Pb is 49.5–51.2 dpm/g with an average sedimentation velocity of 0.8–1.0 cm/yr. The corresponding age-to-depth profile determined by the model, together with the two age markers, is given in Fig. 3.3b. By direct core measurements over the years since 1938, and utilizing ^{137}Cs and ^{210}Pb, McCall et al. (1984) have noted that the increased urbanization around the area since 1940 (with a population doubling time of about 20 yrs) would seem to be in direct one-to-one correspondence with both the increase in sedimentation rate (Fig. 3.3c) and the continuing increase in ^{210}Pb flux to Lake Rockwell. Thus, the influences of various growth stages of urbanization can be tracked for their impact on sediment fill using the ^{210}Pb method.

2. **Marine sediments and multi-radionuclide geochronology**

In most marine situations the luxury of having available historical events as age markers is not always present. In addition, unless the main effort is to understand modern sedimentation

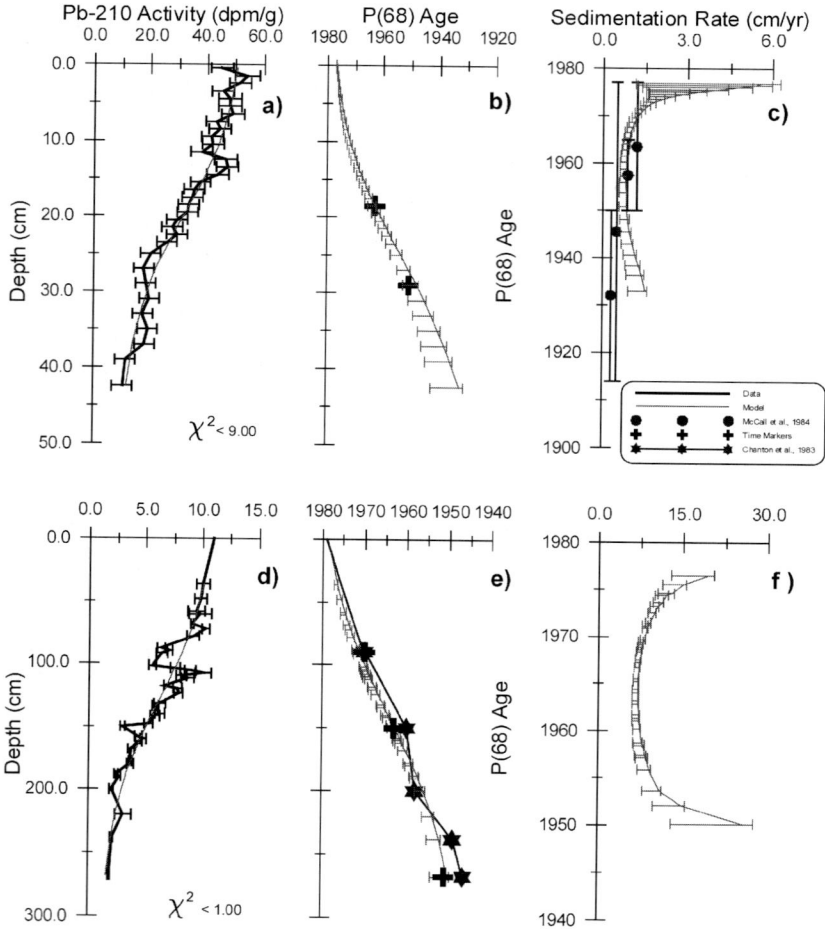

Fig. 3.2. (a–c) Lake Rockwell: (a) Measured ^{210}Pb (dpm/g) versus depth (m) and associated errors (from McCall et al., 1984). The smooth curve represents SIT model $P(68)$ ^{210}Pb activities versus depth ($x^2 < 9.0$). (b) Probability distribution of age-to-depth variations consistent with the smooth curve in Fig. 3.2a. The two crosses represent age markers of the peak in atmospheric ^{137}Cs and the onset of ^{137}Cs injection into the atmosphere at the beginning of the weapons production era. (c) Probability distribution of sediment accumulation rate variations with time. Vertical lines represent sediment accumulation rates determined for several time periods by McCall et al. (1994). Note the uncertainty in sediment accumulation rates for 1970–1978. (d–f) Cape Lookout Bight, NC: (d) Measured ^{210}Pb (dpm/g) versus depth (m) and associated errors (from Chanton et al., 1989). The smooth curve represents SIT model $P(68)$ ^{210}Pb activities versus depth ($x^2 < 1.0$). (e) Probability distribution of age-to-depth variations consistent with the smooth curve in Fig. 3.3d. The crosses represent the time markers described in Table 3.1. The starred curve is the age-to-depth determination of Chanton et al. (1989). (f) Probability distribution of sediment accumulation rates with time.

rate variations, the short (~ 22 yr) half-life of unsupported ^{210}Pb precludes extending ^{210}Pb procedures much further back in time than about 5 half-lives, i.e., of the order of 100 yrs or so. For earlier times other radionuclides must be used with longer half-lives. The rapidity with

(a)

(b)

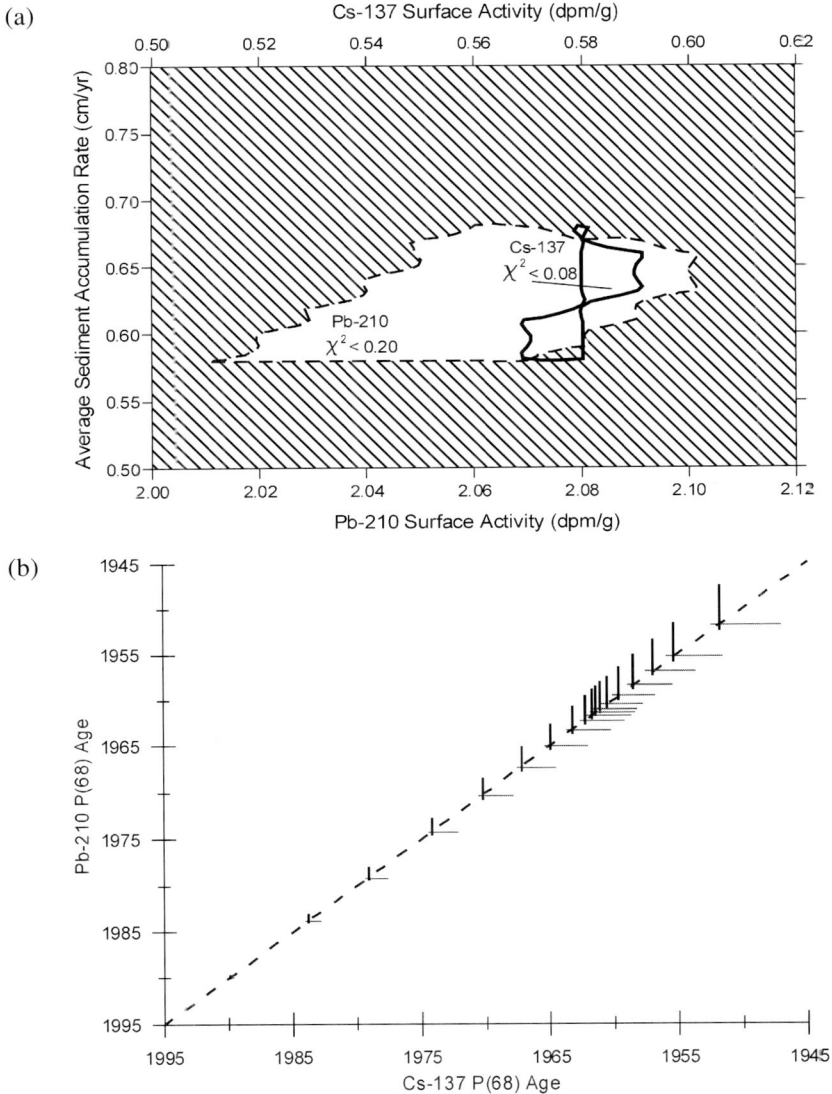

Fig. 3.3. (a) Cross-plot of the best surficial values for ^{137}Cs and ^{210}Pb yielding the minimum mismatch for Lake Rockwell; (b) Cross-plot of the ages determined by both ^{137}Cs and ^{210}Pb for Lake Rockwell. The dashed line at 45° represents consistent ages from both indicators. Note that the interpreted ages sit squarely across that line.

which historical age markers disappear at times much earlier than about 2000 B.C., then forces reliance of radionuclide age-dating methods on consistency between radionuclides of different half-lives. The argument here is that while different radionuclides may show different rates of scavenging to sedimentary particles, corresponding to different depositional fluxes and activities, nevertheless the disentanglement of nuclide profiles with depth into independent age-to-depth behaviors and time-dependent surficial flux behaviors, should lead to the same sedimentation rate profile no matter which radionuclide is used.

In a sense, then, one can use the age dates determined from one radionuclide as markers and require consistency of age dates from a second nuclide, or vice versa. In the best of all possible worlds, consistency should be automatic to within resolution.

Here we consider two different profiles, the first from North Carolina utilizes ^{210}Pb, together with direct knowledge of historical events (Hurricane Ginger in 1971), to disentangle the radionuclide profile. The second, from a core on the Galapagos microplate, uses ^{231}Pa and ^{230}Th profiles to determine a self-consistent age-to-depth for both profiles.

2.1. *Cape Lookout Bight, North Carolina*

The accumulation history in Cape Lookout Bight is complicated by storm deposits of sand and by the supply of silts and clays from multiple sources. Several ^{210}Pb profiles collected from a back-barrier site show exponential decreases in ^{210}Pb in the upper one meter of sediments but variable ^{210}Pb activities below one meter. Different models were employed to analyze different sections of the ^{210}Pb profiles, including the Simple Model and the Constant Specific Activity Model. To define the ^{210}Pb specific activity, Chanton et al. (1989): (i) separated sediment into three size fractions (silt, clay, sand); (ii) assumed that the measured ^{210}Pb activity of the silt-clay fraction had remained constant over the last forty years; and (iii) assumed that sand acted as an inert filler. Verification of the accumulation rates were derived from ^{137}Cs profiles, correlation of a sand layer to Hurricane Ginger, mineralogy on the sand layer, and bathymetric charts. In order to be precise, scuba divers tracked the burial of one sand layer over a period of four years.

To analyze the data profile using the inductive model, the sand component that was removed in the original investigation, was reintroduced giving a profile of uncorrected excess ^{210}Pb concentration with sediment depth. The re-introduction of the sand component resulted in lower excess ^{210}Pb activities in the intervals that contained significant quantities of sand. Three time markers were used to minimize the number of solutions obtained by inductive modelling. These time markers correspond to the initial appearance of ^{137}Cs in the sediments (1951 ± 3 yrs at 268 cm); the ^{137}Cs maximum (1963 ± 3 yrs at 150 cm); and the time of Hurricane Ginger (1971 at 90 cm) (Chanton et al., 1989). The model reproduces the subtle fluctuations in the excess ^{210}Pb data profile and misses only the abrupt fluctuation in activity at 100 cm (Fig. 3.3d). The resulting age-to-depth profile (Fig. 3.3e) satisfies all of the available time information for the three discrete depth horizons. The probability distribution for a surface ^{210}Pb activity is 10.6–11.0 dpm/g and average sediment accumulation rates of 9.2–11.9 cm/yr. With the exception of the very bottom sediment depths, the down-core variation in sediment accumulation rates ranges from 7–12 cm/yr (Fig. 3.3f). Sediment accumulation rates increase substantially to unrealistic rates at the bottom of the core (100's cm/yr) due to the abrupt decrease in excess ^{210}Pb and the lack of additional data at greater depths.

The sediment accumulation rates obtained by employing inductive and conceptual models are similar. The rates determined by conceptual modelling varied between 8.4–11.8 cm/yr with a mean rate of 10.3 ± 1.7 cm/yr as compared to 9.2–11.9 cm/yr determined by inductive modeling. However, using the inductive procedure, variations in sediment accumulation rates over discrete sections of the sediment core were obtained rather than averages with depth.

In addition, information on sediment grain size variations with depth, although extremely useful for the purpose of understanding the dynamics of deposition in this complicated coastal environment, was not critical for the determination of the age-to-depth relationship of the sediment layers by the inductive model, a major advantage in cutting down the amount of work needed to deduce age-to-depth conversions.

2.2. *Core KLH093, Galapagos Microplate*

A 290 cm core (KLH093) was collected in 3259 m water depth during an investigation of the southern part of the Galapagos Microplate at a distance of about 20 km from the active spreading centers (Frank et al., 1994). Core measurements on KLH093 were taken every 5 cm with the exception of the top 40 cm of the core, which was sampled every 2 cm. The measurements represent averages over the corresponding sampling interval. Both ^{230}Th and ^{231}Pa profiles with sediment depth were constructed (Frank et al., 1994) as exhibited in Fig. 3.4a and 3.4d, respectively.

Because ^{231}Pa (half-life = 32,400 yrs) has a much shorter half-life than ^{230}Th (half-life = 75,200 yrs), a greater fractional decay with time (and so depth) is expected for ^{231}Pa, as observed in Fig. 3.5 (a and d). The implication is that over the range with depth of ^{231}Pa where measurement resolution is not becoming a severe problem (i.e., at ^{231}Pa values greater than about 0.5 dpm/g). the rapid variation of ^{231}Pa should provide a sensitive age-to-depth measure.

Indeed, varying the average sedimentation rate from 2.5 through 2 to 1.5 cm/ky, and then performing a least squares fit of the inductive model to the ^{231}Pa profile (as shown by the overlying dashed lines in Fig. 3.5a) permits the construction of the age-to-depth curves shown in Fig. 3.5b. For average sedimentation rates outside the range 1.5–2.5 cm/ky the model fit to the ^{231}Pa profile with depth worsens considerably. The surficial flux of ^{231}Pa then has three extreme end-member values corresponding to 1.5, 2, and 2.5 cm/ky, respectively, as shown in Fig. 3.5c.

Thus, without more information it is not possible to refine further the age-to-depth conversion from ^{231}Pa shown on Fig. 3.5b. One could use the "extra" age markers from δ^{18}O and/or electron spin resonance (ESR) measurements, as given in Frank et al. (1994), in attempts to constrain the age-to-depth conversion. These points are shown in Fig. 3.5b together with nominal statistical error bars associated with calibration against the "standardized" oxygen isotope record of Martinson et al. (1987).

However, as noted by Frank et al. (1994), there is "an enhanced hydrothermal influence during isotope stages 4 and 5 and to a lesser extent during isotope stage 1 in core KLH093" and, in addition, "doubts on the age of the isotope stage transitions have arisen, especially termination II and older stages" (Winograd et al., 1992). In the case of ESR measurements, Mudelsee et al. (1992) have noted that "the results quoted are more or less at the stage of 'relative' dating attempts as the accuracy for a certain accumulated dose determination was very poor and linear extrapolations have been used or some parameters have been adapted to fit the known δ^{18}O ages".

But part of the problem can also arise because of major bioturbation effects of the first sample (δ^{18}O age of 13.6 ± 2.4 kyrs at 55 cm depth) (Frank et al., 1994), so that the true age of undisturbed sediments at 55–58 cm depth must be older than 16 kyrs. Indeed, the EST

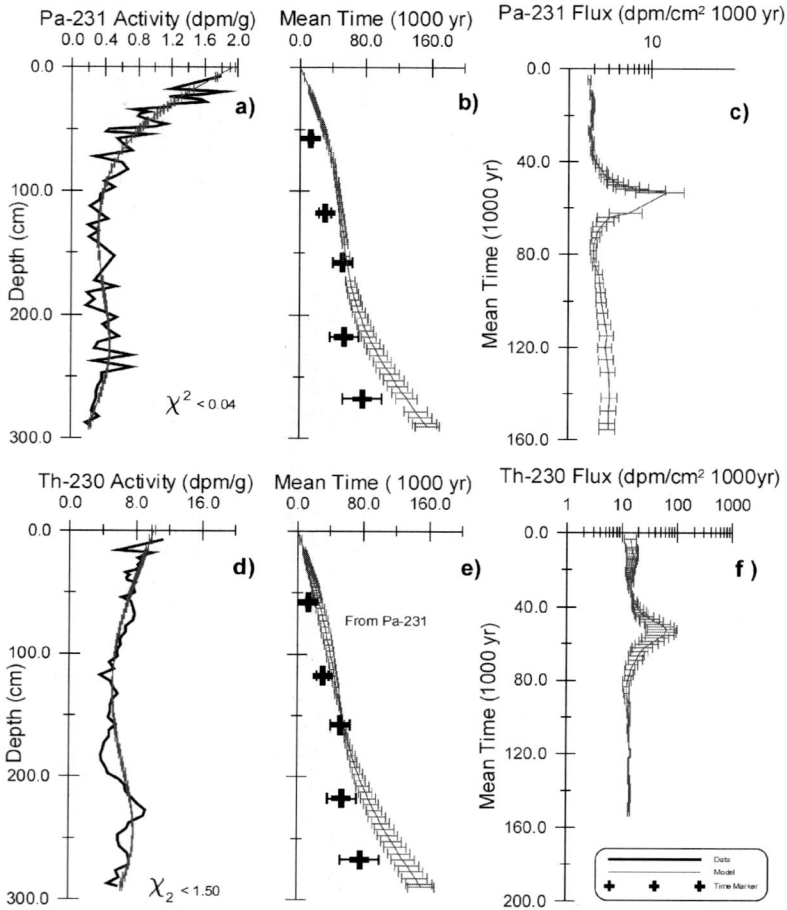

Fig. 3.4. (a–f) Galapagos Microplate, Site KLH093: (a) Measured ^{231}Pa (dpm/g) versus depth (m) (from Frank et al., 1994). The smooth curve represents SIT model $P(68)$ ^{213}Pa activities versus depth, with $P(10)$ and $P(90)$ ranges shown as error bars ($\chi^2 < 0.4$). (b) Probability distribution of age-to-depth variations consistent with the probability distribution in (a). The crosses represent ages determined by electron spin resonance (ESR) and δ^{18}O methods. These sediment ages were *not* used as time markers in the SIT model. (c) Probability distribution of ^{231}Pa fluxes with time (dpm cm^{-2} yr^{-1}). (d) Measured ^{230}Th (dpm/g) versus depth (m) and associated errors (from Frank et al., 1994). The smooth curve represents SIT model $P(68)$ ^{230}Th activities versus depth with $P(10)$ and $P(90)$ ranges shown as error bars ($\chi^2 < 1.5$). (e) Probability distribution of age-to-depth variations consistent with the probability distribution in (d). The crosses represent ages determined by ESR and δ^{18}O methods. These sediment ages were *not* used as time markers in the SIT model. The triangle represents 41 ± 3 ky at 152.5 cm from the ^{231}Pa age-to-depth relationship shown in (b). (f) Probability distribution of ^{230}Th fluxes with time (dpm cm^{-2} yr^{-1}); both ^{231}Pa and ^{230}Th fluxes increase between 50–60 kyr BP, but the uncertainty on the flux estimates is large.

maximum age is at least 17 kyrs at 55–58 cm, and may be considerably larger (up to 50 kyrs) than this nominal estimate (Mudelsee et al., 1992). The ^{231}Pa records a range of ages from 28 to 37 kyrs, depending on the value taken for average sedimentation rate between 1.5 to 2.5 cm/kyr.

(a)

(b)

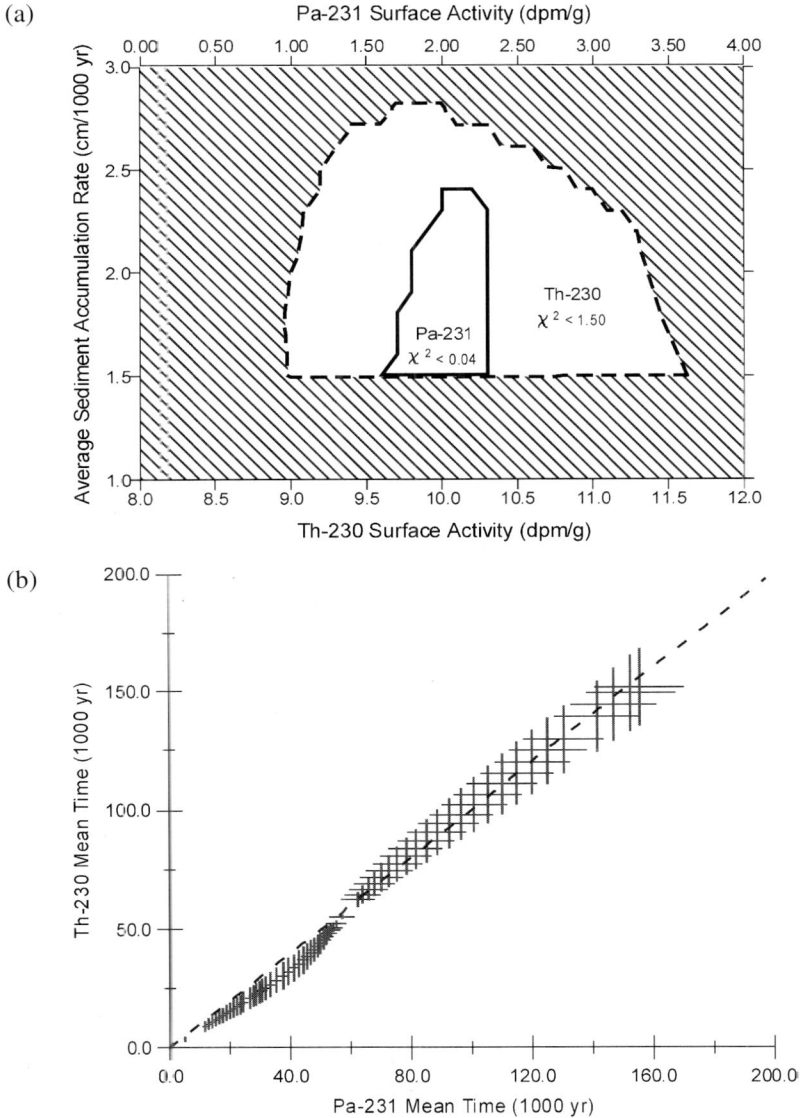

Fig. 3.5. (a) Intersection of ^{231}Pa and ^{230}Th average sediment accumulation rates for site KLH093 at the Galapagos Microplate (1.5–2.4 cm/kyr). The "flat barrier" at 1.5 cm/kyr occurs because sediment ages are constrained to be younger than 200,000 yrs (approximately six times the half-life of ^{231}Pa) in the SIT model. (b) Cross-plot of $P(68)$ ages obtained from Fig. 3.4 (b and e) of ^{231}Pa and ^{230}Th, respectively, with $P(10)$ and $P(90)$ ranges shown as error bars. The dashed line is one of $45°$ which passes through nearly all of the datum values.

For the ^{230}Th profile with depth, as shown in Fig. 3.4d, again the three cases of average sedimentation of 1.5, 2 and 2.5 cm/ky were used in conjunction with the inductive model, yielding the age-to-depth curves shown in Fig. 3.4e, with corresponding surficial flux variations given in Fig. 3.4f. In this case, because of the slower decay of ^{230}Th compared to ^{231}Pa, the variations of age-to-depth have a different pattern of behavior with depth (compare Fig. 3.4b and 3.4e) and can be arranged to encompass nearly all of the bottom four age markers from δ^{18}O and ESR measurements (to within $\pm 1\sigma$ of the nominal statistical errors on δ^{18}O and ESR).

The differences between age-to-depth curves derived solely from ^{231}Pa and ^{230}Th profiles, each taken on its own, can now be used in combination by demanding that an age-to-depth curve from both ^{231}Pa and ^{230}Th must be the same to within resolution. Thus age-to-depth markers from ^{231}Pa are used to constrain ^{230}Th ages and vice versa. The result is a combined curve for age-to-depth which is not only consistent with both ^{231}Pa and ^{230}Th, but in which the average sedimentation rate is also better constrained in order to be consistent with both profiles.

Plotted on Fig. 3.5a are the age-to-depth curves most consistent with both ^{230}Th and ^{231}Pa taken together, with ranges that would be allowed if an average sedimentation rate range of 1.5 to 2.5 cm/ky were to be invoked. Also plotted on Fig. 3.5a are the δ^{18}O ages and ESR ages with estimated ranges of error, showing the systematic lower ages (higher sedimentation rates) obtained by these methods (see Frank et al., 1994; Mudelsee et al., 1992).

In order to see if any systematic bias is present, at each depth the mean age from ^{230}Th versus that from ^{231}Pa was cross-plotted as shown in Fig. 3.5b, together with error bar ranges. Also drawn is a line of 45° slope, representing equal age determination. To within statistical uncertainty, the straight line passes squarely through all of the points used.

The result is that the age-to-depth conversion is consistent with an average sedimentation of 2 cm/kyr but with variations around that value, with a relative high in sedimentation rate from about 40 to 65 ky BP and relative lows to either side. Correspondingly, the surficial fluxes of ^{230}Th and ^{231}Pa both show increases over the same time period of 40–65 ky BP, as shown by the probability curves of Fig. 3.4 (c and f), corresponding to isotope stage 3 for which there is a diminution in scavenging of ^{230}Th relative to all other isotope stages (Frank et al., 1994).

There is uncertainty in the magnitude of the radionuclide fluxes due essentially to the *rate* of depth-to-age conversion, i.e., sedimentation rate. While it is the case that accumulated sedimentation with time is well determined, the derivative (sedimentation rate) is less well constrained.

The advantage of the inductive model in this case is that it enables rapid disentanglement of age-to-depth effects in the presence of both surficial flux and activity variations of radionuclides. Indeed, Frank (1994, private communication) has pointed out that "the flux of ^{230}Th into the sediments was probably not constant due to the hydrothermic influence. This, for example, prevents the calculation of a ^{230}Th constant flux model, which we usually apply to determine high resolution sedimentation rates in sediment cores".

The application here shows how both ^{230}Th and ^{231}Pa profiles are capable of being used in conjunction to determine a high resolution age-to-depth profile when both ^{230}Th and ^{231}Pa fluxes can vary with time. In addition, and in agreement with Frank et al. (1994), Winograd et al. (1992), and Mudelsee et al. (1992), the δ^{18}O and ESR ages do not agree with the more carefully crafted radionuclide ages, either because of bioturbation, intrinsic measurement uncertainty (dominantly ESR), or hydrothermal effects influencing the nuclide stage age determination against a standard curve (Martinson et al., 1987) that does not include such an effect.

3. Creel Bay and Main Bay, Devil's Lake, North Dakota

Alexander (1996) has supplied both excess ^{210}Pb and also ^{137}Cs data for the above two areas to see what can be determined using the SIT procedure in a blind test model, i.e., without further information being made available.

The results from both cores are indicative of the current strength of the SIT method and also provide suggestions for further needed enhancements.

3.1. *General observations*

As a general observation, the data were relatively straightforward to model with little difficulty being encountered for either core. The results are relatively robust because no time-markers or other restrictions were invoked. As a consequence, the model results are based strictly on the model fit to the data for minimum χ^2 values. The error bars reported on each figure are based on the cumulative probability distribution for each variable, with the upper (lower) error bar limits representing the 90% (10%) cumulative probability limits. The central datum marker point represents the 68% probability value for the variable.

For each core six figures are provided. The first figure provides contours of χ^2 values for combinations of model parameters for the surface excess ^{210}Pb activity value and the *average* sediment accumulation rate for the core. Basically the figure identifies which model parameter combinations fit the data within the minimum χ^2 contours. For the central values of the parameters above, the second figure of the sextuplet shows the superposed depth variations of observed excess ^{210}Pb and model predictions. In this way one can see immediately how well, or poorly, the model reproduces the data. Using the disentanglement aspect of the SIT model, the third figure of the sextuplet provides the relationship between current depth of a parcel of sediment and when the sediment was deposited, i.e., the age of the sediment. The error bars displayed are again based on the 10 and 90% cumulative probabilities and provide a measure of the uncertainty of age determination at each depth.

The disentanglement aspect of SIT is also used to generate the fourth figure of each sextuplet where the variation in sediment accumulation rate with time is shown; while the fifth figure presents the same information but with the sediment rate viewed with respect to sediment depth rather than time. Finally, the sixth figure of each sextuplet presents variations of the contaminant ^{137}Cs activity with time, showing the "spiking" around the early 1960s as a result of nuclear bomb testing.

3.2. *Creel Bay*

The excess ^{210}Pb profile with depth was extremely easy to model with SIT, basically because the profile is quite smooth with only minor data fluctuations. A one-term SIT model is more than adequate to fit the data to within a χ^2 value of less than about 0.1.

The results indicate that both the surficial ^{210}Pb and the average sediment rate are sharply confined (Fig. 3.6), and the model fit to the data is of extremely high quality (Fig. 3.7). The corresponding age versus depth profile (Fig. 3.8) is extremely well constrained at the sediment surface and gradually becomes less well constrained with increasing depth. Such a pattern is, by now, expected because the excess ^{210}Pb concentrations are extremely small at depth in the

J. Carroll, I. Lerche

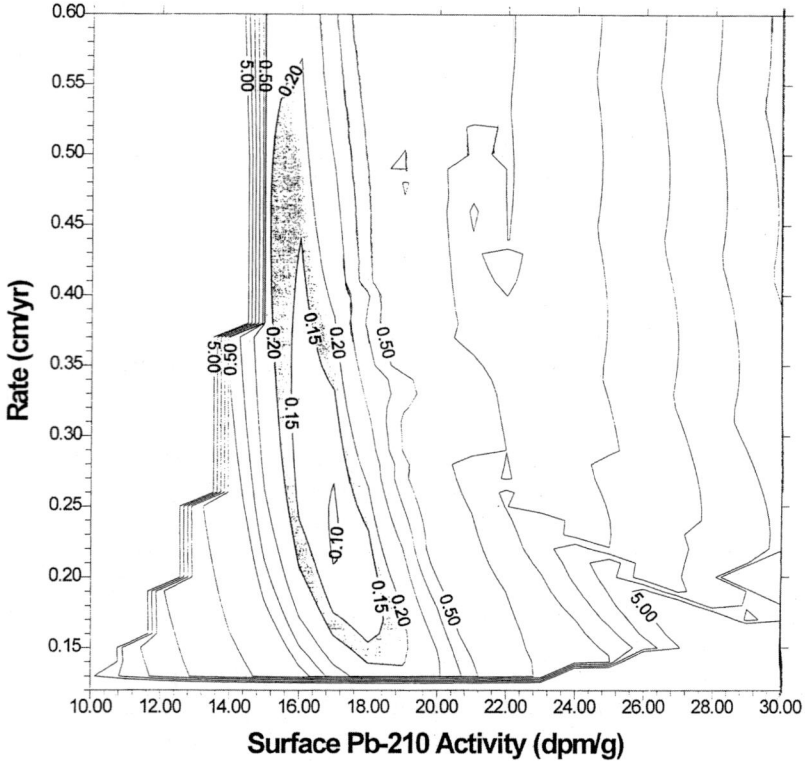

Fig. 3.6. Contours of χ^2 mismatch in a two-dimensional plot of surficial value for excess ^{210}Pb and average sediment accumulation rate for Creel Bay.

core and because the ages of the deeper sediments are dependent on the ages of the younger overlying sediments. The most probable ages for the deeper section of the core lie close to the 10% cumulative probability values, and so their uncertainty rests mainly on the determinations of the upper age limits for each depth interval. Based on the fit, it would appear that the bottom of the core is no older than 1914 but could be as young as 1930.

Considering the sediment accumulation rate, one recognizes that the rate is calculated for each depth interval, making for a more volatile behavior. When the sediment age does not change much from interval to interval, the rate determination will be very uncertain as a result of the small change in the sediment age with depth. In general, the results show a rate decrease with time, most rapidly until about 1960, and more gradually thereafter. The sedimentation rate decreases from about 0.45 cm/yr to about 0.15cm/yr (Figs 3.9 and 3.10). For the ^{137}Cs data (Fig. 3.11), there is a clear maximum extracted by the SIT method at about 1960, and the activity matches rather well with the model ages determined from the excess ^{210}Pb data. However, also observable are finite concentrations of ^{137}Cs activity extending further to the base of the core, indicative of diffusive or advective transport of ^{137}Cs downward, or indicative of the level of resolution of the age determinations.

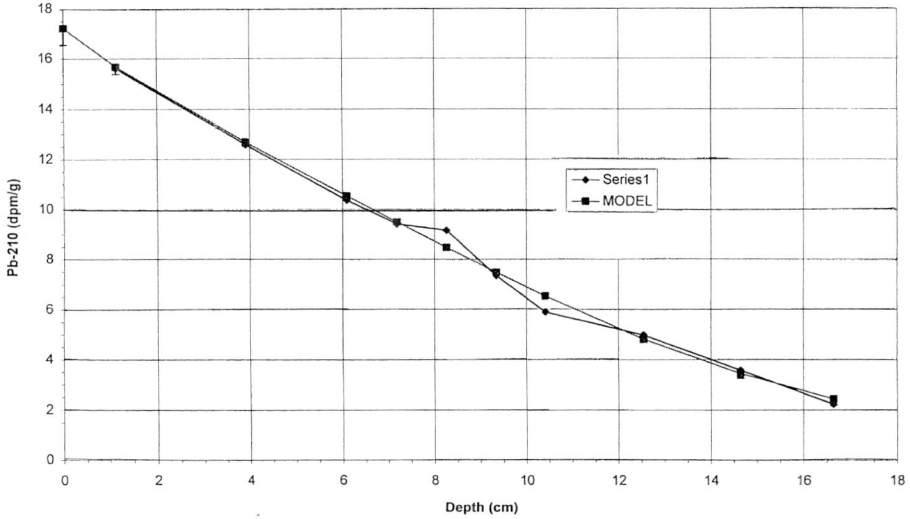

Fig. 3.7. Excess ^{210}Pb observed and SIT model predicted variations with depth for Creel Bay.

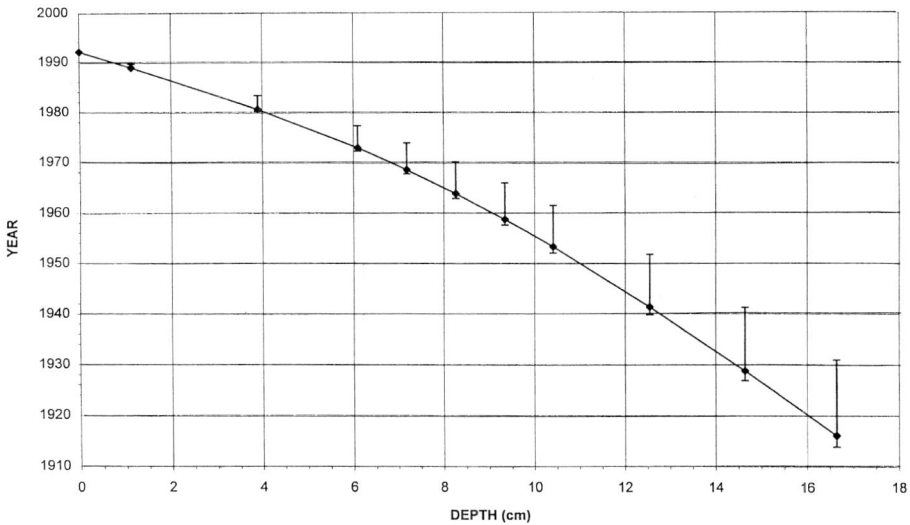

Fig. 3.8. Age to depth conversion deduced from the SIT model for Creel Bay ^{210}Pb. Error bars indicate uncertainty (see text for definition).

3.3. *Main Bay*

Of the radionuclide data provided, the Main Bay data required a more complex evaluation than just the two-term SIT model used for Creel Bay, basically caused by the fluctuations in the surficial value for excess ^{210}Pb. Precisely how the fluctuations are interpreted by the SIT

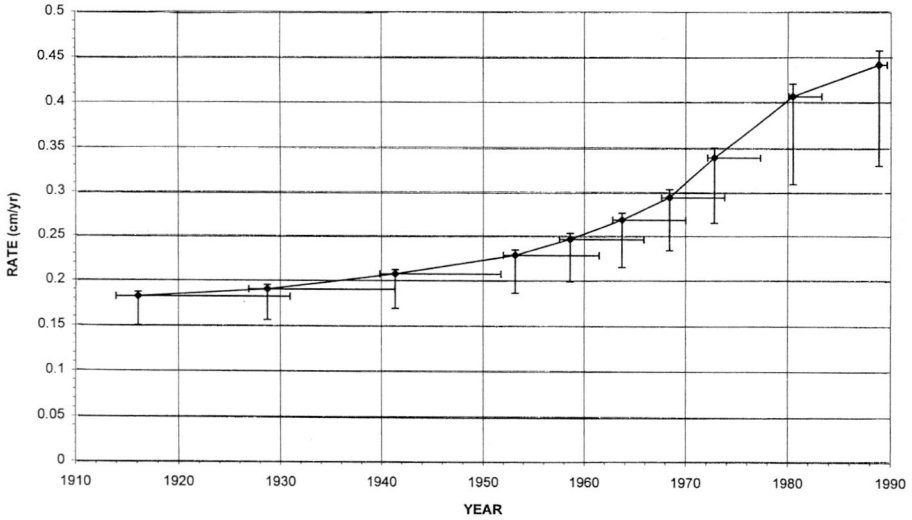

Fig. 3.9. Sediment accumulation rate with time deduced from ^{210}Pb data, Creel Bay.

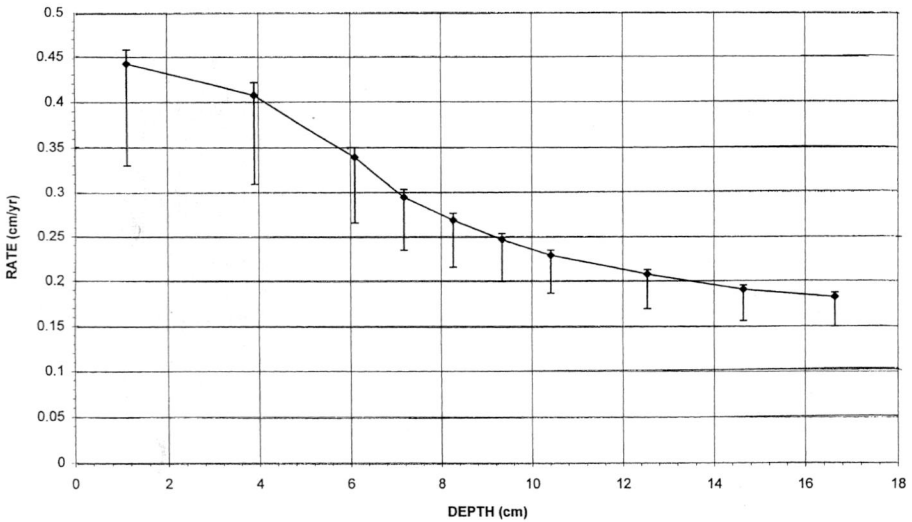

Fig. 3.10. As for Fig. 3.9 but with depth replacing time, Creel Bay data.

model has major implications for the rest of the core due to the dependence of sediment ages with depth.

Accordingly, the uncertainty in the average sediment accumulation rate is larger for Main Bay than for Creel Bay (Fig. 3.12). The overall fit to the data is still quite good (a χ^2 value of less than 0.45) but not as good as for Creel Bay. The SIT model reproduces the data trend with depth in an acceptable manner (Fig. 3.13). Near the surface, the model suggest that

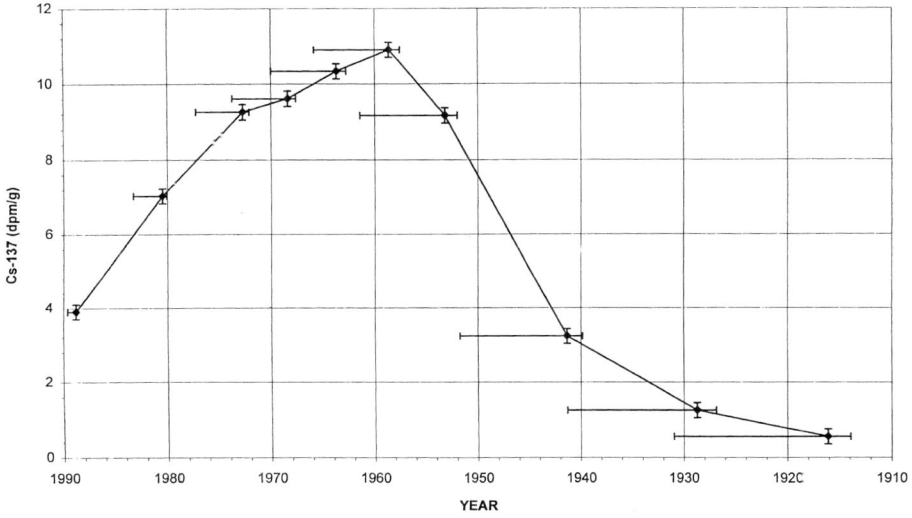

Fig. 3.11. Activity of ^{137}Cs versus time for Creel Bay data.

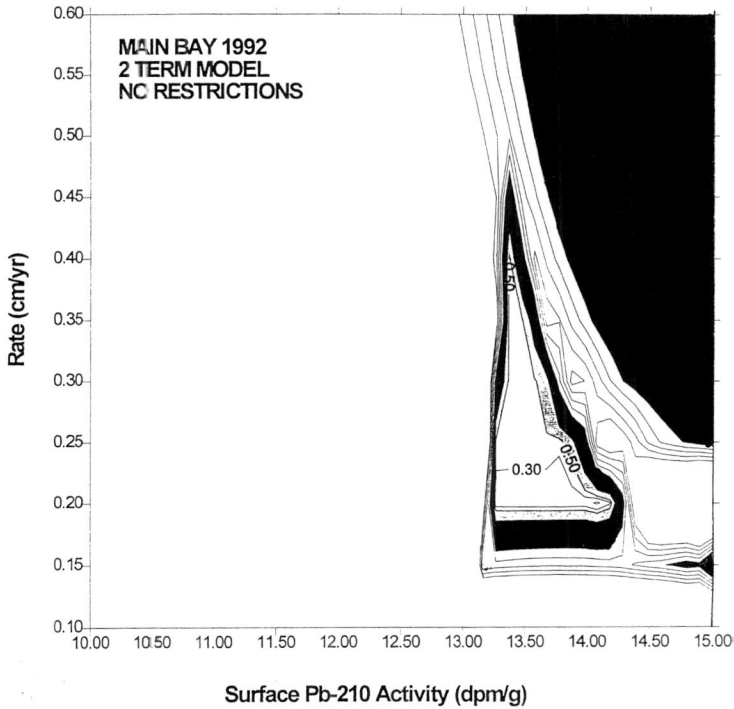

Fig. 3.12. As for Fig. 3.6 but for data from Main Bay.

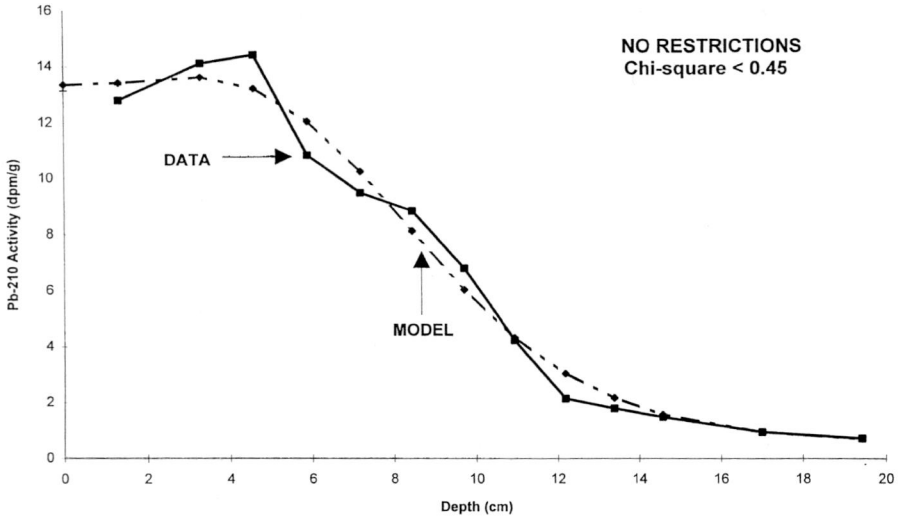

Fig. 3.13. As for Fig. 3.7 but for data from Main Bay.

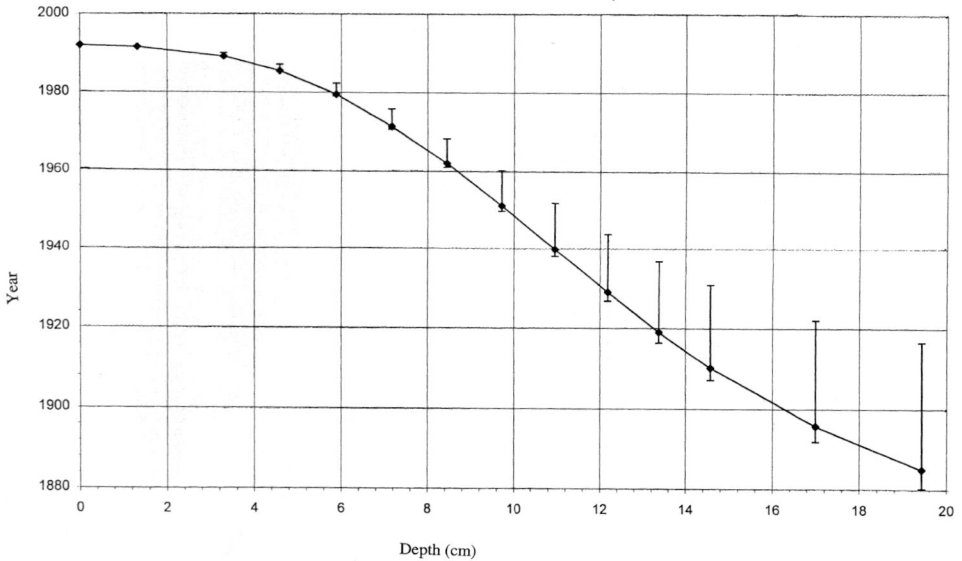

Fig. 3.14. As for Fig. 3.8 but for data from Main Bay.

little change occurred in excess ^{210}Pb, which is indicative of rapid sediment accumulation (perhaps a dumping or mixing event). As was the case for Creel Bay, the ages of recent sediments are well constrained with the ages of deeper sediments progressively less well-constrained (Fig. 3.14). The nominal sediment accumulation rate in the very first interval below the sediment surface has the formal value 200^{+24}_{-160} cm/yr, which is a direct consequence

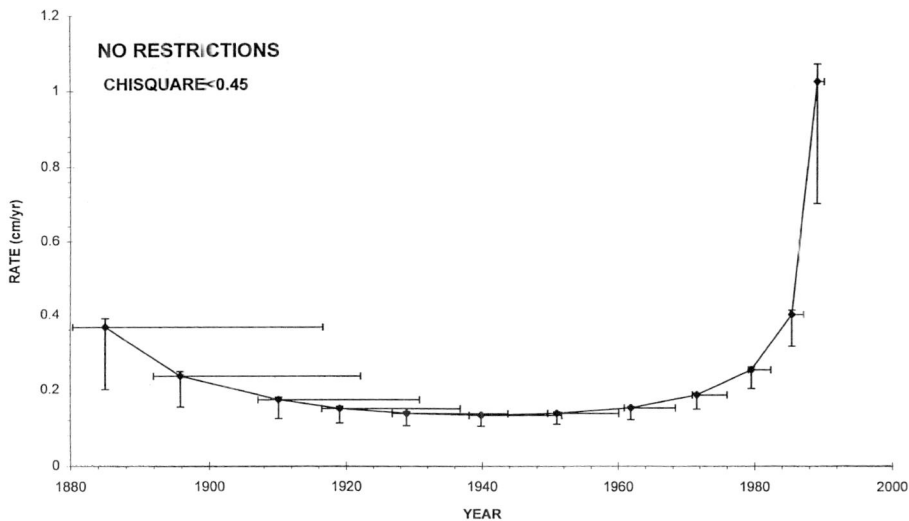

Fig. 3.15. As for Fig. 3.9 but for data from Main Bay.

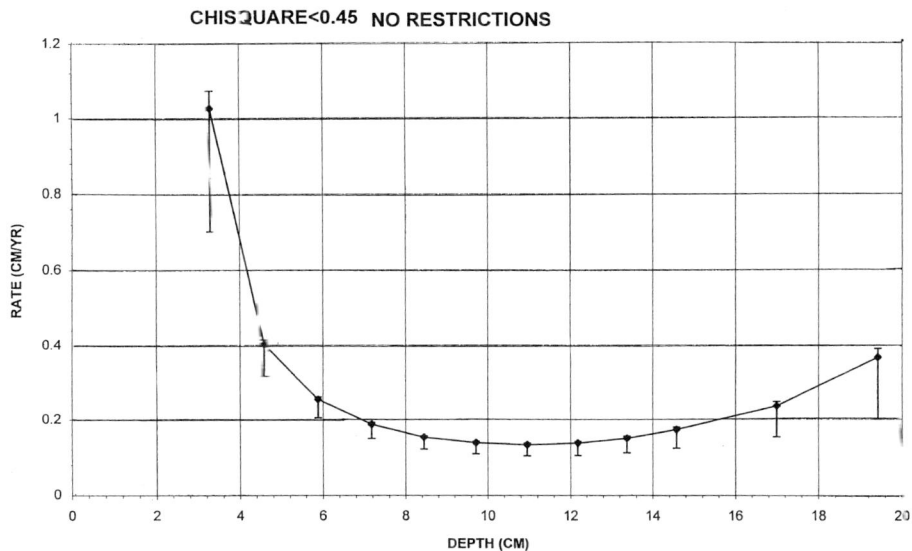

Fig. 3.16. As for Fig. 3.10 but for data from Main Bay.

of little change in the first interval (Figs 3.15 and 3.16) and also a consequence of the large uncertainty in the surficial value for excess ^{210}Pb. Once below that first interval, however, the sediment accumulation rates promptly revert to "rational" values of less than about 0.2 cm/yr, indicating that the uncertainty is confined to the nearest surface layer. There is the possibility of disturbance in the surficial layers providing the anomaly in the first interval.

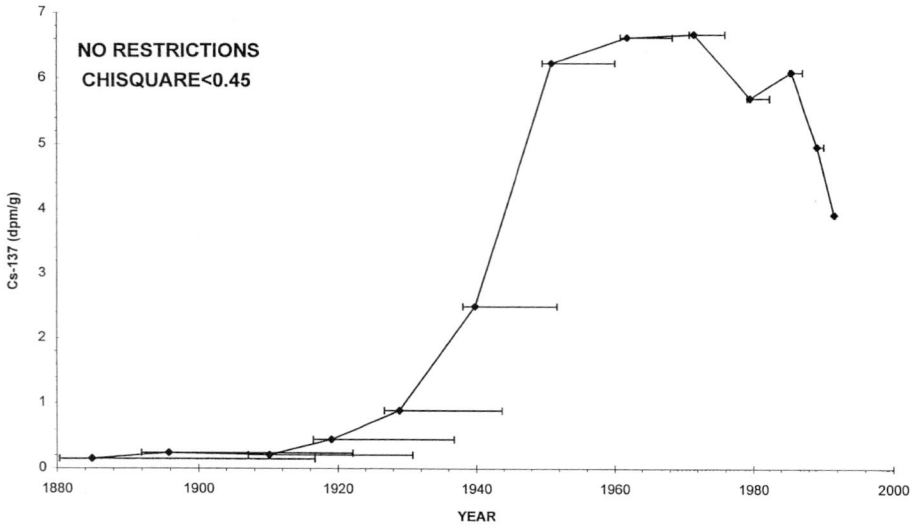

Fig. 3.17. As for Fig. 3.11 but for data from Main Bay.

The age at the base of the core is only a little older than that at Creel Bay, ranging from 1880 through to 1917, corresponding perhaps to the fact that the Main Bay core extends slightly deeper into the sediments than that from Creel Bay.

The ^{137}Cs profile (Fig. 3.17) is more difficult to interpret for the Main Bay data, basically because there is no sharply delineated contamination peak, but the highest values of ^{137}Cs activity do correspond to the early 1960s according to the model determined depth to age conversion. Again, however, Cs activity is found deeper in the core than would have been predicted based solely on nuclear bomb effects, suggesting that diffusion/advection are again acting as transporting processes, just as for Creel Bay.

3.4. *Summary*

The results from the two cores complement each other quite well. There is the same general trend observed in both with site to site variations in sedimentation processes. In addition, despite *not* using any time markers to constrain the system, nevertheless the resulting ^{137}Cs profiles peak *precisely* at the time of maximum nuclear bomb activity, with the time determined solely from the excess ^{210}Pb disentanglement and not using the ^{137}Cs. This fact provides strong underpinning support for the accuracy of the SIT procedure.

4. Discussion and conclusions

The purpose of this chapter has been to demonstrate how signal theory approaches are used to interpret single site measurements of radionuclides in a variety of diverse aquatic systems. Sedimentation histories and flux effects with time were recovered for the systems and compared with results based on applying classic conceptual model methodologies. For the cases of all the lake situations it was demonstrated that the inductive model procedure was able to accurately unscramble effects in agreement with known historical events. In the case of a shallow marine core. it was shown that inductive modelling can rather easily confirm direct measurements in a very short period of time and with less need for massive scientific infrastructure. In the case of the deep sea core, the use of two different radionuclide profiles (^{230}Th and ^{231}Pa) with depth, enabled both to be used in conjunction to maximal advantage in extracting a consistent age-to-depth behavior as well as surficial flux variations with time. These illustrative examples demonstrate the variety of conditions under which the assumptions of inductive models are not violated. When applying inductive models both surficial flux and sedimentation rates can vary with time and single or multiple radionuclide profiles can be used to disentangle sedimentation and flux effects, either in the presence and/or in the absence of historical age markers.

Appendix A. Resolution and uncertainty for disentanglement procedures

This appendix discusses the procedures used for assessing resolution and uncertainty on results of disentanglement of sedimentation and source effects from radionuclide profiles with depth using the SIT model.

As expressed in detail earlier, radionuclide profile, $P(x)$, with sediment depth, x, is represented by the model dependence

$$P(x) = P_0 \exp\left[-T(x) + S(x)\right], \tag{A.1}$$

with

$$T(x) = Bx + \sum_{n=1}^{N} \frac{a_n}{n\pi} \sin(n\pi x/x_m), \tag{A.2}$$

$$S(x) = \sum_{n=1}^{N} \frac{b_n}{n\pi}\left[1 - \cos(n\pi x/x_m)\right], \tag{A.3}$$

where x_m is the greatest depth of measurement; $B = \lambda/V$, where λ is the decay rate of the radionuclide and V is the average sedimentation velocity between $0 \leqslant x \leqslant x_m$; and where P_0 is the present day surface value of the nuclide. The term $T(x)$ represents time-to-depth behavior and the term $\exp(S(x))$ accounts for the temporal flux behavior. The coefficients $\{a_n\}$ and $\{b_n\}$ are uniquely determined by matching the predicted behavior (A.1) to observations once P_0 and V are specified for a given number of terms N. The factor $T(x)$ is always positive and increasing with increasing depth, representing greater ages at greater depths.

As P_0 and V are both varied, account is kept of the mean square mismatch (MSR) between predicted and observed radionuclide profiles with depth so that a contour plot of MSR in the two-parameter space of P_0 and V provides the range of values most consistent (lowest MSR) with the observations. Figures 3.18 and 3.19 present contours of MSR for the ^{210}Pb data profiles from Lake Nômmejärv, Lake Kurtna, Lake Rockwell Reservoir, and Cape Lookout Bight, respectively. Figure 3.20 gives the contours of MSR for site KLH093, Galapagos

Lake Nômmejärv

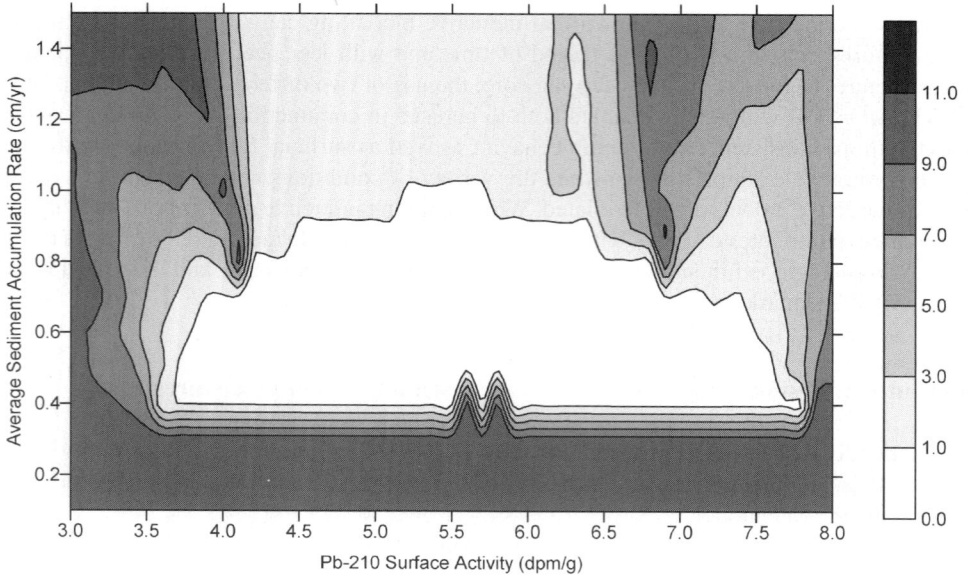

Lake Kurtna

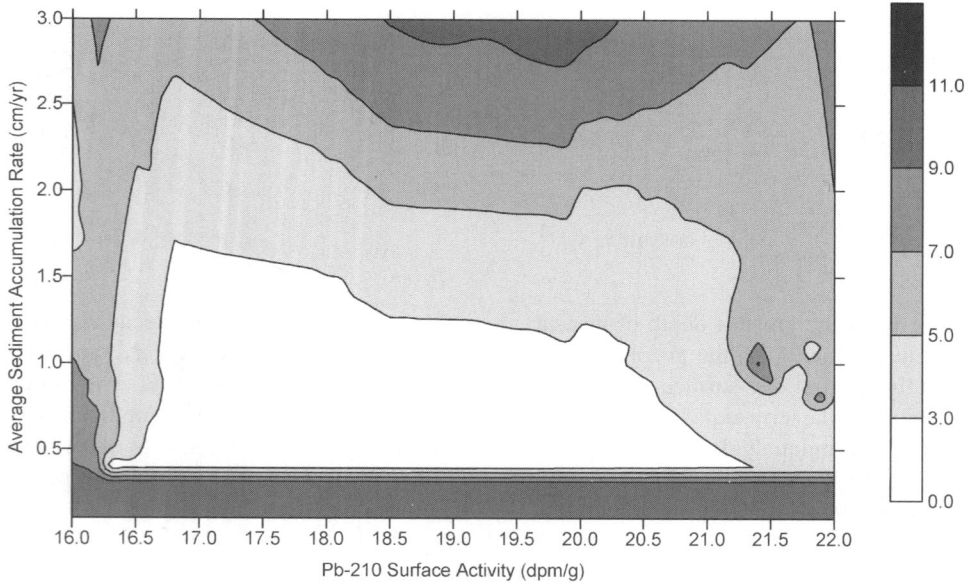

Fig. 3.18. Plot of contours of constant MSR as average sediment velocity, V, and the surface ^{210}Pb radionuclide value, P_0 (dpm/g) are varied, for the data from Lake Nômmejärv (top) and Lake Kurtna (bottom). No age markers were used to constrain the ranges for V or P_0 but the maximum age of the sediment core must be less than 134 yrs (six times the half-life of ^{210}Pb).

Lake Rockwell

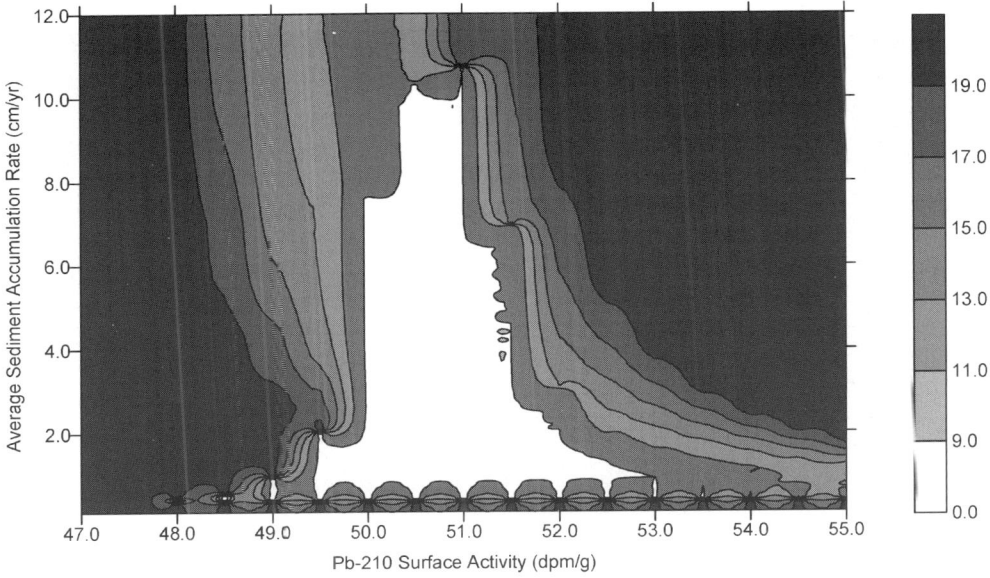

Cape Lookout Bight

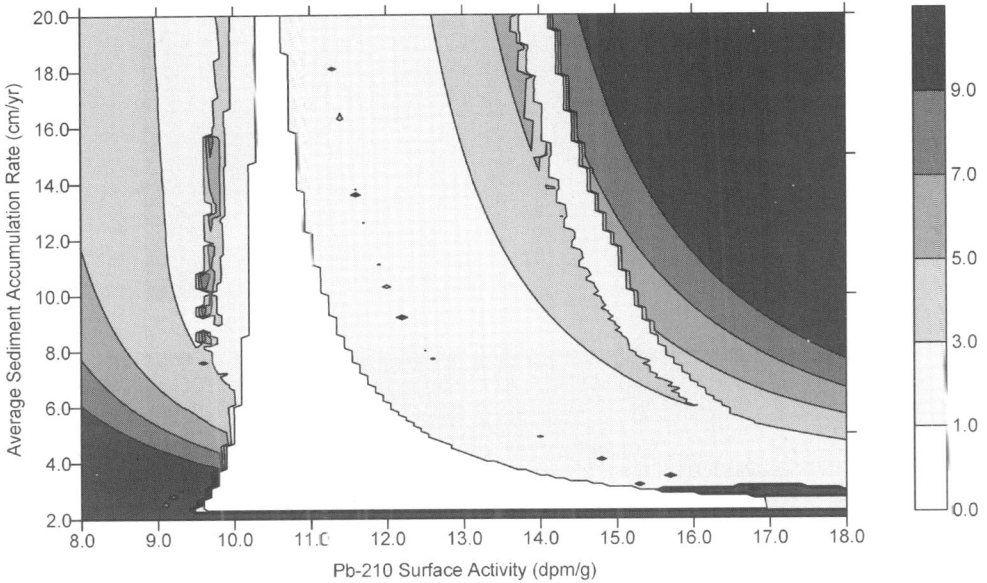

Fig. 3.19. As for Fig. 3.18 but with ^{210}Pb data for Lake Rockwell (top) and Cape Lookout Bight (bottom).

Galapagos Microplate

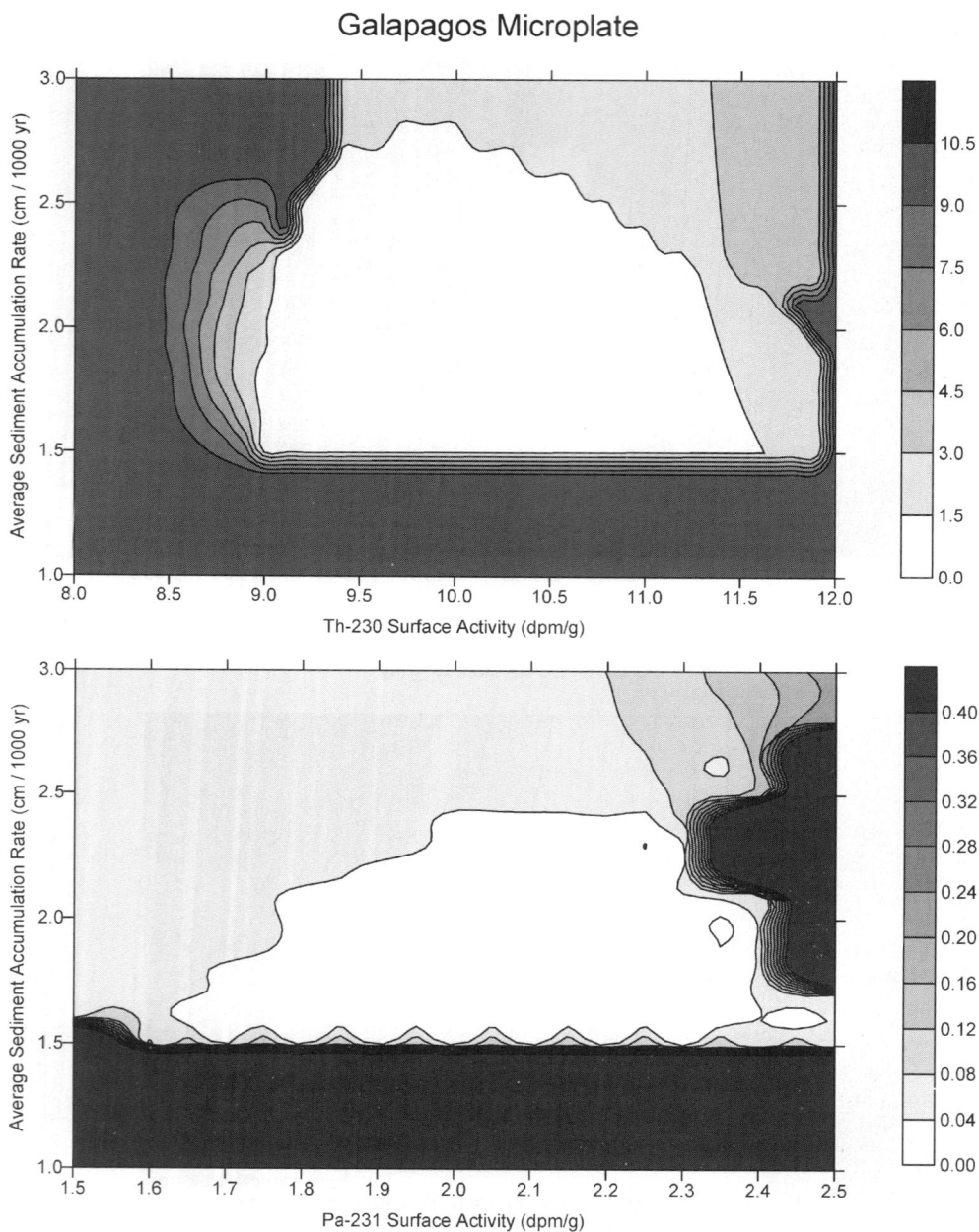

Fig. 3.20. Plot of contours of constant MSR as average sediment velocity, V, and surface ^{230}Th (top panel) and surface ^{231}Pa (bottom panel) vary (dpm/g) for the Galapogos case studied in text.

Microplate for ^{231}Pa and ^{230}Th data profiles, respectively, with each profile considered to be independent of the other, while Fig. 3.6 (text) presents the overlay of MSR contours for ^{231}Pa and ^{230}Th from site KLH093, so that a common average sedimentation velocity, consistent with both profiles, can be extracted. In general, there are finite ranges in both V and P_0 values which will provide equally acceptable fits to the observations, in the sense of providing an MSR that is within the resolution of the quality, quantity and sampling frequency distribution of the data available.

The minimum, maximum and mid-point values of the parameters P_0 and V are then determined satisfying the lowest MSR criterion. Curves of age-to-depth can then readily be drawn for each choice of V(min), V(max) and V(mid-point); each choice yields different coefficient values for the sets $\{a_n\}$ and $\{b_n\}$, thereby providing a non-uniform age-to-depth conversion, e.g., doubling V does not halve ages at a given depth. For each of the three values of V, there are three corresponding ages at a given depth T(min), T(max) and T(mid-point) but these ages are non-uniformly distributed and, in general, T(mid-point) $\neq \frac{1}{2}(T(\text{min}) + T(\text{max}))$, not does it have to be the case that T(min) corresponds to V(min) at a given depth. But the ages obtained do characterise what corresponds to acceptable fits (lowest risk) to the radionuclide with depth profile.

From the three ages at each depth, Simpson's rule is used to calculate an expected average age $E_1(T)$, and the variance, $\sigma(T)^2$, around the average from

$$E_1(T) = \frac{1}{3}\left(T(\text{min}) + T(\text{max}) + T(\text{mid-point})\right) \tag{A.4}$$

and

$$
\begin{aligned}
\sigma(T)^2 &= \left(E_2(T) - E_1(T)^2\right) \\
&= \frac{1}{18}\left\{\left[T(\text{mid-point}) - \frac{1}{2}(T(\text{max}) + T(\text{min}))\right]^2 \right. \\
&\quad \left. + \frac{3}{4}(T(\text{max}) - T(\text{min}))^2\right\}.
\end{aligned}
\tag{A.5}
$$

For a distribution of ages which is cumulatively log-normally distributed or approximately so (Feller, 1957), with $\mu^2 = \ln(E_2(T)/E_1(T)^2)$, the cumulative probability of 68% ($P(68)$) occurs at about $E_1(T)$, that of 50% ($P(50)$) at $E_{50} \equiv E_1(T)\exp(-\mu^2/2)$, that of 16% ($P(16)$) at $E_{50}(T)\exp(-\mu) < E_1(T)$, and that of 84% ($P(84)$) at $E_{50}(T)\exp(\mu) > E_{50}(T)$.

On log probability paper, a plot can then be made and the $P(10)$ and $P(90)$ values read off (Fig. 3.21).

The age distribution at each depth is then taken to be represented by the expected average value at $P(68)$ viz. $E_1(T)$, with uncertainty around the $P(68)$ value described through the $P(10)$ and $P(90)$ values; and precisely this sort of error uncertainty is plotted on the age-to-depth curves given in the text for each case history.

In addition, for each of the values of V and P_0, the temporal variation of radionuclide flux can be plotted and, at each time (or depth), a distribution of nuclide fluxes can be drawn. Following similar arguments to those above for the age-distribution, mean expected surface

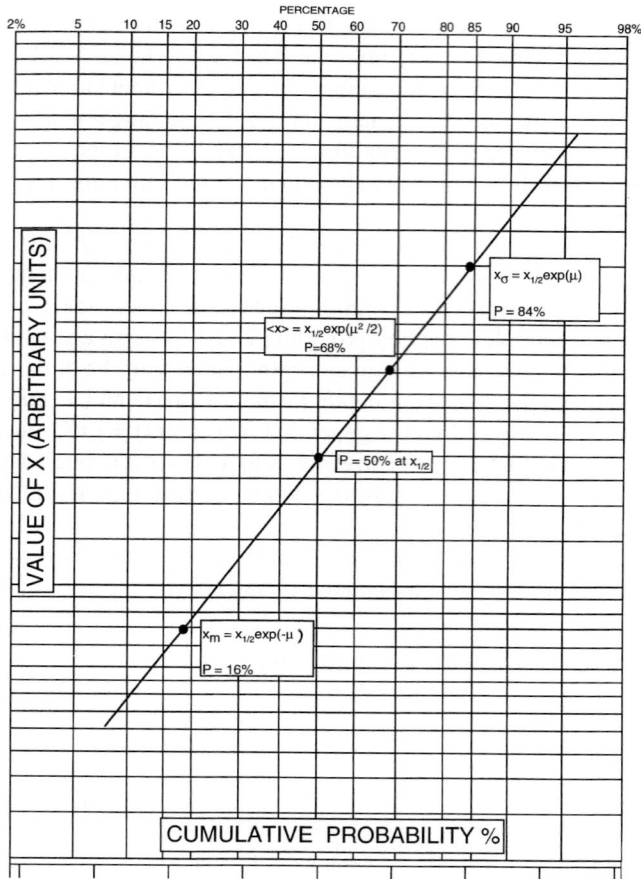

Fig. 3.21. Cumulative lognormal probability distribution with special points marked.

flux values with time at the $P(68)$ cumulative probability value, as well as uncertainties at the $P(10)$ and $P(90)$ cumulative probability values, can be plotted to represent measures of the uncertainty. It is these measures that are drawn on the figures for depositional flux with time given in the text for each case history.

Chapter 4

Multiple site studies

While radionuclides are widely used in geological studies, predominantly to ascribe ages to strata and to provide an assessment of sedimentation rates, the difficulties in handling lateral variations of sediment transport, deposition, erosion and total accumulation are well-known.

The aim of the present chapter is to use the SIT procedure to disentangle effects due to sediment variations with time (or depth) from intrinsic source variations with time (or depth) using observed radionuclide variations with depth across a basin.

Basically, without a suite of sedimentary cores taken at different locations in a geological area of interest, it is not possible to determine to any significant resolution precisely how sediments are transported.

Fortunately, in several cases of significant geological interest, multiple sediment cores have been taken from several wells so that it is possible to provide some idea, from radionuclide measurements in the core material, of lateral and vertical variations with time of sedimentary accumulation.

Three such cases are considered here to illustrate not only the diversity of behaviors describable, but also how extraneous information can be used to provide qualitative and semi-quantitative limits on patterns of behavior, both vertically and laterally. Consider each case in turn.

1. Amazon Delta test case

1.1. *Background*

The Amazon River is the world's largest river. There have been many studies of aspects of sediment transport and of deposition and accumulation processes on the Amazon's broad deltaic environment (Kuehl et al., 1986; Gibbs, 1976; Meade et al., 1985; Nittrouer et al., 1986; DeMaster et al., 1986). Seismic reflection profiles reveal that the inner shelf mud deposit (< 70 m water depth) is a sub-aqueous delta (Figueiredo et al., 1972; Nittrouer et al., 1986). The delta contains topset, foreset, and bottomset strata characteristic of classic river deltas.

The Amazon River sediment discharge is estimated to be $1.1 \pm 1 \times 10^{12}$ kg yr^{-1} (Meade et al., 1985) and is composed of about 85–95% silt and clay-sized sediment (Gibbs, 1967; Meade, 1985). The transport of sediments is under the influence of several, roughly equally dominant, agents such as surface waves, tidal currents and the North Brazilian Coast Current (Nittrouer et al., 1986), and a northwestward trend for sediment transport is inferred.

Previous studies showed that radionuclide profiles vary even within a very small horizontal distance (Kuehl et al., 1986). Kuehl et al. (1986) describe 4 types of ^{210}Pb sediment profiles observed on the Amazon shelf:

(1) a moderate accumulation profile (≈ 1 cm/yr) which typically exhibits a zone of constant ^{210}Pb followed by a zone of exponential decay and a zone of constant ^{210}Pb;
(2) a rapid accumulation profile (≈ 10 cm/yr) showing a slow decrease in ^{210}Pb with depth;
(3) a profile of variable ^{210}Pb activity throughout the length of the core; and
(4) no excess ^{210}Pb present.

Profile type 3 is a common feature in rapidly depositing deltaic sedimentary environments, e.g., the Ganges River Delta (Kuehl et al., 1989), and it is the most difficult type from which to determine ^{210}Pb ages with depth. Techniques that have been designed are based on the premise that changes in ^{210}Pb content are due to changes in the supply of ^{210}Pb to the seafloor; for example, as a result of changes in the specific activity of sediments (Robbins, 1978; Appleby & Oldfield, 1978) or as a result of sediment grain size variations (Chanton et al., 1989).

Even this constraint proves limiting in areas such as the Amazon Delta. The advantage of the SIT technique is that no a priori assumptions need be made about the importance or causes of potential changes in specific activity, and we do not assume sedimentation rates are constant through time.

Five examples of profile type 3 from the Amazon shelf are given in Fig. 4.2 (a–e) (Kuehl, 1985). Core collection positions are shown on Fig. 4.1 (Kuehl et al., 1986). All data sets show a mixing property of decay and source variation reflected by the highly variable distributions of excess ^{210}Pb activities with depth. Kuehl (1985) was unable to determine sedimentation rates in these locations because of variable ^{210}Pb distributions. Bioturbation effects were not observed in the cores (Kuehl et al., 1986) and are not determined by the procedure. If a separate kinetic model of bioturbation were to be developed it could be incorporated within the SIT technique exploited here, but this point is not addressed any further in this chapter.

1.2. *Data and predictions*

1.2.1. *Measurements and predictions*
The comparison of measurements and predictions of excess ^{210}Pb activity (dpm/g) are shown in Fig. 4.3 (a–e), respectively. The predicted variations of excess ^{210}Pb activity (dpm/gm) with depth catch the varying trend of the measurements quite accurately in each data set. Taking a measurement error to be 5–10% of the radionuclide activity, the predictions and measurements are almost identical for most of the cases.

1.2.2. *Sensitivity analysis*
To illustrate the stability of the model and the sensitivity of the parameters, we also used the linear search method on several of the data sets. Figures 4.4 (a–d), 4.5 (a–d) present the MSR (root mean square residual) vs. P_0, B, N, and w, respectively, for the data sets. The analyses reveal that (1) the resolution for parameter determination is relatively stable within the search range for all four of the parameters; (2) different parameters have different sensitivities in the modeling, as can be seen from the shapes and scales on the individual figures.

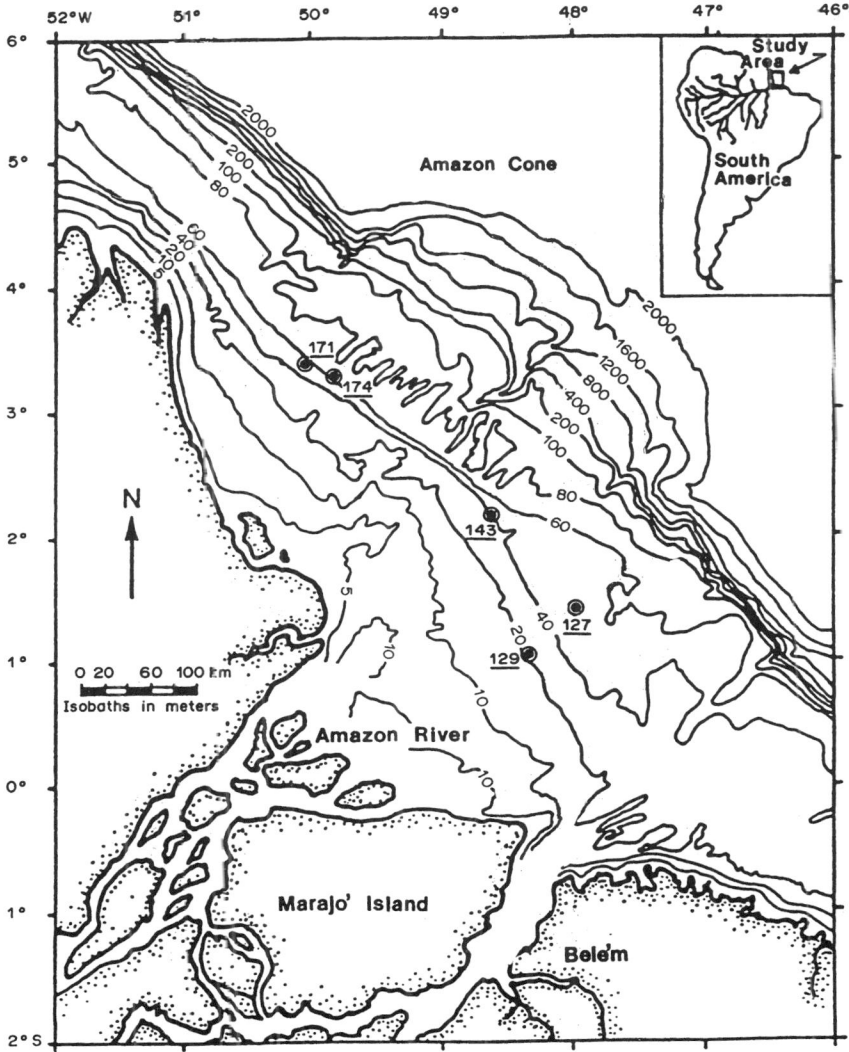

Fig. 4.1. The outskirts of the Amazon basin. Sampling positions are also located on the map (Kuehl et al., 1986).

1.2.3. *Separating decay and source variations*

Using the detailed procedure given in Chapter 2, Fig. 4.6 (a–e) presents the source variation vs. depth, while Fig. 4.7 (a–e) shows the source variation vs. time, for each of the five data sets. In order to compare the relative influence of radioactive decay and source variation on the excess ^{210}Pb activity we superimposed the radioactive decay vs. depth and vs. time. By comparing the decay curve and source curve of a data set, the relative contributions were obtained for the two parts to the excess ^{210}Pb activity with depth/time. It is also noticeable that for these data sets, even though some samples are collected in close proximity, the excess ^{210}Pb activity

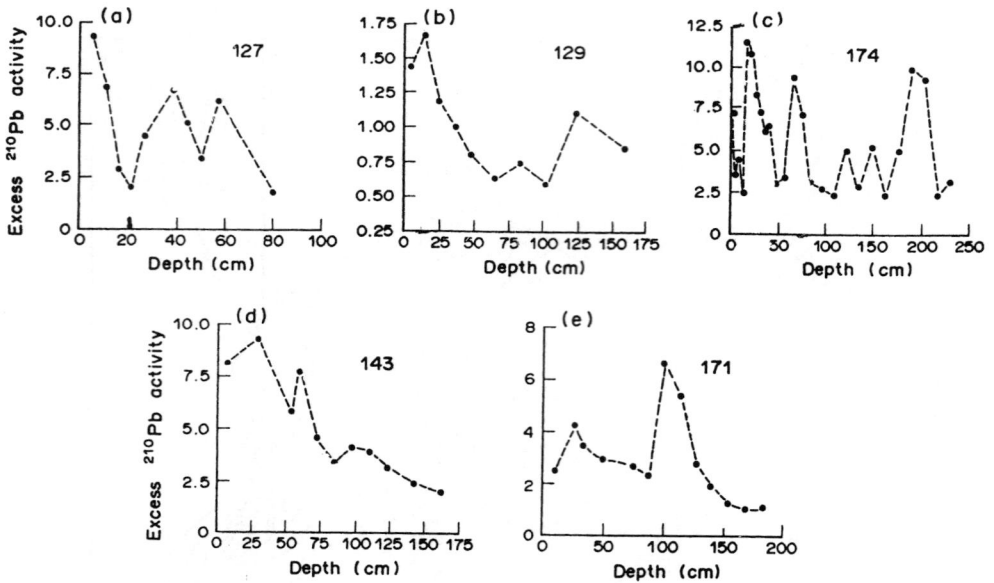

Fig. 4.2. Excess ^{210}Pb activity (dpm/g) profiles collected from the Amazon basin from five coring locations: (a) KC127; (b) KC129; (c) KC174; (d) KC143, and (e) KC171.

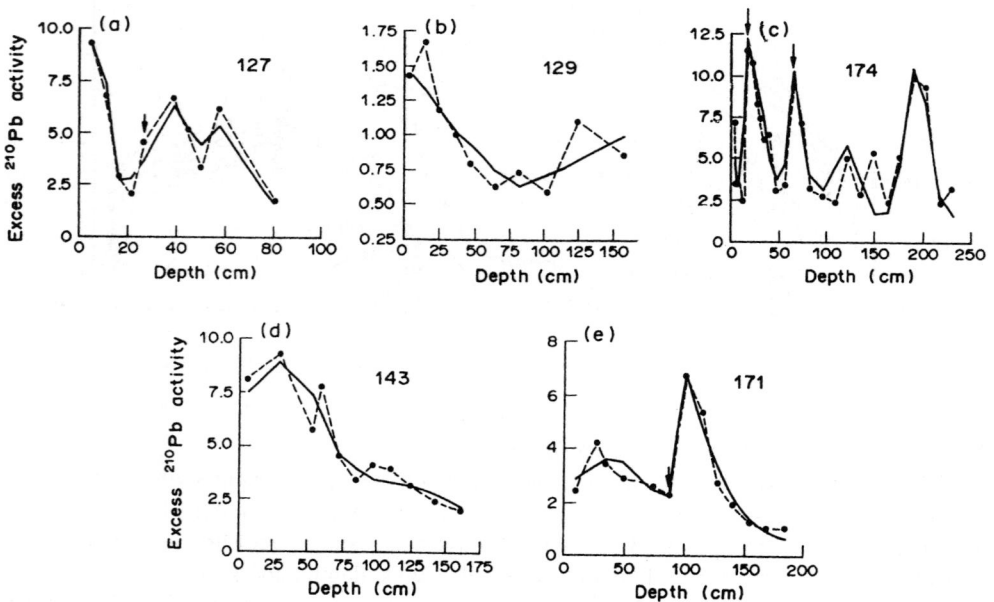

Fig. 4.3. Comparison of measured and predicted excess ^{210}Pb activities (dpm/g) for each data set allowing for a 5–10% measurement error. Vertical arrows mark partitioning points as explained in Chapter 2.

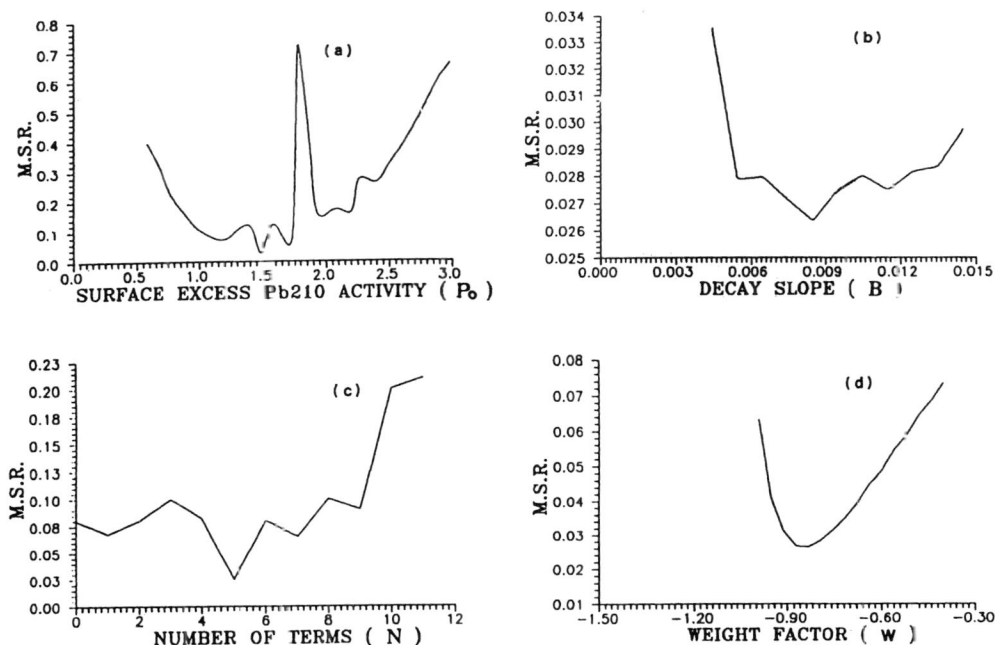

Fig. 4.4. Sensitivity analysis results for data set KC129: (a) MSR vs. surface excess ^{210}Pb activity (dpm/g); (b) MSR vs. radioactive decay slope (B); (c) MSR vs. number of terms (N); (d) MSR vs. weight factor (w). Comparing the relative changes of MSR when varying each of the parameters individually provides an assessment of the parameter resolution. The minimum MSR values give the parameter values used in the predicted curves of Fig. 4.3.

profiles, (and thus the source variation and decay for these data sets) are quite different. The implications are that: (a) the sediment accumulation processes in deltaic environments are very complex and vary on the smallest spatial sampling scale; (b) the nuclide concentrations are not evenly distributed spatially. After separating the contributions of radioactive decay and source variation in the excess ^{210}Pb profile, it is possible to determine quantitatively the time to depth conversion, as well as the sediment accumulation rate variation, in the sedimentary series represented by each data set.

1.2.4. *Time to depth conversion*

Figure 4.8 (a–e) shows the age variation of sediments using the predicted excess ^{210}Pb activity profile. Superimposing the five data curves on one plot (Fig. 4.9) permits a quick visual comparison between the data sets.

Except for samples from KC127, which has a low slope of time vs. depth, all the other data sets (KC129, KC143, KC171, KC174) show relatively similar slopes. For most, the age vs. depth curves are relatively smooth. However, a noticeable change is seen in KC174 at approx. 75 cm depth, where the slope has a significant shift, indicating changes in the sedimentation process.

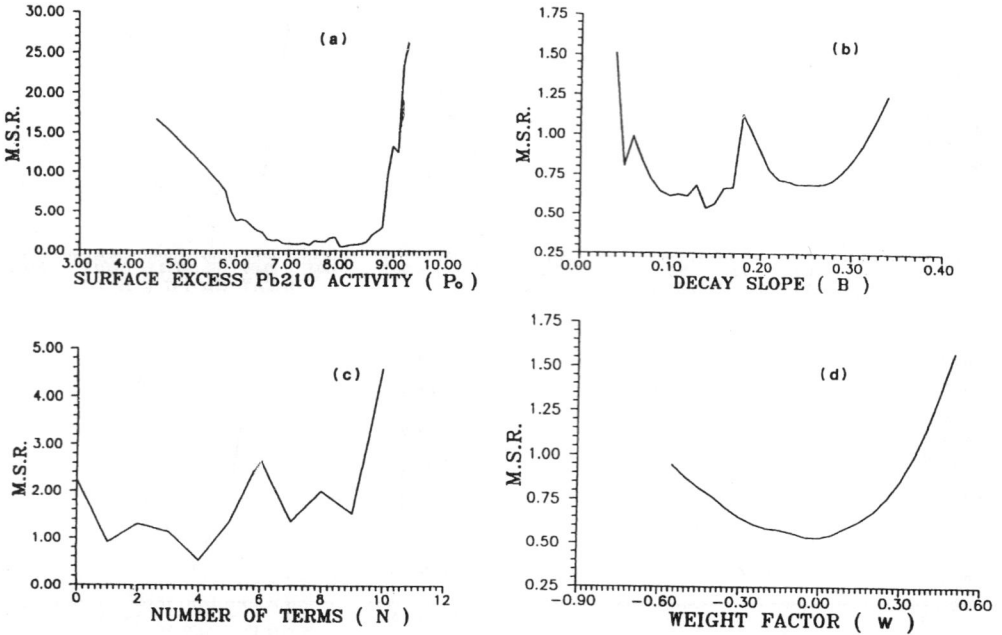

Fig. 4.5. Sensitivity analysis results for data set KC143, as for Fig. 4.4.

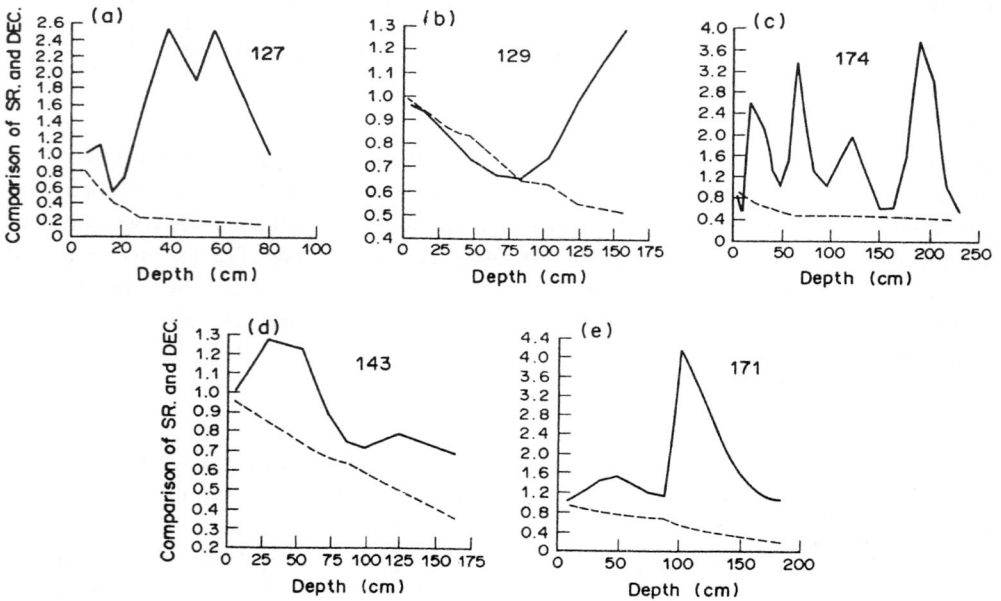

Fig. 4.6. Plots of source variation and intrinsic decay curves vs. depth showing the separation of source variation and radioactive decay process in each nuclide profile. The variations of excess ^{210}Pb activity are mainly contributed by source variation.

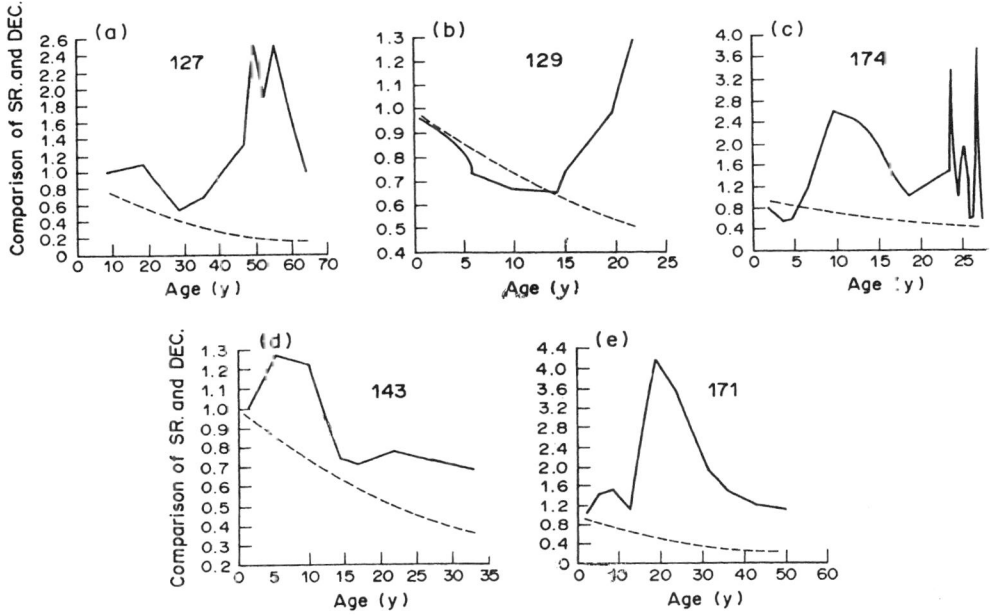

Fig. 4.7. Source variations and decay variation vs. ages of the sediments.

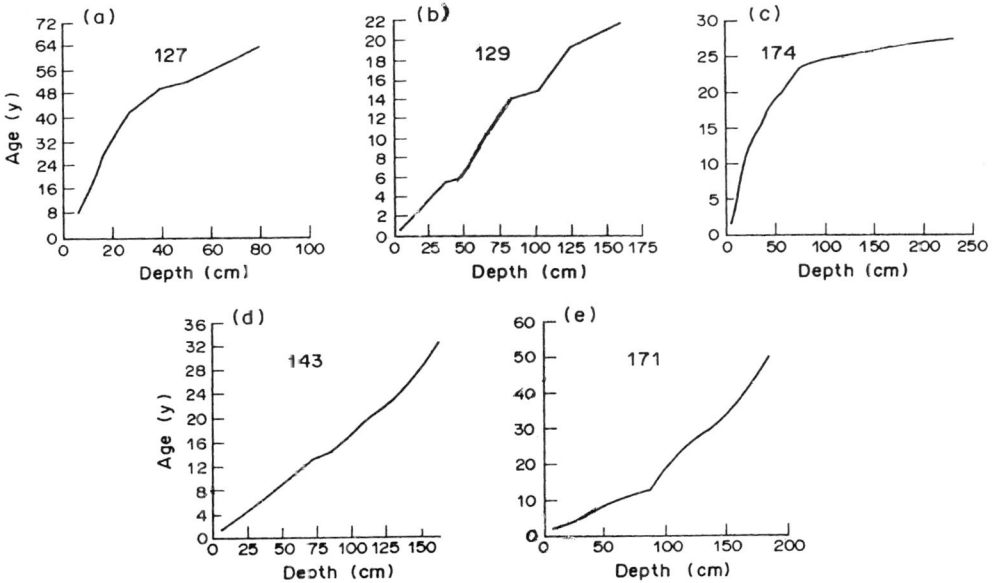

Fig. 4.8. Age vs. depth plots. In Figs 4.8a (KC127) and 4.8e, the slopes of the curves have obvious changes, indicating variations in the sediment accumulation process.

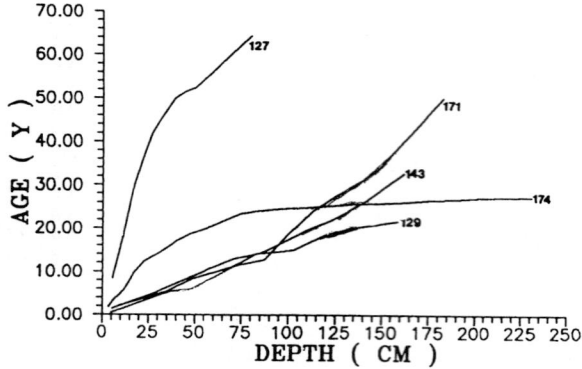

Fig. 4.9. Plot of age vs. depth for all five sediment cores showing three different types of time-depth conversion trends represented by (a) KC127; (b) KC129, KC143, and KC171; (c) KC174.

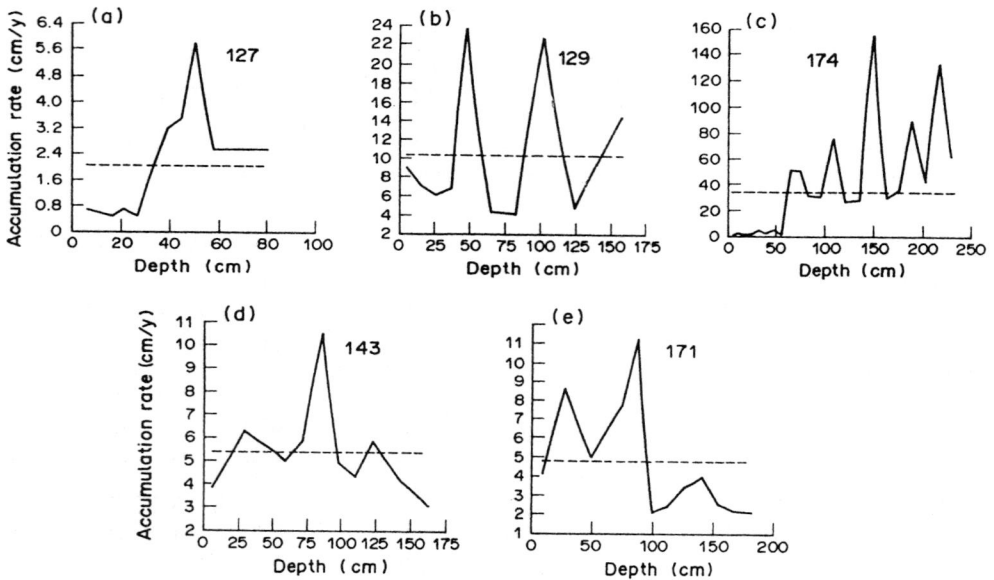

Fig. 4.10. Sediment accumulation rate vs. depth plot for each data set. Solid lines represent the predicted accumulation rate variation with depth, dashed lines show the average accumulation rate.

1.2.5. *Accumulation rate*

Figure 4.10 (a–e) illustrate the sediment accumulation rate versus *depth* for each of the five sediment cores. Figure 4.11 (a, b) shows the sediment accumulation rate vs. *age* of sediments. In order to enhance the fluctuations of accumulation rate around the average values, the dashed lines in Figs 4.10 and 4.11 show average accumulation.

Sediment cores analyzed by Kuehl et al. (1986) using a constant sedimentation model gave results of 0.1–10 cm/yr average accumulation on the Amazon Shelf. From the ^{210}Pb profiles

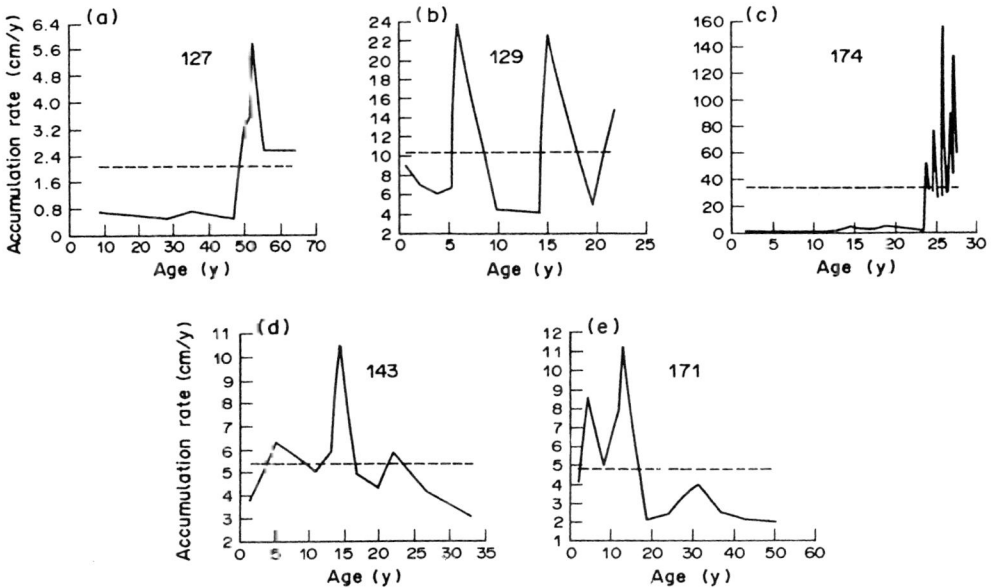

Fig. 4.11. Plots of sediment accumulation rate vs. time. To be noted are periodic variations of sediment accumulation rates in different time intervals and the changes of patterns of accumulation rate with time.

analyzed here, the variability of accumulation rates yields values sometimes much higher than the averages reported by Kuehl et al. (1986). It is apparent that the sediment accumulation rate can sometimes become very high in a short time period, as shown in Figs 4.10c, 4.11c, 4.10e and 4.11e for locations KC174 and KC171. These very high accumulation rates can correspond to abrupt changes of sedimentation environments, such as massive flooding or long-term seasonal trends. At KC174 the accumulation rates vary from 110 cm/yr, prior to 25 yrs ago, to several cm/yr in the last 25 yrs. This rapid shift may be related to the sampling position on a foreset region, with the inference that the characteristic shift may be due to quick movement of the foreset.

When plotting the sediment accumulation rates of all the data sets on one figure (Fig. 4.12), there are periodic variations for the accumulation rate, in about 3–9, 11–18, and 23–28 yrs intervals. This periodic variation of sediment accumulation rate in the whole basin may be related to changes in sea level or to massive regional shifts in sediment transport or deposition.

1.3. *Conclusions*

Based on the disentanglement procedure, and on the resolution and sensitivity of the results, we conclude:

(1) The results show that although a radionuclide profile in a sediment series may be very variable and have a complex distribution with depth, it is possible to apply the methods given in Chapter 2 to separate contributions from the source variation and the intrinsic radioactive decay in the nuclide measurements. A quantitative, and more precise, inter-

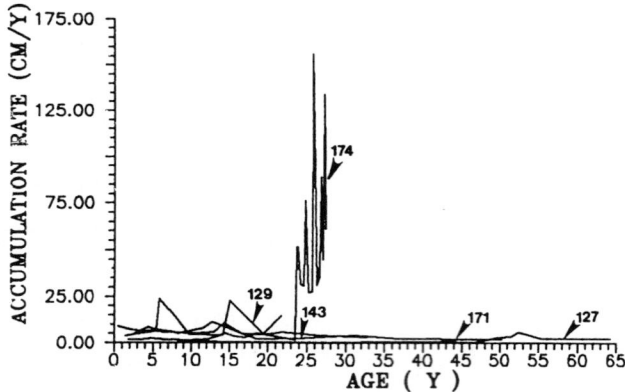

Fig. 4.12. Superimposed sediment accumulation rate curves for all five locations. There appears to be regional periodic variations of accumulation rate with time.

pretation for time to depth conversion can then be made, as can assessments of source variation with time and accumulation rate with time. In addition the stability of parameter determinations can be addressed quantitatively, as can the location-dependent behaviors relative to location sampling position in a basin.

(2) The case history study in the Amazon basin indicates that the sedimentation, radionuclide mixing and concentration processes are quite complex, with sediment accumulation rate and source variation changing markedly within short lateral distances. The sediment accumulation rate seems to have some sort of periodic variations for most of the data sets.

2. $^{87}Sr/^{86}Sr$ sediment ages in stratigraphy studies

Depth to time conversions based on $^{87}Sr/^{86}Sr$ nuclide ratios in sediments have been used to determine depositional histories in various marine sedimentary environments. Here we show how the SIT method can detect samples that have been contaminated, usually by diagenesis or sediment reworking. Because sample contamination is often otherwise difficult to detect, the method helps to avoid incorrect age determinations and the misinterpretation of depositional histories.

2.1. Background

$^{87}Sr/^{86}Sr$ (herein referred to as Sr) versus age curves have been developed for use in interpreting global paleo-oceanography (DePaulo, 1986; Hess et al., 1986; DePaulo & Ingram, 1985; Palmer & Elderfield, 1985; Burke, 1982). Investigators have applied these curves to interpret the history of marine depositional environments (Rundberg & Smalley, 1989; Hurst, 1986). This application represents a powerful extension of biostratigraphy for dating stratigraphic sequences. Mass spectrometry can detect changes in Sr at the level of a part in a million representing changes in time on the order of 0.5 million years.

One requirement for using a universal Sr to age calibration curve is an absolute agreement on Sr values among laboratories. Measurement accuracy is difficult to achieve because of sys-

Table 4.1
^{86}Sr/^{87}Sr measurements

N	Hmc	Reference
42	0.709180	Burke et al. (1982)
19	0.709174	DePaolo & Ingram (1985)
22	0.709215	Palmer & Elderfield (1985)
9	0.709228	Hess et al. (1986)
Average value	0.709199±15	

tematic biases among different laboratories' measurements. As an alternative to laboratory intercalibration, researchers normalize Sr values from Sr vs. age curves relative to an estimate of Sr for Hmc (Holocene marine carbonate). Hmc values are determined from recent samples or taken as the average among reported values from the literature. However reported Hmc values vary significantly among laboratories (Table 4.1) by over 50 ppm. Multiple measurements from a particular laboratory exhibit a similar range of variability (see, for example, Hess et al., 1986). Accordingly, allowance for variations in Hmc must be incorporated in the depth to age conversion. Such variations increase the uncertainties associated with the determined ages to about ±1.5 million years (Palmer & Elderfield, 1985).

To refine the conversion from depth to time for Sr measurements, improvements are needed in determining pristine samples that are free from secondary distortions to the age determinations. Diagenetic alteration, and sample displacement from the original depositional horizon by caving or sediment reworking processes, are the main mechanisms which compromise sample integrity. At present few criteria have been developed to determine when sample alteration or reworking has occurred. Typically samples are examined for evidence of diagenesis before Sr measurement and evaluated for internal consistency of Sr values after measurement.

We have used a modified version of the SIT algorithm (see Chapter 2) to predict the shape of the Sr vs. depth profile from the data by minimizing the least square error between measured and predicted Sr values. We have previously shown that in curves where depth coverage is good our predicted Sr values are within sample measurement error (Carroll & Lerche, 1989).

Rundberg & Smalley (1989) use Sr from a small number of samples to interpret depositional history along two parallel transects in the northern North Sea. We were curious as to how sensitive their Sr ages were to an Hmc value they determined by averaging values from the literature. They used an Hmc value of 0.709200. From our least squares model procedure we can determine an Hmc value for each of the Sr profiles of Rundberg & Smalley (1989), and then investigate whether our model detects any contamination for a given data set.

Data profiles are fit by a function of the form:

$$Sr \equiv \, ^{87}Sr/^{86}Sr$$

$$= A \exp\left[Bz + \sum_{n=1}^{m} \frac{a(n)}{n\pi} \sin(n\pi z) + \sum_{n=1}^{m} \frac{b(n)}{n\pi} (\cos(n\pi z) - 1) \right], \tag{4.1}$$

where z is depth (normalized to unity at the depth of the deepest sample), and where A, B, $a(n)$, $b(n)$ are variable parameters determined from the data as is m, the best number of terms to use to fit the data. At $z = 0$ (the sediment–water interface), the exponential term becomes 1 and the model parameter A thus represents the best estimate of the Hmc Sr value for the given data. The only boundary condition is that the resulting parameters defining the equation must provide the best fit to the data in a least squares sense. The model is quite versatile and has the capability to account for nonlinearities (i.e., departures from a simple exponential decrease with increasing depth) in the data provided enough data are present to resolve such distortions (Carroll & Lerche, 1989).

2.2. Data and predictions

We applied the model to the Rundberg & Smalley (1989) Sr profiles for stations H, J, K, and L which represent a transect across the North Sea basin from continent to ocean. Stations E and D are located along a second transect across the basin (Fig. 4.10). In Fig. 4.13, we have plotted the data, the model best fit, and also the model fit based on the Hmc value

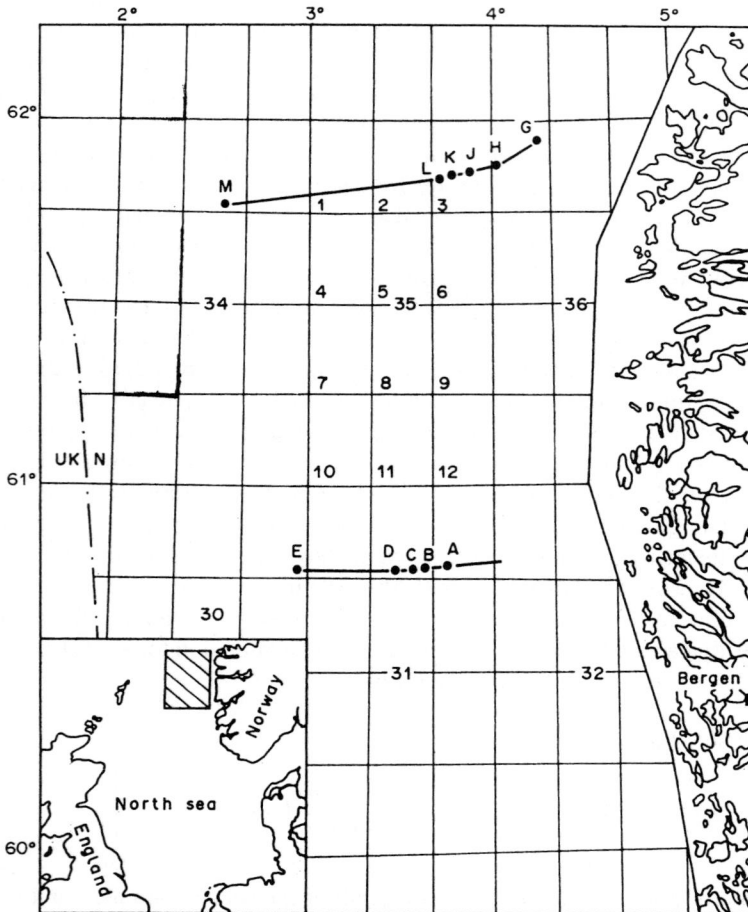

Fig. 4.13. Station locations from Rundberg & Smalley (1989).

Table 4.2
Model results

Core	A (Hmc)	B	Chi-square (arbitrary units)	Comments
J	0.708820	0.79	16	
	0.709200	1.05	31	Rundberg & Smalley (1989), Hmc
H	0.708684	0.60	35	
	0.709200	0.93	77	Rundberg & Smalley (1989), Hmc
K	0.713592	2.13	166	All data
	0.708982	0.72	1	Shallowest point omitted
	0.709200	0.84	3	Rundberg & Smalley (1989), Hmc
L	0.710180	9.23	262	
	0.709200	0.40	929	Rundberg & Smalley (1989), Hmc
E	0.709733	1.01	8	
	0.709200	0.71	19	Rundberg & Smalley (1989), Hmc
D	0.713500	1.46	23	
	0.709200	0.09	46	Rundberg & Smalley (1989), Hmc

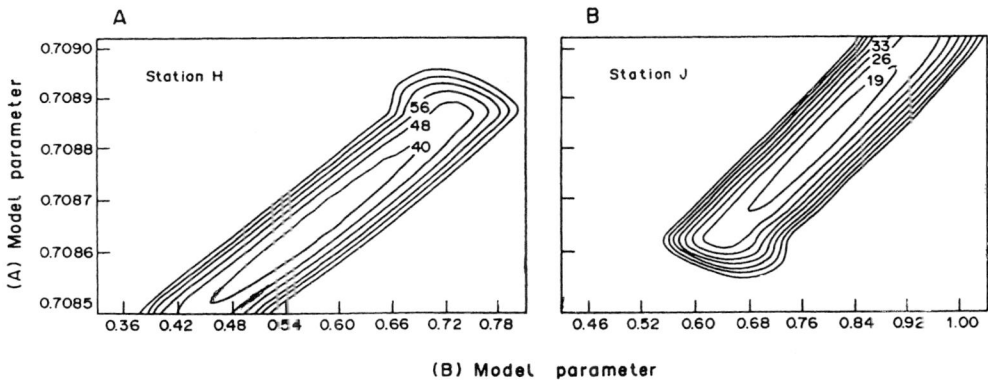

Fig. 4.14. Contour of chi-square for A vs. B. Stations H and J contours are sharp resulting in well resolved model parameters.

used by Rundberg & Smalley (1989) during their analyses (0.709200). A summary of model parameters and chi-square values for each case is given in Table 4.2. In all four cases, the best fit to the data occurred when $n = 0$. For this situation the model equation simplifies to an exponential of the form:

$$^{87}\mathrm{Sr}/^{86}\mathrm{Sr} = A\exp(Bz). \tag{4.2}$$

Profiles H and J had the largest number of data points (five and six, respectively) and showed highly significant chi-square values. The average difference between measured and predicted values was 0.000042 ± 28. However the data for profiles H and J predict Hmc values considerably lower than the value 0.709200 used by Rundberg & Smalley for their age determinations. The model determined values were ($H = 0.708684$; $J = 0.708820$). As shown in Fig. 4.14, contours of chi-square for A vs. B are extremely sharp so that the resulting Hmc values are well resolved.

For profile K we conducted two model runs. The first included all data while the second excluded the shallowest sample. Rundberg & Smalley (1989) suspected the shallow Sr sample to be contaminated by sediment caving. In the first case (all data), the model best fit Hmc value was unrealistically high (0.713592). However, in the second case, the model Hmc value was 0.708982, which is similar to the values determined for profiles H and J and resulted in a significantly lower chi-square. This result also suggests the shallowest sample was contaminated and could not provide a realistic age. For profile L, the best fit Hmc equals 0.71018. This high Hmc value and poor fit is likely due to a contamination problem, however the sparsity of data points (four) precludes a definitive determination. Like profiles K and L, Sr ratios for various depths in profiles D and E are significantly larger than model predicted values and Hmc predictions are unusually high ($D = 0.713500$; $E = 0.709733$). However, profile D contains only three datum points, and profile E contains 8 datum points but these are tightly clustered into two groups of 4 each, leading to very little resolution capability with depth (Fig. 4.15).

2.3. Discussion

The results obtained here suggest caution when interpreting the ages determined for several of the samples. In those cases where chi-square values were determined to be low, i.e., H, J, K, the model determined best Hmc values are considerably lower than the range reported on measured Hmc values and confirmed by this model with many tens of datum values (Carroll & Lerche, 1989). The reason for this behavior is likely due to a combination of poor resolution in the upper strata of the sediment cores and contamination in the lower sections. In such circumstances it is, perhaps, better to estimate the surface value as the global average of Hmc, accepting the associated increase in error for the age estimates and the loss of information on spatial variations. The results also indicate that core L likely has significant contamination making the ages derived for this profile suspect.

2.4. Conclusions

In summary, the use of Sr ratios as age markers is practical because measurement precision is high. However, this high precision in turn necessitates careful screening of samples since accurate age determinations are sensitive to small losses of sample quality. While sample inspection techniques are a necessary precursor to measurement, such do not guarantee accurate results. We have demonstrated how a model based on a least squares error minimization routine provides a sensitive technique for evaluating sample quality within the context of the sedimentary environment.

3. Eastern Arabian Sea

Somayajulu et al. (1999) have investigated eight cores from the active continental margin of the eastern Arabian Sea to the west of the Indian subcontinent. Accelerator mass spectrometry

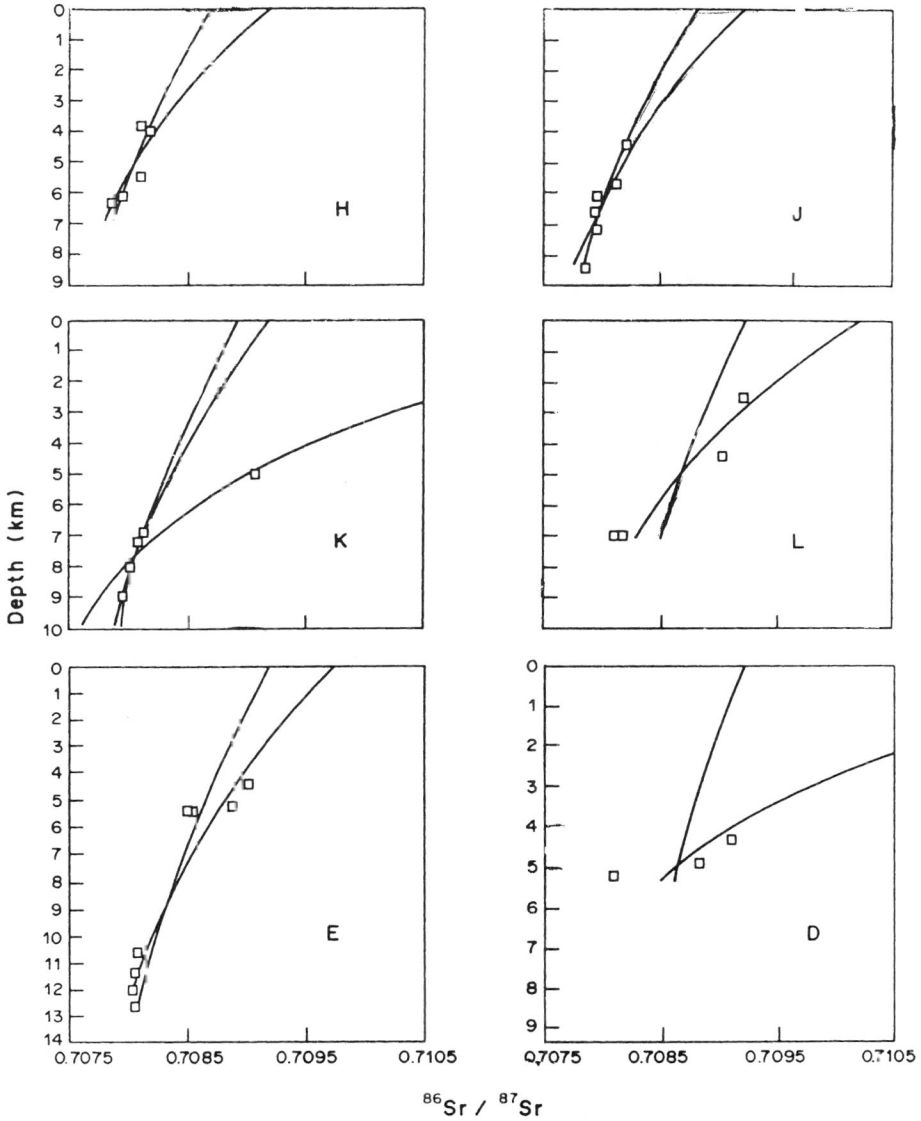

Fig. 4.15. Model predicted $^{87}Sr/^{86}Sr$ profiles vs. depth for stations H, J, K, L, E, and D. See text for further details.

was performed on planktonic foraminifera for ^{14}C, yielding long-term (of order 1000 yrs) average sediment accumulation rates ranging between 4×10^{-3} to 1.3×10^{-1} cm/yr. For the shorter term (< 100 yrs), accumulation rate data were obtained for both excess ^{210}Pb and ^{137}Cs. Only in five of the cores was ^{137}Cs observed, but unsupported ^{210}Pb was measured in all eight cores. A detailed investigation of all core results is given by Somayajulu et al. (1999).

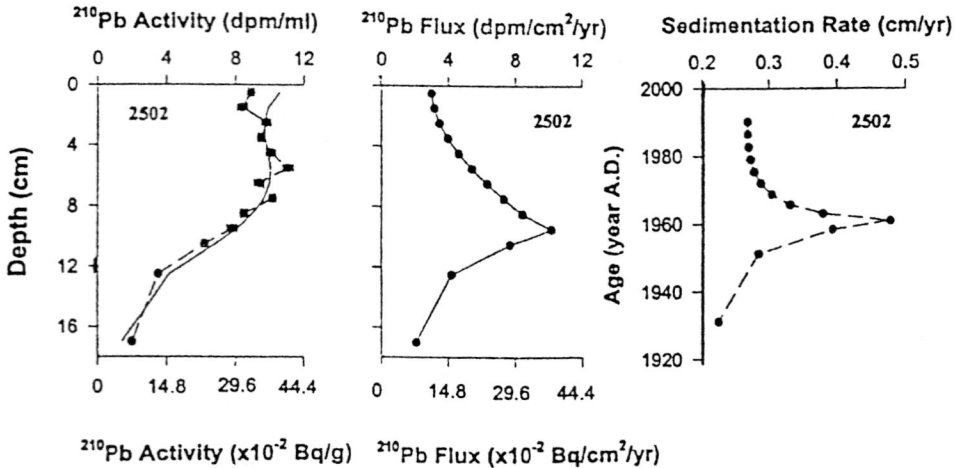

Fig. 4.16. Site 2502, eastern Arabian Sea. Left panel: variation of excess ^{210}Pb with depth, solid line is SIT model fit; Central panel: flux variation of excess ^{210}Pb with depth or time; Right panel: sediment accumulation rate variations with time (adapted from Somayajulu et al., 1999).

Here we concentrate on just the one core from site 2502 because not only is it representative of the general procedure used but it also corresponds to the analysis performed in a similar manner for the North Dakota data (see Chapter 3).

Shown in Fig. 4.16 (left panel) is the variation of excess ^{210}Pb with depth, together with a solid line representing the model fit from the SIT procedure. Except for the very near surface values of excess ^{210}Pb, the fit is remarkably good. The center panel of Fig. 4.16 shows the surficial flux variation of excess ^{210}Pb, showing a peak around a depth of 9 cm, which corresponds to an age of 1960. The corresponding sedimentation rate is also variable with time (right panel of Fig. 4.16), indicating a peak of about 0.48 cm/yr around 1960 and decreasing to about 0.22 cm/yr at the base of the core (dated as 1930) and also diminishing systematically near the top of the core to about 0.27 cm/yr.

When one looks at the average rate of sediment accumulation over the eight wells on the long time-scale (as determined from ^{14}C) there is a variation of over a factor 30 between sites for the average alone. It is, therefore, not surprising that sedimentation accumulation rates also vary on the short time-scale. Perhaps it is remarkable that the sediment accumulation rate at site 2502 varies by only a factor 2. Viewed in this light, and because site 2502 is in a very active environment, it is most likely that both the excess ^{210}Pb flux and the sediment rate vary with time; in our opinion, it would be even more remarkable if they did not vary!

The ^{137}Cs curve with depth (Fig. 4.17) for site 2502 shows peak activity at 9 cm depth, corresponding precisely to the early 1960 age estimated from the SIT method applied to the excess ^{210}Pb profile, and so confirming the statistical sharpness of the estimates made.

In short: the SIT procedure disentangles flux and sediment variations in a manner consistent with all knowledge from multiple radionuclides.

$$^{137}Cs \text{ [dpm/cm}^3\text{]}$$

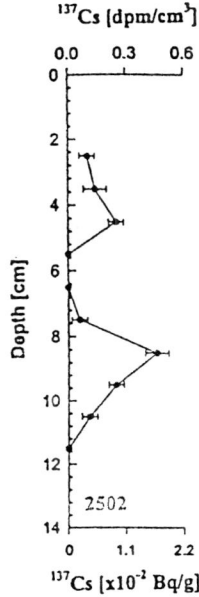

Fig. 4.17. Site 2502, eastern Arabian Sea. Variation of ^{137}Cs with depth showing peak at 9 cm depth, corresponding to about year 1960 (from Somayajulu et al., 1999).

Appendix A

The following is a list of symbols used in the equations:

$A_0 = $ initial ^{210}Pb excess activity (dpm/g),
$A_x = $ excess ^{210}Pb activity (dpm/g) for depth x in the sediments,
$x = $ depth of sample (cm),
$V = $ velocity of sedimentation (cm/yr),
$\lambda = $ decay constant of ^{210}Pb (yr^{-1}),
$AR = $ sediment bulk accumulation rate,
C_{org} accum rate $= $ organic accumulation rate (mg C/cm^2/yr),
$C_{org} = \%$ organic carbon.

Equations:

$$SA = \left[\ln(T/I)\right]/[T/I - 1], \tag{A.1}$$

$$A_0 = A_x \exp(\lambda x / V), \tag{A.2}$$

$$C_{org}\text{accum rate} = AR(C_{org}), \tag{A.3}$$

$SA = $ self-absorption correction as specified by Cutshall et al. (1988). The attenuated beam intensity in counts per minute (cpm) (T) was measured for each sample. The unattenuated beam was determined for the detector, $I = 101$ cpm for ^{210}Pb.

Chapter 5

Sediment mixing

1. Introduction

The process of bioturbation results from the burrowing and feeding of benthic macrofauna so that particles are mixed and redistributed within a column of sediment or, alternatively, they are resuspended and redeposited or washed away. Mixing disturbs the sedimentary record of accumulation, erasing the signature of time given by the decrease in radionuclide concentration with sediment depth. The only way to recover information on time is then to apply a model that describes the effect of sediment mixing. Bioturbation is most often expressed mathematically using various formulations of the general sedimentary diagenetic equations (Berner, 1980). These models may be subdivided into two broad classes: (1) particle transport models and (2) diffusion-reaction models. In this chapter our focus is on the first class of models whereby tracers are strongly bound to particle phases; in Chapter 6, our focus will be on the second class of models whereby tracers are transferable between particles and pore water.

Some mixing models are not predicated on a mechanistic model of the mixing process itself, but rather have been shown to produce a high degree of correlation with depth profiles of tracers gathered from a number of sedimentary systems. In these models bioturbation is typically expressed mathematically as a diffusive process. Alternatively, a variety of model formulations exist whereby bioturbation is described according to the individual feeding and burrowing strategies employed by certain organisms. Collectively, these processes of non-diffusive bioturbation are called non-local mixing. Deposit feeders, such as tubificid worms for example, ingest sediment from a range of depths and excrete it at the surface. Other organisms feed in a conveyor-belt fashion, ingesting sediments over a range of depths while depositing feces at the sediment surface or directly behind them. Alternatively, large macrobenthic organisms such as deep burrowing crustaceans, dig large holes that are subsequently back-filled with surficial material. The result is that sedimentary material is exchanged over distances equal to or greater than the scale over which the concentration of a tracer changes substantially. Interestingly, on a worldwide basis, the depth of mixing is restricted to a narrow surficial zone with a mean of 9.8 ± 4.5 cm (Boudreau, 1998). This depth zone does not correlate with biological mixing intensity or water depth, indicating that no single controlling factor has yet been identified on the basis of observational data.

Similarly, non-biological, or physical mixing of sediments caused by periodic processes (e.g., bottom currents, waves and tides) and episodic events (e.g., storms and floods, turbidites) can exchange material among different depth horizons at a single location or remove material from one location depositing it elsewhere. These processes can have a profound effect on the distribution of radionuclides, and thus the determination of sedimentation rates, especially in environments where deposition amounts are low relative to sediment mixing thicknesses. The result is that there is no unique age at each depth, but rather a mixture of material of different ages. Unless one can account in a quantitative fashion for the mixing processes, inferred sedimentation rates could be significantly in error. To determine which pattern is caused by physical transport and which is due to intrinsic concentration variation, when coupled with sediment mixing possibilities, is not always clear-cut, as the illustrations given at the end of this chapter will demonstrate.

A number of different conceptual models exist describing sediment mixing effects caused by both physical and biological processes valid under restricted assumptions of temporal or spatial steady state or of constant and invariant mixing. Later in this volume we will describe how these models can be used, in conjunction with the signal theory model presented in Chapter 2, to determine sediment accumulation rates in environments where mixing occurs. But first, we review some of the general types and forms of the different methods used for describing the effects of sediment mixing on sedimentation rate determinations.

2. Basic model

Based on the general theory of diagenetic reactions in sediments, some of the earliest formulations express bioturbation mathematically as a diffusive process (Berger & Heath, 1968; Guinasso & Schink, 1975; Robbins & Edgington, 1975; Benninger et al., 1979). The general theory is that the steady-state distribution of a solid species in a constant porosity sediment is subject to first-order decay and negligible advection. Mixing is described by the mixing function which is the concentration profile of a conservative tracer that results from a unit impulse falling on the sediment surface at a given time, as a function of depth and time. In reality, mixing need not be instantaneous but the rate of sediment mixing must be much faster than the sedimentation rate so that what is normally an exponential decrease in radionuclide activity becomes uniform activity within the upper layer of sediment (Mélières et al., 1991). The process is governed by the conservation equation

$$D_B \, d^2C/dx^2 - \lambda C = \text{Flux divergence} \tag{5.1a}$$

with depth (x), concentration (C) of the tracer in the solid phase, D_B the mixing coefficient $(cm^2 \, sec^{-1})$, and a rate constant (λ) (sec^{-1}). The tracer distribution is then described in terms of a sedimentation rate (SR) as follows:

$$\partial C/\partial t = \partial(D_B \partial C/\partial x)/\partial x - SR \partial C/\partial x - \lambda C. \tag{5.1b}$$

Values of D_B range typically from 10^{-2} to 10^2 $cm^2 \, yr^{-1}$ for fine-grained marine environments (Matisoff, 1982; Dellapenna et al., 1998). This model is employed when the mixing depth is clearly delineated in the upper surface layer of the tracer profile with an abrupt transition from constant to exponentially decreasing radionuclide activity with increasing depth. Bioturbation acts only within a bioturbation zone near the sediment–water interface and the mixing is rapid relative to the tracer half-life. This simplified assumption is often invoked when no additional information is available to correct for mixing effects. When, in addition the sedimentation rate is constant, the radionuclide activity will decline exponentially below the mixed depth. In practice, the zone of transition from mixed to unmixed sediments is normally chosen as the depth where the nuclide activity ceases to be constant. This approach is one of the limiting cases of more general model formulations based on the advection-diffusion equation. In Chapter 6, a numerical scheme is developed to extract the intrinsic depositional behavior from a biologically mixed tracer profile.

In the more general case, the mixing or biodiffusion coefficient is depth dependent. It is often described with a Gaussian shape and a maximum value of D_0 at the surface. D_B is constant down to a depth L, the mixing layer thickness, and zero for greater depths. For convenience, a dimensionless number (G) is used to express mixing in terms of a single coefficient, $G = x/(SRL)$.

The maximum of the mixing function is displaced from the initial input location in the sediment by a distance varying from 0 (corresponding to the mixing case with $G = 0$) to L ($G = \infty$), respectively. When mixing is fast, the sediments are immediately homogenized and $G = \infty$. This is the end-member case described previously corresponding to rapid mixing in the surface layer. Early applications of this modelling approach confined the mixing process to one or two mixed layers of defined thickness (e.g., Guinasso & Schink, 1975; Benninger et al., 1979; Christensen, 1982; Anderson et al., 1988; Boudreau, 1986; Smith et al., 1995). Alternatively, Olsen et al. (1981) divided the entire sedimentary column into a number of homogeneous sub-layers of equal thickness and employed numerical solutions using a mixing coefficient that decreased exponentially between successive sub-layers.

For these models to be valid, it is essential to obtain accurate information on the flux of the chosen tracer(s) to the sediments. In the case of ^{210}Pb, one can sometimes invoke the additional simplifying solution that the flux of ^{210}Pb is invariant with time over the entire length of a sediment core, or at least over a limited number of discrete depth zones as was described earlier in Chapter 2. For other tracers (e.g., ^{137}Cs, ^{239,240}Pu, ^{60}Co) the input functions are time-dependent and must be fully described (Olsen et al., 1981; Smith et al., 1986).

The mechanisms and rates of mixing are not necessarily independent of the tracer (Smith et al., 1993). This phenomenon, known as age-dependent mixing, is associated with the idea that recent sediment (fresh), food-rich particles are ingested and distributed at higher rates by organisms feeding at the sediment–water interface than are older, food-poor particles (Lauerman et al., 1997). Thus, it is additionally important to match the characteristic time scales of mixing tracers to modelled reactants.

A wide variety of examples are available in the literature demonstrating various permutations on the general approaches given above and/or site-specific applications of these models as previously formulated. For example, Hancock & Hunter (1999) use the depth dependent formulation of Christensen (1982), in conjunction with profiles for two tracers (^{228}Th and ^{210}Pb), to constrain mixing coefficients and interpret sediment accumulation rates in Port Phillip Bay, Australia. Henderson et al. (1999) describe variations in bioturbation with water depth on marine slopes applying the approach of Anderson et al. (1988) to ^{210}Pb profiles from the Little Bahamas Bank. In a model combining molecular diffusion and bio-irrigation processes, Schlüter et al. (2000) were able to characterise the seasonal variability in bio-irrigation within coastal sediments of Kiel Bight (Baltic Sea).

3. Organism-specific models

The development of mathematical representations of biological mixing processes moved toward describing specific feeding and burrowing behaviors of benthic organisms (Boudreau, 1986; Boudreau & Imboden, 1987). These include the conveyor-belt model developed by Robbins (1986) describing the benthic organisms that feed by ingesting sediments over a range of depths while depositing gut contents primarily on the sediment surface. This conveyor-belt model has also been employed inversely to simulate the subsurface egestion of surficial sediments (plus tracer) at discrete depths below the sediment–water interface (Smith et al., 1986/87).

A more extreme mode of sediment replacement is produced by organisms such as deep burrowing crustaceans that create deep burrow networks that may either fill with surficial material progressively when vacated or catastrophically during storms (Tedesco & Aller, 1997). Unlike the cases of bio-advection described previously, there is no net movement of the sediment at depth. Due to the injection of surface material at depth, this process leads to an effective "regeneration" of any radionuclide previously deposited. Gardner et al. (1987) depict the "regeneration" effect as being described by a term of the form

$$\left(\frac{dA}{dz}\right)_{bio} = K(A_0 - A)\left(\frac{dt}{dz}\right)\exp(-z/L). \tag{5.1c}$$

Here A is the specific radionuclide activity (usually measured in dpm/g) at sediment depth z, A_0 is the present-day surficial ($z = 0$) activity, K is the rate constant (units of time^{-1}) describing the overall bioturbation rate; dt/dz is the "slowness" at sediment depth z (which is $1/V$, where V is the sedimentation velocity) and the scale-depth L in the exponential factor, $\exp(-z/L)$, describes how greater sedimentary depths are less susceptible to bioturbation effects.

4. Physical turbation

In the case of physical sedimentary erosion and replacement by turbidites, allocthonous deposition brings with it radionuclide activity from other locations, which may be higher or lower than that originally deposited. Here we consider the effect of physical turbidite transport (in both erosion and depositional aspects) in the profile of activity with sedimentary depth. A physical turbation model can replace previously deposited sediments with other sediments from different locales, or can just erode previously deposited sediments without replacement, or can leave previously deposited sediments alone and superpose sediments from different locales. In all three cases one obtains completely different variations of radionuclides with depth than previously deposited sediments would suggest. We illustrate the patterns available in all three cases with pictorial representations.

As shown in Fig. 5.1, two sites (A and B) are subject to sedimentation with radionuclide activity with depth of sediment. Prior to any further disturbance nuclide activities at sites A and B are as shown in Fig. 5.1a. Now consider that the sediments at site B are physically unstable and undergo turbidite transport (perhaps in the manner detailed in Bagirov & Lerche,

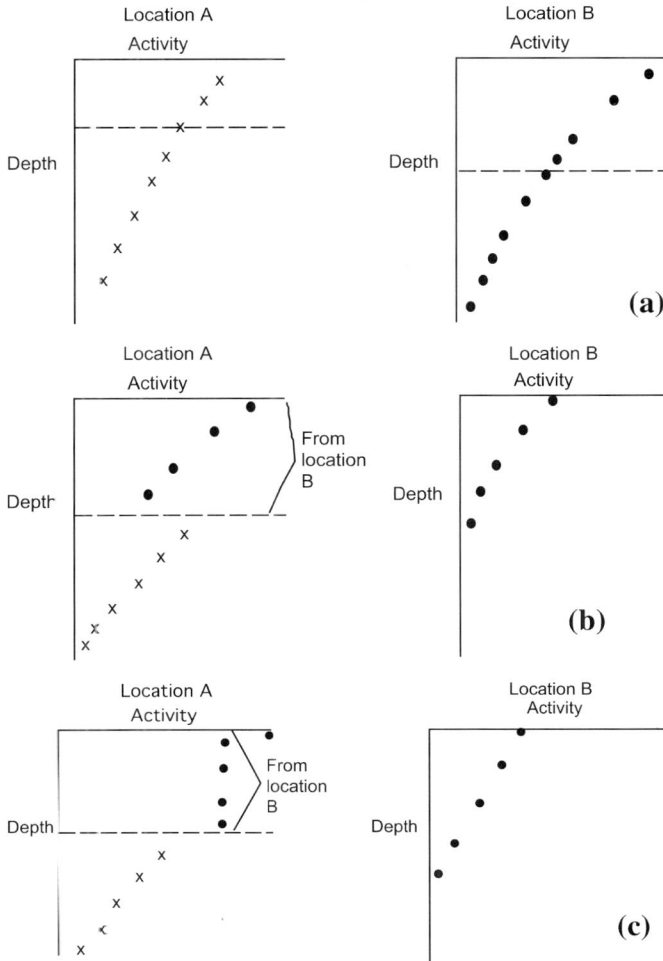

Fig. 5.1. Illustrative depictions of the influence of turbidite erosion and redeposition from one location to another, both in the absence and presence of mixing during transport, followed by later sediment deposition.

1998). Take the top sediments at location B (denoted by a dashed line on Fig. 5.1a) to be transported wholesale to location A, where the sediments shown by a dashed line are eroded by the turbidite action. If the sediments from location B are deposited at location A without physical mixing (i.e., as a slab of material) then the subsequent activity profile at location A is as shown in Fig. 5.1b, showing a reversal of activity with depth. At location B, the corresponding new surficial activity is lowered due to removal of the top sediments.

If homogeneous mixing were to occur of the turbidite sediments from location B prior to their resettlement at location A, then the resultant activity profiles at locations A and B would appear as shown in Fig. 5.1c, so that a reversal of activity at depth does *not* take place.

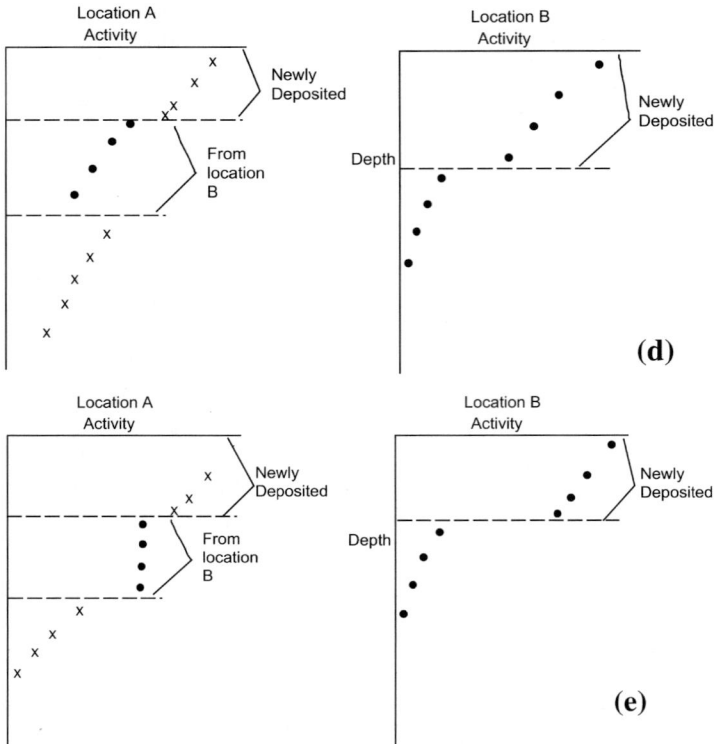

Fig. 5.1. (*Continued*).

In general, the result at location A will lie between the two extremes depicted in Fig. 5.1 (b and c).

If further deposition now takes place at both locations A and B, then the later modification of activity with depth of Fig. 5.1 (b and c) is shown in Fig. 5.1 (d and e), respectively.

Clearly, by repeating the process of turbidite deposition and/or erosion in variable amounts, one can construct a large variety of profiles of activity with depth due to physical transport. Even in the absence of erosion at location A, by superposing turbidite contributions from different locales, in different thicknesses, and with different activity profiles (partially or totally mixed, or unmixed, during transport, redistribution and resettlement) one can construct exceedingly complex patterns of activity with depth at given locations.

5. A sedimentation model with bioturbation

We demonstrate how the regeneration model can be incorporated into the inductive model developed in Chapter 2.

The equation describing the variation of radionuclide activity under the two conditions of radioactive decay plus surficial variation at deposition is given in Chapter 2. The activity equation written in the general form is:

$$\frac{dA}{dz} = A(-\lambda \, dt/dz + dS/dz), \tag{5.2}$$

where λ is the radioactive decay rate.

If one now adds the bioturbation (or physical turbation) effect in the form prescribed by Gardner et al. (1987), then the general activity equation with depth is

$$\frac{dA}{dz} = A(-\lambda \, dt/dz + dS/dz) + K\frac{dt}{dz}(A_0 - A)\exp(-z/L). \tag{5.3}$$

5.1. *Simple models*

In the cases examined by Gardner et al. (1989) the further assumptions were made that: (a) the velocity, dz/dt, was constant at a value V; (b) there were no surficial activity variations ($dS/dz = 0$). In that particular case equation (5.3) reduces to

$$V\frac{dA}{dz} = -A\left[\lambda + K\exp(-z/L)\right] + K A_0 \exp(-z/L). \tag{5.4a}$$

Note that dA/dz can be zero only when

$$A = A_0 K \exp(-z/L)/\left[\lambda + K \exp(-z/L)\right] < A_0. \tag{5.4b}$$

Alternatively, at the depth locations $\{z_i\}$ (with corresponding observed values $\{A_i\}$), where dA/dz is observed to change sign, then

$$A_i\lambda = K(A_0 - A_i)\exp(-z_i/L). \tag{5.4c}$$

Thus three, or more, turning points for $A(z)$ can be used alone to determine A_0, K and L.

In addition, note that at any such turning point the value of A_i *must* be less than A_0. Further, the ratio of two turning point values (A_1/A_2) occurring at z_1 and z_2 (with $z_1 < z_2$) must be in the ratio

$$A_1/A_2 = \left(K + \lambda \exp(z_2/L)\right)/\left(K + \lambda \exp(z_1/L)\right) > 1. \tag{5.4d}$$

Thus the turning point values must follow the sequential magnitude ordering $A_0 > A_1 > A_2 > A_3 \cdots$. If this sequential ordering is *not* obeyed then some cause other than the model for turbation *must* be responsible for the observed variations – or the model is incorrect.

But, because $A \to 0$ as $z \to \infty$, dA/dz must have a negative slope at large z. This fact, combined with the requirements on any slope change directions listed above, means that $A(z)$ does *not* increase at any depth. Effectively, the turbation model delays the radioactive decrease with depth, but does not allow an increase of $A(z)$ at any depth over the surface value A_C. Any profile of activity showing an increase must then be due to other causes, such as surficial variations due to concentration variations at deposition and/or physical turbidite replacement (which, of course, masquerades as a concentration variation at deposition).

In addition, note that $A(z) \leqslant A_0$, everywhere because if $A(z)$ were to exceed A_0 then dA/dz would be negative, forcing $A(z)$ to be less than A_0. And, on $z = 0$ note that $dA/dz < 0$, so that A initially decreases from its surficial value of A_0.

The unknown components in equation (5.4) are V, K, L and A_0 (the decay timescale constant, λ, is known for each radioactive species). Gardner et al. (1989) used a trial-and-error procedure to determine the values of V, K, L and A_0 that best fit their observations of $A(z)$ (see later in this chapter). However, there is a more quantitative way to proceed. Note that equation (5.4) is linear in V, K and the product $(A_0 K)$, in the sense that doubling each leaves the equation unchanged. Note also that equation (5.4) is non-linear in the scale depth, L. Let the data measurements of activity be $A_1, A_2 \ldots$ occurring at the increasing depth values $z_1 \leqslant z_2 < z_3 \cdots$.

Suppose then, that one were to multiply equation (5.4) by A^n and then integrate from $z = z_1$ (where $A = A_1$) – to the last data point at $z = z_m$ (where $A = A_m$). Then, because V is assumed constant over the total depth range of measurements one obtains the set of linear moment equations

$$\frac{V}{(n+1)} \{ A_1^{n+1} - A_m^{n+1} \} = \lambda I_{n+1} + K J_{n+1} - (K A_0) J_n, \tag{5.5}$$

where

$$I_{n+1} = \int_{z_1}^{z_m} A^{n+1} \, dz; \qquad J_{n+1} = \int_{z_1}^{z_m} A^{n+1} \exp(-z/L) \, dz. \tag{5.6}$$

For brevity, let $y_{n+1} = (n+1)^{-1}(A_1^{n+1} - A_m^{n+1})$ and set $K A_0 = X$. Note that y_{n+1} is observed, and that V, K and X are to be determined. Then consider the quadratic function

$$Q = \sum_{n=0}^{N} \{ V y_{n+1} - K J_{n+1} + X J_n - \lambda I_{n+1} \}^2, \tag{5.7}$$

where one includes the first N moments.

A minimum of Q with respect to V, K and X occurs when

$$V \sum_{n=0}^{N} y_{n+1}^2 - K \sum_{n=0}^{N} J_{n+1} y_{n+1} + X \sum_{n=0}^{N} J_n y_{n+1} = \lambda \sum_{n=0}^{N} I_{n+1} y_{n+1}, \tag{5.8a}$$

$$V \sum_{n=0}^{\infty} y_{n+1} J_{n+1} - K \sum_{n=0}^{N} J_{n+1}^2 + X \sum_{n=0}^{N} J_n J_{n+1} = \lambda \sum_{n=0}^{N} I_{n+1} J_{n+1}, \tag{5.8b}$$

$$V \sum_{n=0}^{\infty} y_{n+1} J_n - K \sum_{n=0}^{\infty} J_{n+1} J_n + X \sum_{n=0}^{N} J_n^2 = \lambda \sum_{n=0}^{N} I_{n+1} J_n. \tag{5.8c}$$

For each value of the scale-length, L, the linear (in V, K and X) equations (5.8) can be solved to obtain values of V, K and X. Then insertion of these values into equation (5.7) produces a value for Q. As L is systematically increased one repeats the procedure, thereby obtaining a curve of Q versus L. Hence one can find trivially graphically the minimum Q and the corresponding L value. Then use of this L value in equations (5.8) a final time produces the best least squares fits for V, K and X, and so for $A_0 \equiv X/K$.

Hence, under the constraints imposed by Gardner et al. (1987) one has the best fits to a given data field. A numerical illustration is given later.

In addition, one can also use the quadratic procedure given above to determine if *constant* values for V, K, X and L really do provide the best fit to the data. The procedure operates as follows. Over the depth range z_i to z_j (with $z_i < z_j$) where corresponding activity measurement values are A_i, $A_{i+1}, \ldots, A_j$, suppose one had integrated equation (5.4) only over z_i to z_j. Then, replacement of equation (5.5) would be

$$\frac{V}{(n+1)}\left[A_i^{n+1} - A_j^{n+1}\right] = \lambda I_{n+1} + K J_{n+1} - (K A_0) J_n, \tag{5.9}$$

with

$$I_{n+1} = \int_{z_i}^{z_j} A^{n+1}\,dz; \qquad J_{n+1} = \int_{z_i}^{z_j} A^{n+1}\exp(-z/L)\,dz. \tag{5.10}$$

Then, following the same procedure as just outlined one can obtain values for V, K, X and L that minimize misfit to the activity data field in $z_j \geqslant z \geqslant z_i$. By varying the range of z_i and z_j, one obtains a suite of values for each of the four parameters. If, to within resolution error of the data, all values are identical to those obtained using the total data field, then indeed that particular data field is adequately described by constant values throughout. But, should it transpire that the corresponding values are variable outside the range of uncertainty incorporated in the data field, then the intrinsic assumptions of constant values for V, K, L and (possibly) A_0 need to be modified, as does the assumption of no intrinsic source depositional variations.

5.2. *Velocity and source variations*

As shown in detail in Chapter 2, in the absence of a turbation effect, one can determine the steady and variable components of sedimentation rate and of source variations. The addition of a turbidite mixing component, as in equation (5.3), complicates the technical determination of sedimentation rate and source variation because there are also the parameters of the turbation effect to determine (viz. K, L) and, of greater concern, is that the turbation effect (last term on the right-hand side of equation (5.3)) is directly coupled with the slowness, dt/dz, so that, if the sedimentation rate is *not* constant, then it becomes more difficult to disentangle the two components.

The general sense of argument for disentanglement proceeds as for Chapter 2. First the time to depth conversion is represented through

$$t = Bz + \sum_{n=1}^{N} z_{\max} a_n/(n\pi)\sin(n\pi z/z_{\max}), \tag{5.11a}$$

where B is the average slowness over the depth range $z_{max} \geqslant z \geqslant 0$, and z_{max} is the maximum depth of measurement. The parameters B, $\{a_n\}$ are to be determined. Equally the source variation, S, is described through

$$S(z) = \sum_{n=1}^{N} z_{max} b_n/(n\pi)\left[1 - \cos(n\pi z/z_{max})\right], \tag{5.11b}$$

with the coefficients $\{b_n\}$ to be determined.
    Then

$$\frac{dt}{dz} = B + \sum_{n=1}^{N} a_n \cos(n\pi z/z_{max}) \tag{5.12a}$$

and

$$\frac{dS}{dz} = \sum_{n=1}^{N} b_n \sin(n\pi z/z_{max}) \tag{5.12b}$$

so that the average slowness is indeed

$$z_{max}^{-1} \int_0^{z_{max}} dt/dz \, dz = B. \tag{5.13}$$

Now reconsider equation (5.3) written in the form

$$\frac{dA}{dz} = A\left[-\lambda \, dt/dz + dS/dz - K \exp(-z/L) \, dt/dz\right] + K A_0 \exp(-z/L)\frac{dt}{dz}. \tag{5.14}$$

Unlike the simple case of the previous section, equation (5.14) is no longer linear in all parameters, so the least squares control function, so effective for equations with linear parameter dependence, is no longer an appropriate vehicle for efficiently disentangling and determining the various parameters.

    Instead, we follow a variation of the procedure used in Chapter 2.

    Thus: multiply equation (5.14) by $A(z) \exp(i\pi kz/z_{max})$ where k is an integer, and then integrate over $z_{max} \geqslant z \geqslant 0$ to obtain

$$-\frac{1}{2}A_0^2(-1)^k - \frac{1}{2}ik\pi \int_0^1 A(y)^2 \exp(ik\pi y) \, dy$$

$$= \int_0^1 A(y)^2 \exp(ik\pi y)$$

$$\times \left\{-(\lambda + K \exp(-z_{max}y/L))\left(B + \sum_{n=1}^{N} a_n \cos n\pi y\right) + \sum_{n=1}^{N} b_n \sin n\pi y\right\} dy$$

$$+ K A_0 \int_0^1 A(y) \exp(ik\pi y) \exp(-z_{max}y/L) \times \left[B + \sum a_n \cos(n\pi y)\right] dy, \tag{5.15}$$

where $y = z/z_{max}$.

Then note that if $K = 0$ one could follow precisely the procedure laid down in Chapter 2 to obtain both the source variation and the time-to-depth conversion. And, further, if K, A_0 and L were to be given then, once again, one could follow the procedure of Chapter 2. The variation required of the previously given procedure can then be implemented by combining equation (5.15) with a different form of equation (5.14). To do so, it is convenient to note that a general solution to equation (5.14) can be written in the form

$$A(z) = A_0 \exp(-\tau(z))$$

$$+ \exp(-\tau) \int_0^t K(A_0 - A(z')) \exp(-z'/L) \, dt/dz' \exp(\tau')a(\tau')^{-1} \, d\tau' \quad (5.16)$$

where

$$\tau(z) = \int_0^z a(z') \, dz' \quad (5.17a)$$

with

$$a(z) = \lambda \frac{dt}{dz} - \frac{dS(z)}{dz}. \quad (5.17b)$$

Thus, if a source variation, $S(z)$ is provided and also a time-to-depth conversion, dt/dz, then $a(z)$ is known. For a set of measurements $\{A_i\}$ at $\{z_i\}$ $(i = 1, \ldots, M)$ one can then write the corresponding least squares control function for equation (5.16) in the form

$$Q = \sum_{i=1}^M \left\{ A_i - A_0 \exp(-t_i) \right.$$

$$\left. - \exp(-t_i) \int_0^{t_i} K(A_0 - A(z')) \exp(-z'/L)a(z')^{-1}(dt'/dz') \, dt' \right\}^2. \quad (5.18)$$

Equation (5.18) has a minimum with respect to A_0 when

$$\sum_{i=1}^M \left\{ A_i - A_0 \exp(-t_i) - \exp(-t_i) \int_0^{t_i} K(A_0 - A(z')) \exp(-z'/L)a(z')^{-1} \frac{dt'}{dz'} \, dt' \right\}$$

$$\times \exp(-t_i) \left[1 + K \int_0^{t_i} \exp(-z'/L)a(z')^{-1} \frac{dt'}{dz'} \, dt' \right] = 0 \quad (5.19a)$$

and a minimum with respect to K when

$$\sum_{i=1}^M \left\{ A_i - A_0 \exp(-t_i) - \exp(-t_i)K \int_0^{t_i} [A_0 - A(z')] \exp(-z'/L)a(z')^{-1} \frac{dt'}{dz'} \, dt' \right\}$$

$$\times \exp(-t_i) \int_0^{t_i} [A_0 - A(z')] \exp(-z'/L)a(z')^{-1} \frac{dt'}{dz'} \, d\tau' = 0, \quad (5.19b)$$

where $t_i = \tau(z_i)$.

We recognize that equations (5.19) represent a pair of quadratic equations for determining A_0 and K. Hence they can be solved analytically, albeit tediously. The solutions can be inserted into the control function (5.18) and then L varied. Thus one produces the best fit A_0, K and L for a given set of source and time-to-depth conversions. These can then be inserted into equation (5.15) and the procedure of Chapter 2 followed to evaluate updated source and time-to-depth functions – which are then used in equations (5.19a) and (5.19b). Thus, the iterative loop can be followed through to convergence.

A procedure for parameter determination which is less direct, but which has greater flexibility, is as follows. For both the real and imaginary parts of equation (5.15), for each k value one writes the *difference* between left and right sides as $R_k + iI_k$, where R and I depend on all parameters involved. Then construct the control function

$$Q = \sum_{k=0}^{P} (R_k^2 + I_k^2)$$ (5.20)

and remember that Q depends on all parameters involved, as well as on the data field $\{A_i\}$.

Then proceed as follows. For each parameter choose an initial, but arbitrary, minimum and maximum. Denote by p the pth parameter. Then write

$$p = p_{min} + s(p_{max} - p_{min}),$$ (5.21)

where $0 \leqslant s \leqslant 1$.

Then one follows precisely the general non-linear iteration scheme given in Chapter 2 to determine the parameter values. In this way the distinction between linear versus non-linear dependence of parameters, so crucial in other procedures, is completely subsumed by the general non-linear procedure.

6. Applications and limitations

The limitation of the bioturbation model behavior for the cases of (i) constant sedimentation; and, simultaneously, (ii) constant input concentration at the sediment surface, is clearly that the activity *must* decrease with increasing depth. The idea is to delay the decrease relative to the exponential behavior that would otherwise occur. The converse side is that any variation of activity that shows an increase with depth *cannot* be accounted for by the bioturbation model on its own but requires a source variation with time.

In addition, just because activity decreases with depth in a non-exponential manner does *not* imply that bioturbation is the only cause nor, indeed, that it is even a cause. The reason for this latter statement is that under the two limitations above there are only 4 parameters to determine completely the profile of activity with depth (viz. velocity, V, scale-length, L, surface activity, A_0, and bioturbation rate constant, K). There is no guarantee that a given profile will allow a fit by the four parameters and the functional form (5.4) without significant disagreement with the observations.

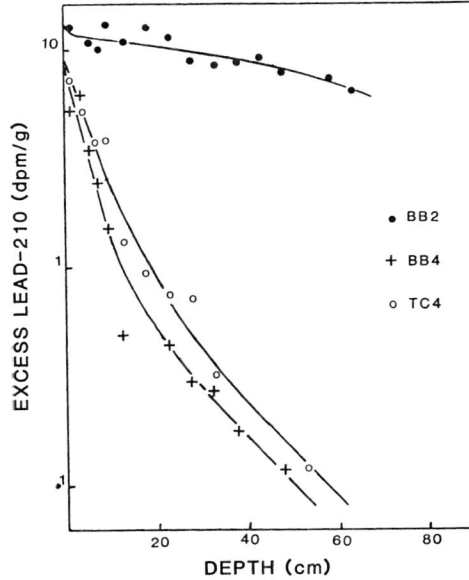

Fig. 5.2. Bioturbation model fits to salt marsh data of ^{210}Pb for three profiles at North Inlet, SC (after Gardner et al., 1987).

6.1. *North Inlet, SC*

Gardner et al. (1987) applied the bioturbation model to three different profiles of excess ^{210}Pb taken from salt marsh sediments in North Inlet, SC. Reproduced in Fig. 5.2 are their data and their fits of equation (5.4) to the data. To be noted from Fig. 5.2 is that an overall decrease of activity with depth is simulated for all three sedimentary profiles. In addition, in the case of profile BB2 there is a mismatch of theory against the data of around 10–20%, particularly at the shallower depths, while the mismatch to profiles BB4 and TC4 is considerably better at around 1–5%.

In the cases of profiles BB4 and TC4, the bioturbation rate coefficient K, as reported by Gardner et al. (1987), is small (3.6×10^{-3} and 4.3×10^{-3} yrs^{-1}, respectively) so that the dominant component to the profile shapes is pure decay ($\lambda \approx 3 \times 10^{-2}$ yr^{-1}) with, at best, only a small admixture of bioturbation. However in the case of profile BB2, the bioturbation coefficient is around 0.23 yr^{-1}, so that throughout the total profile it is bioturbation that controls the behavior rather than decay. The explanation put forward by Gardner et al. (1987) for this major difference between the studied sites is that site BB2 is at a highly burrowed creek bank, while sites BB4 and TC4 are in the marsh interior and are not so susceptible to fiddler crab burrowing due to the ecological preference for the crabs to be at creek bank locations.

6.2. *Lake Baikal, Russia*

Edgington et al. (1991) measured profiles of both excess ^{210}Pb and ^{137}Cs at ten sites in Lake Baikal. The data they reported are presented in Fig. 5.3, after removal of superposed lines

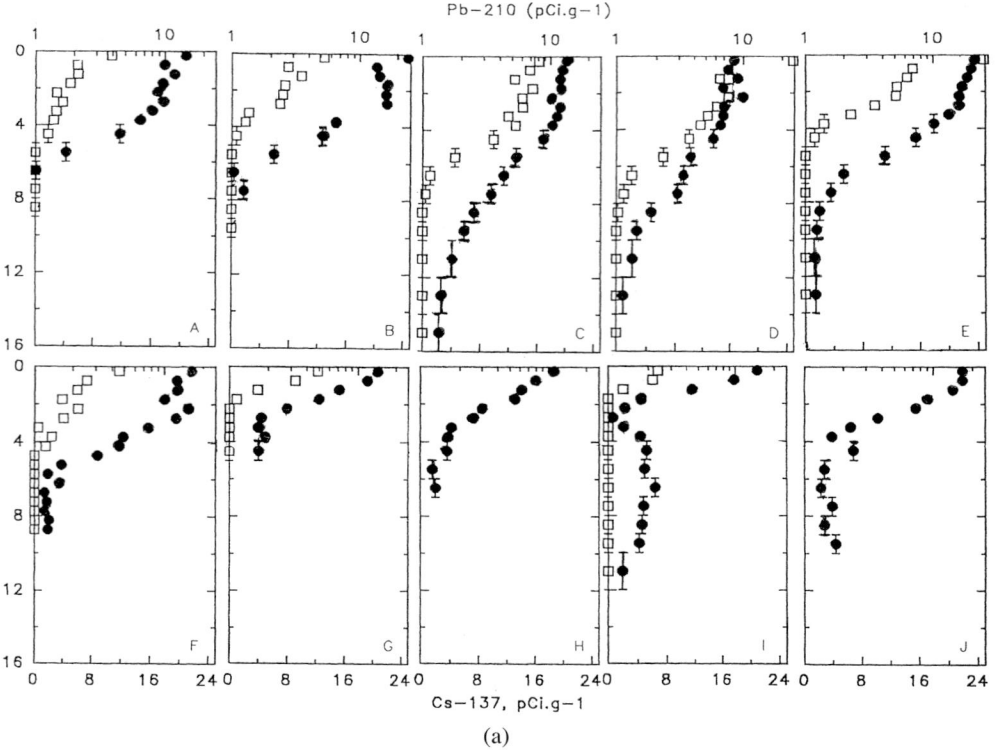

Fig. 5.3. (a) Ten profiles of ^{210}Pb and of ^{137}Cs for Lake Baikal sediments. Upper logarithmic abscissa is concentration of ^{210}Pb (pCi/g), lower linear abscissa is concentration of ^{137}Cs (pCi/g). Ordinate is sediment depth in cm (after Edgington et al., 1991).

from the original paper. There is a claim (Edgington et al., 1991) that "Each ^{210}Pb profile shows a layer of constant activity (surficial mixing) followed by an exponential decrease in concentration with depth to a constant value maintained by in situ decay of ^{226}Ra". In fact, direct inspection of the data profiles for ^{210}Pb shows that this claim is overstated. For instance, station A shows first a decrease {from about 15 to 10 in units of pCi g^{-1} (where 1 pCi = 37 mBq = 2.22 dpm)} followed by an increase (from 10 to around 12); station B shows first a decrease (from about 22 to around 12) followed by an increase (to around 14); and similarly for other stations. In fact none of the profiles show a *constant* near surficial activity with increasing depth; there is either a decay of activity or a reversal near the surface. Further, not all ^{210}Pb profiles are, thereafter, exponentially decreasing – as can be seen by inspection of the profiles for stations *I* and *J* for instance.

The changes in radionuclide concentration with depth are attributed to redistribution of sediments by mixing, a longer residence time of ^{137}Cs in the water than in other lakes, and/or from post-depositional mobility by molecular-scale diffusion (Edgington et al., 1991). However, bioturbation of deposited radionuclide concentrations is *not* necessarily the reason for the observed variations, but rather changes in radionuclide composition due to source varia-

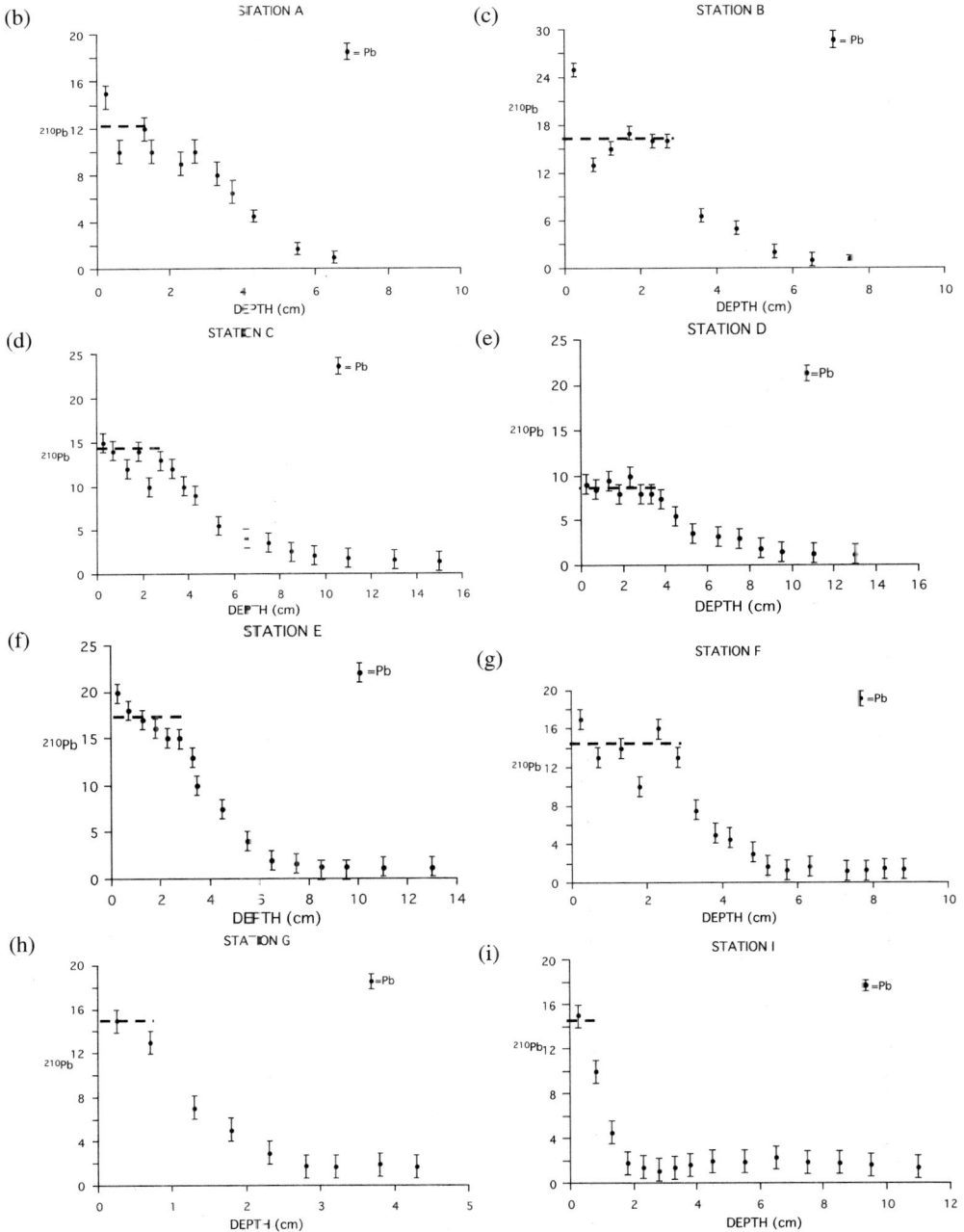

Fig. 5.3. (*Continued*) (b)–(i) Eight profiles on a *linear* scale of ^{210}Pb for Lake Baikal sediments (stations A–G and I). [137]Cs was not measured at stations H and J and therefore these are not included. Abscissa is sediment depth in cm; ordinate is excess ^{210}Pb in pCi/g. The sub-figures were produced by reading directly from the logarithmic figures of Edgington et al. (1991) due to inability to obtain the actual data. Note that 1 pCi = 37 mBq = 2.22 dpm.

tions with time as will be shown (Chapter 7) to be the case for two other profiles from Lake Baikal.

The point of including this example in the bioturbation chapter is to point out that there is a great danger in not using all of the radionuclide data to evaluate causes of observed profiles. In addition, plotting on a logarithmic scale tends to obscure the fact that variations of up to almost factors of 2 occur in the shallowest sedimentary depths, which of itself can lead to serious misinterpretations if a "constant" mixing model is used (Fig. 5.3). The dotted line drawn on each station profile is the "constant" value claimed to provide evidence of shallow mixing; the error bars have been added to provide an assessment of likely uncertainty. It can be clearly seen that the constant value fails to satisfy the shallow ^{210}Pb data in all but one or two cases – and the misfits are up to factors of around 50–100%. It is also important to note that while ^{137}Cs production peaked on a worldwide scale in 1962/64, it does *not* follow that on the local scale of Lake Baikal that such a peak will be observed. The Chinese nuclear testing at Lop Nor was much later than 1964, and the prevailing winds would enhance the ^{137}Cs input from at least the Selenga River into very recent sediments.

Chapter 6

Biological mixing coefficients

Biological mixing models are conveniently subdivided into two broad classes: (1) particle transport models whereby tracers are strongly bound to particle phases and reaction-diffusion models. Reaction-diffusion models are normally used to evaluate the effects of organisms on chemical diagenesis and sediment–water exchange processes (e.g., Aller, 1980a, 1980b; Soetaert et al., 1996b; Schlüter et al., 2000). The basis of these models is the basic equations for the conservation of both solids and porewater as given by Berner (1980).

When using reaction-diffusion models additional complications arise involving the coefficients used to describe chemical adsorption and diffusion processes for mixtures of sediment and water. These sometimes lead to the inability of radionuclide activity with depth profiles to be statistically sharp in age determinations of sediments. We consider in turn three of these processes, simple linear adsorption, microscale diffusive processes and bulk diffusive processes. We conclude this chapter with a simple technique for distinguishing between depositional concentration behavior and rapid mixing under the assumption of steady, constant, sedimentation.

1. Simple linear adsorption

Due to the relative partitioning of radioactive species between water and sedimentary particles, deposited sediments have variable activity per gm due to both intrinsic variations, modifications during burial influx, and modifications after deposition. The deposition rate of sedimentary particles versus the residence time of radioactivity in the overlying water column determines precisely the level of depositional activity at the seabed surface. Fine-grained particles, which spend a long time in the water column compared to coarse-grained materials, presumably reach a closer equilibrium balance (partitioned activity) of radioactive exchange with the surrounding water milieu. After burial, radionuclides bound to sedimentary particles may exchange as a result of sediment diagenesis. Conservation of mass dictates that exchanges resulting in changes in particulate concentration must be balanced by an equal and opposite change in the dissolved concentration of the radionuclide. Therefore, basic information on the sorption potential of radionuclides onto particulate matter becomes a prerequisite for the use of reaction-diffusion models.

A practical representation of the sorption process uses equilibrium distribution coefficients (K_ds), where $K_d = $ Ased/Aw, where Ased and Aw are the radionuclide activitiy of particles (Bq/g sed) and filtered seawater, or porewater (Bq/g water), respectively. Equilibrium K_ds are dependent upon the relative influence of the chemical properties of the radionuclides, particle

97

characteristics and concentrations, and the moderating influence of the oceanic environment (Wepener et al., 2000; Duursma & Carroll, 1996; Smith & Comans, 1996; Dai & Martin, 1995; Honeyman et al., 1988; Li, 1981).

In reaction-diffusion models, the exchange process is expressed in terms of mass conservation. The rate of tracer transfer into the particulate (dissolved) phase in a given volume of water (sediment) is equal to the rate of tracer transfer out of the particulate (dissolved) phase (Robbins, 1986).

A practical representation of the sorption process in heterogeneous systems is to use a water partition coefficient, P_w, for each radionuclide. In turn, this partition coefficient is tied to the concentration of particles in the system, C, and to K_d.

Assuming for the moment that the K_d value is known (see, however, later in this chapter) then, if the particle concentration, C, is also precisely known, it follows that the water partition coefficient is

$$P_w = (1 + K_d C)^{-1},$$

(6.1a)

so that the sedimentary particle activity partition, P_p, is

$$P_p = 1 - P_w = K_d C / (1 + K_d C).$$

(6.1b)

This formulation is useful when tracer activities are expressed in terms of the composition of whole sediment (when porewater and solids are not separated experimentally). The equation of mass conservation between porewater and deposited sediments can then be expressed in terms of the partition coefficient as given by Robbins (1986, his equation (5)).

Similarly, the above formulation is useful when tracer activities are expressed in terms of the composition of unfiltered water (when suspended solids and water are not separated experimentally). In this case, the equation of mass conservation between suspended sediments and water is as given by Robbins (1986, his equation (9)).

Thus the formulation of equation (6.1) is useful in determining the partitioning of tracers between suspended sediment and water for models that require estimates of the influx of tracer from the water column to the seabed.

A problem that arises in attempting to use P_w (or P_p) to assess the likely changes in sedimentary activity brought about (either prior to or after deposition) is that no precise value is available to evaluate the water (or sediment) partition coefficient. Instead one must allow for the measured (or inferred) concentration variability. Second, even if the concentration were to be precisely known, there is still considerable uncertainty in determinations of the K_d values. Thus the partition coefficient will also have a dynamical range. Part of the task of this section of the chapter is to show, by illustration, how one goes about determining this range and how it can be tied to the cumulative probability for estimating the likelihood of activity at deposition in excess of a specified amount per gm. To illustrate the way one calculates the corresponding probabilities for partition coefficients proceed as follows.

The approach is to make an "educated guess" for the average and range of values of S and K_d. We show how to compute the probability distribution of $P_w(C)$ from knowledge of the individual probability distributions of K_d and C, and we illustrate with an example the differences between assuming that P_w can be computed solely from the average values of K

and C versus from the exact probability distribution. As we shall see the difference can be quite substantial.

Clearly, what is needed to understand the process of potential sedimentary particle pick-up (or stripping) of radioactive activity relative to that in the water column is an estimate of the water partition coefficient, P_w, for each radionuclide. In turn, this partition coefficient is tied to the water concentration of each radionuclide and to a coefficient, K_d, which measures the relative affinity of sediment and water for reaching an equilibrium balance in a radionuclide when full solubility can be reached.

There is still considerable uncertainty in determinations of the K_d values for particular radionuclides and, it would seem, that K_d values are, at the least, pressure, temperature, and water salinity-dependent (Duursma & Carroll, 1996) so that some range of values is always present. This factor is important because it provides an idea of the fraction of radioactive concentration that is likely to be original to the sedimentary particles, and also the fraction that is likely "allochthonous" due to a combination of sorption from the particles into riverine or standing waters in the basin and sorption by the sedimentary particles from the waters during their journey to final deposition.

To illustrate the way one calculates the corresponding probabilities for partition coefficients proceed as follows.

The actual values of $\bar{K}_d$ and C are poorly known in most areas of the ocean, and are variable spatially and temporally anyway, as is especially true for the Russian Arctic, where data are sparse because access has been limited until recently. Nonetheless, risk assessors and radionuclide transport modelers must decide on suitable values for these parameters. A number of assessments were carried out to determine the potential risk to human health and the environment posed by nuclear wastes discharged by the former Soviet Union into the Kara Sea.

As a result of the paucity of data for the Kara Sea, the approach to date has been to make an "educated guess" for the average value of C in the region. But sediment concentrations vary from a few milligrams per liter of seawater in the open Kara Sea to as high as hundreds of milligrams per liter near the seabed and near the mixing zone of the Ob and Yenisi Rivers. For K_d, the approach has been to use average values for the various radionuclides of interest taken from IAEA (1985); but the values for individual radionuclides range over several orders of magnitude. Alternatively, it is sometimes assumed that none of the radioactivity sorbs onto sediment particles, i.e., all of the radioactivity remains dissolved in seawater (i.e., $p_c(S) = 0$ and $p_c(W) = 1$), where S stands for sediments and W for water.

We show how to compute the probability distribution of $p_c(S)$ from knowledge of the individual probability distributions of K ($= K_d$) and C, and we illustrate with an example the differences between assuming that p_c can be computed solely from the average values of K and C versus from the exact probability distribution.

1.1. *General mathematical development*

For brevity write the partition coefficient $p_c(S)$ as λ with, from equation (6.1),

$$\lambda = u/(1 + u), \tag{6.2}$$

where $u = KC$.

Let the probability distribution of finding a value of K in the range K to $K + dK$ be $p_1(K) \, dK$; for C let it be $p_2(C) \, dC$, with normalizations

$$\int_0^\infty p_1(K) \, dK = 1 = \int_0^\infty p_2(C) \, dC. \tag{6.3}$$

Then the probability distribution of finding a value of u ($\equiv KC$) in the range u to $u + du$ is

$$p(u) \, du = \left[\iint dK \, dC \, \delta(u - KC) p_1(K) p_2(C) \right] du, \tag{6.4}$$

where $\delta(x)$ is the Dirac δ-function of argument x, where the range of the integrals is $0 \leqslant K \leqslant \infty$, and $0 \leqslant C \leqslant \infty$ and where equation (6.4) has the automatic normalization

$$\int_0^\infty p(u) \, du = 1. \tag{6.5}$$

The range of u is $0 \leqslant u \leqslant \infty$, so that $0 \leqslant \lambda \leqslant 1$, together with $u = \lambda/(1 - \lambda)$, and $du = d\lambda/(1 - \lambda)^2$. Hence the probability distribution of finding a value of λ in the range λ to $\lambda + d\lambda$ is given by

$$p(\lambda) \, d\lambda = (1 - \lambda)^{-2} \, d\lambda \int_0^\infty dK \int_0^\infty dC \, p_1(K) p_2(C) \delta\big[\lambda(1 - \lambda)^{-1} - KC\big], \tag{6.6}$$

with

$$\int_0^1 p(\lambda) \, d\lambda = 1. \tag{6.7}$$

Use the property of the δ-function that

$$\delta\big(KC - \lambda(1 - \lambda)^{-1}\big) = \frac{1}{C} \delta\big[K - \lambda C^{-1}(1 - \lambda)^{-1}\big] \tag{6.8}$$

to re-write equation (6.7) in the form

$$p(\lambda) \, d\lambda = d\lambda(1 - \lambda)^{-2} \int_0^\infty C^{-1} p_2(C) p_1\big(\lambda C^{-1}(1 - \lambda)^{-1}\big) \, dC \tag{6.9}$$

which expresses directly the probability distribution of the partition coefficient in terms of the probability distributions for the sediment/water distribution coefficient and the sediment concentration.

1.2. An illustrative example

We illustrate the approach for ^{137}Cs, which is one of the primary radionuclides listed in inventories of nuclear waste in the Kara Sea (White Book, 1993). Combining information from

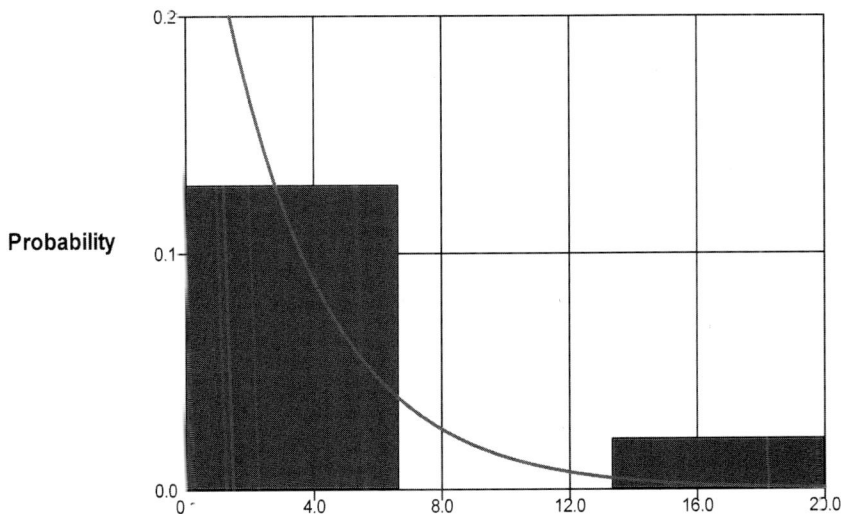

Fig. 6.1. Idealized probability function for the sediment/water distribution coefficient (K) of ^{137}Cs. The range of K is from IAEA (1985). The form of the exponential function is based on the clustering of measurements at the lower limit of the distribution, as shown in the underlying bar graph.

IAEA (1985) with empirical measurements (Schell & Sibley, 1982; Sholkovitz et al., 1983; Sholkovitz & Mann, 1984), take the sediment/water distribution coefficient, K, to be approximately exponentially distributed in $0 \leqslant K \leqslant \infty$ (Fig. 6.1, see also Duursma & Carroll, 1996), with

$$p_1(K) = K_*^{-1} \exp(-K/K_*), \tag{6.10}$$

where K_* is a positive constant. Equation (6.10) has unit normalization and mean value $\langle K \rangle = K_*$.

For our illustrative case of the Kara Sea, we must rely on a generalized understanding of sediment dynamics in coastal seas to define the probability distribution for sediment concentration. As previously mentioned, we expect sediment concentration to vary from a few milligrams to hundreds of milligrams per liter of seawater. Most values will be near the lower limit of the range (e.g., ≈ 5 mg/l). Therefore, take the sediment concentration, C, to be approximately a gamma distribution with parameters 2 and $1/C_*$ and with expected value $\langle C \rangle = 2C_*$, i.e.,

$$p_2(C) = CC_*^{-2} \exp(-C/C_*). \tag{6.11}$$

If the average values of $\langle K \rangle$ and $\langle C \rangle$ were to be used in the partition coefficient (6.11), one obtains

$$\lambda_{\text{av}} = 2u_*/(1 + 2u_*) \tag{6.12}$$

where $u_* = K_* C_*$. It is λ_{av} which has to be compared against the exact average value obtained from the probability distribution for $p(\lambda)$.

Using expressions (6.10) and (6.11) reduces equation (6.9) to the form

$$p(\lambda)\,d\lambda = K_*^{-1} C_*^{-2}\,d\lambda (1 - \lambda)^{-2} \int_0^\infty \exp\{-C/C_* - K_*^{-1}\lambda(1 - \lambda)^{-1}C^{-1}\}\,dC.$$

$$(6.13)$$

The integral over C in equation (6.13) can be performed in terms of $K_1(x)$, the modified Bessel function of the second kind of order unity and of argument x (Gradshteyn & Ryzhik, 1965), yielding

$$p(\lambda)\,d\lambda = 2 u_*^{-3/2}\lambda^{1/2}(1 - \lambda)^{-5/2} K_1\big(2 u_*^{-1/2}\lambda^{1/2}(1 - \lambda)^{-1/2}\big)\,d\lambda.$$

$$(6.14)$$

The cumulative probability $P(l)$, of obtaining a value less than or equal to 1, is given by

$$P(l) = \int_0^l p(\lambda)\,d\lambda \equiv \frac{1}{2}\int_0^{q(l)} x^2 K_1(x)\,dx,$$

$$(6.15)$$

where

$$q(\lambda) = 2 u_*^{-1/2}\lambda^{1/2}(1 - \lambda)^{-1/2}$$

$$(6.16a)$$

so that

$$\lambda = q^2(u_*/4)\big(1 + q^2 u_*/4\big)^{-1} = q^2 u_* \big(4 + q^2 u_*\big)^{-1}.$$

$$(6.16b)$$

Now use the well known Bessel function property that

$$\int_0^a x^2 K_1(x)\,dx = 2 - a^2 K_2(a)$$

$$(6.17)$$

to write the cumulative probability, $P(\lambda)$, of obtaining a partition coefficient less than or equal to λ, as

$$P(\lambda) = 1 - 1/2 q(\lambda)^2 K_2\big(q(\lambda)\big).$$

$$(6.18)$$

It also follows that the exact average value of λ, $\langle\lambda\rangle$, can then be written as

$$\langle\lambda\rangle = \frac{u_*}{8}\int_0^\infty x^4 K_1(x)\big(1 + u_* x^2/4\big)^{-1}\,dx.$$

$$(6.19)$$

A plot of the right-hand side of equation (6.18) is given in Fig. 6.2, showing that the cumulative probability $P(\lambda)$ is almost a linear function of $q(\lambda)$ throughout the range $0 \leqslant q \leqslant 3.5$. Indeed, an accurate approximation in $0 \leqslant q \leqslant 3.5$ is that

$$P(\lambda) = mq(\lambda),$$

$$(6.20)$$

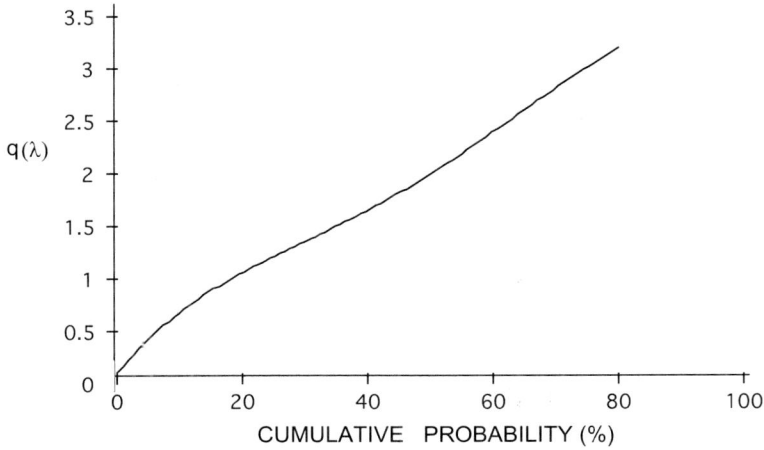

Fig. 6.2. Cumulative probability plot of $q(\lambda)$ taken from equation (6.18).

where $m \approx 0.3$ for $P(\lambda)$ fractional $(0 \leqslant P(\lambda) \leqslant 1)$. Then

$$P(\lambda) = 0.60 u_*^{-1/2} \lambda^{1/2} (1 - \lambda)^{-1/2}. \tag{6.21}$$

The value, λ_p, corresponding to a cumulative probability of P, is then given approximately by

$$\lambda_p = (0.36)^{-1} P^2 u_* \left(1 + P^2 u_*/(0.36)\right)^{-1}. \tag{6.22}$$

For a cumulative probability of 50% (P = 0.5) the median value of λ is

$$\lambda_{0.5} = (1.44)^{-1} u_* \left(1 + u_*/(1.44)\right)^{-1} \tag{6.23a}$$

while the 2/3 chance (P = 0.67) value is

$$\lambda_{0.67} = (0.81)^{-1} u_* \left(1 + u_*/(0.81)\right)^{-1}. \tag{6.23b}$$

At 10 and 90% cumulative probabilities, the corresponding λ values are

$$\lambda_{0.1} = (36)^{-1} u_* \left(1 - u_*/(36)\right)^{-1} \tag{6.23c}$$

and

$$\lambda_{0.9} = (0.44)^{-1} u_* \left(1 + u_*/(0.44)\right)^{-1}. \tag{6.23d}$$

The approximate value, λ_{av}, for the partition coefficient obtained by replacing K and C by their average values, respectively, is given in equation (6.12), so that by equating $P^2 m^{-2}/4$ to

2 the value λ_{av} corresponds to a cumulative probability of $P_{av} = 2^{3/2}m$, which yields $P \approx 0.84$ for $m \approx 0.30$.

Thus: in the computation of the partition coefficient there is about an 84% chance that the correct average value is less than that obtained by using the average values of K and C to compute an approximate λ_{av}.

If one takes the 67% cumulative probability value, $\lambda_{0.67}$, as representative of the mean value then

$$\lambda_{0.67}/\lambda_{av} = (1.6)^{-1}[1 + 1.2u_*]^{-1}[1 + 2u_*]. \tag{6.24}$$

For $u_* \ll 1$, we have

$$\lambda_{0.67}/\lambda_{av} = (1.6)^{-1}\{1 + 0.8u_*\} < 1 \tag{6.25a}$$

while for $u_* \gg 1$

$$\lambda_{0.67}/\lambda_{av} = 1 - 1/(3u_*) < 1. \tag{6.25b}$$

Thus λ_{av} always exceeds $\lambda_{0.67}$. For large values of $u_* \gg 1$, there is but little difference in $\lambda_{0.67}$ and λ_{av}; however, for $u_* \ll 1$, $\lambda_{0.67}$ is about 2/3 of λ_{av}, i.e., λ_{av} overestimates the true $\lambda_{0.67}$ for sediments by about 50%. The converse is true for the partition coefficient for water $(p_s(W) = 1 - \lambda)$. In that case, at $\lambda_{0.67}$, we have

$$p_{0.67}(W) = 1 - \lambda_{0.67} = (1 + 2u_*)^{-1} \tag{6.26a}$$

while at $\lambda = \lambda_{av}$, we have

$$p_{av}(W) = 1 - \lambda_{av} = (1 + 2u_*)^{-1}. \tag{6.26b}$$

In this case, at small values of $u_* \ll 1$, $p_{0.67} \approx p_{av}$, but at high values of $u_* \gg 1$,

$$p_{0.67}/p_{av} \approx 1.67 \tag{6.27a}$$

so that using p_{av} is an *underestimate* of $p_{0.67}$ by a factor 1.67.

From the preceding evaluation we see that a sediment partition coefficient for ^{137}Cs will be *overestimated* if based on the average value of sediment concentration in the Kara Sea and on the ^{137}Cs distribution coefficient. Conversely, a reasonable approximation of the water partition coefficient will be obtained. Because the distribution coefficient, K, for ^{137}Cs is quite low, most of the ^{137}Cs released into the Kara Sea will remain dissolved in seawater, implying the current use of average values of K and C by radionuclide transport modelers and risk assessors is appropriate for ^{137}Cs. The approach here also allows for the determination of the range of uncertainty for the partition coefficient because a measure of the volatility, v, of, say, the 50% chance value is provided by Lerche (1992)

$$v = |(\lambda_{90} - \lambda_{10})/\lambda_{50}|. \tag{6.27b}$$

For $u_* \ll 1$, the volatility is $v = 3.2$, so that there is considerable uncertainty on the *sediment* partition coefficient and, correspondingly, very little uncertainty on the water partition coefficient. At high $u_* \gg 1$, the corresponding volatilities are reversed.

In short: the estimates of partition coefficients for sediments and water imply that *both* cannot be calculated simultaneously with accuracy using the replacement value λ_{av} obtained with the average values for both K and C.

When it is remembered that the partition coefficient is used for sorption determinations, a 50–67% error can have major consequences for assessing total radionuclide leakage from a dumpsite and its sorption onto sediments or into water.

For the radionuclide ^{137}Cs, the error for the water partition coefficient is negligible while the sediment partition coefficient is underestimated by 37%. However, in this case it is more important to accurately predict the water partition coefficient. We need not be as concerned about the error in the sediment partition coefficient because ^{137}Cs has a very low affinity for particles. Thus the risk of human and ecological health effects resulting from the discharge of ^{137}Cs into the Kara Sea will be primarily from exposure to contaminated seawater.

2. Diffusion

Radionuclides bound to sedimentary particles after deposition can be redistributed only by: (i) complete removal of the sediments and deposition elsewhere (turbidites); (ii) in situ vertical displacement of sediments post-deposition; (iii) unbinding of the radionuclides from the sedimentary particles and water flushing, thereby placing a radionuclide elsewhere and importing a different concentration to be bound.

The usual assumptions made are that: (a) a steady-state diffusion approach is adequate; (b) a vertical diffusion approximation is adequate; (c) bioturbation and/or radionuclide sorption and desorption can both be handled by a diffusion process with different choices of diffusion coefficient dependence with sediment depth, z. Based on these assumptions, the corresponding diffusion equation for variations of concentration, c, with sediment depth, z, is customarily (Christensen, 1982; DeMaster & Cochran, 1982; Officer, 1982) written in the form

$$\partial(Vc - D\partial c/\partial z)/\partial z + \Lambda c = 0, \tag{6.28}$$

where c is the concentration activity of the radionuclide at depth z; Λ is the radioactive decay constant, V is the sedimentation velocity, and D is the diffusion coefficient which can be depth-dependent.

3. Microscale diffusion

The parameter D in equation (6.28) represents the diffusion of dissolved constituents in pore fluids subject to microscale diffusive transport. This transport corresponds to transport on scales comparable to or smaller than the spaces between sediment particles (Berner, 1980). The coefficient D can be separated further into two parts, where

$$D = D_s + D_i.$$

Here D_s is the coefficient for molecular diffusion and D_i is an irrigation coefficient that accompanies particle mixing.

Biodiffusion of a dissolved constituent associated with mixing has one important distinction from molecular diffusion in that the flux of a dissolved constituent during mixing can result from a gradient in porosity (water content). The effect of porosity gradients on mixed sediment profiles has long been recognized from a theoretical standpoint (Berner, 1980). The bioturbational mixing of both solids and porewater can result in three different outcomes: (1) homogeneous distribution of porosity; inhomogeneous distribution of particle types relative to one another; (2) inhomogeneous distribution of porosity; homogeneous distribution of particle types relative to one another; (3) inhomogeneous distribution of both porosity and particle types (see Fig. 1 of Mulsow et al., 1998). An investigation of this phenomenon along the eastern Canadian margin showed that porosity effects could be important at roughly one half of the investigated sites (Mulsow et al., 1998). Nonetheless, many models assume D_i is small relative to D_s so that the effect of porosity gradients can be ignored. If $D_i = 0$, then

$$D = D_s = D_0(\phi/\phi_s)^{n-1},$$

where n is approximately equal to 2 (Berner, 1980), ϕ is the porosity, and D_0 is the value of D at the sediment surface where the porosity is ϕ_s.

4. Bulk diffusion

The determination of a bioturbation coefficient in the form of a diffusion coefficient (D_b) has been handled in bioturbation models using different choices of diffusion coefficient dependence with sediment depth, z (Schlüter et al., 2000). The corresponding diffusion equation for variations of concentration, c, with sediment depth, z, is then precisely equation (6.28).

A general analytic solution to equation (6.28) for arbitrary functional forms for $D(z)$ is not available. However, from a numerical perspective this is not a major problem as we now show in two parts: first by taking $D(z)$ to be a constant and then by taking $D(z)$ to be piecewise constant – which is what a numerical scheme for solution of equation (6.28) uses anyway.

4.1. Constant diffusion coefficient with depth

For D constant, the general solution to equation (6.28) can be written in the form

$$c(z) = \exp\big((Vz)/(2D)\big) \times \big[c_1 \exp(\alpha z) + c_2 \exp(-\alpha z)\big], \tag{6.29}$$

where

$$2\alpha = \big[(V/D)^2 + 4\Lambda/D\big]^{1/2}, \tag{6.30}$$

and where c_1 and c_2 are two arbitrary constants to be determined by the boundary conditions on $z = 0$. The two boundary conditions on $z = 0$ are:

(i) the concentration be given, $c = c_0$,
(ii) the net surficial flux, $F_0 = V c_0 - D \frac{\partial c}{\partial z}$, be given.

Then the constants c_1 and c_2 are given by

$$c_1 = \frac{1}{2aD} [F_0 - c_0(V/2 - \alpha D)], \tag{6.31a}$$

$$c_2 = \frac{1}{2aD} [-F_0 + c_0(V/2 + \alpha D)]. \tag{6.31b}$$

However, $\alpha > 0$, so that if the term involving c_1 were to be present then the concentration would increase exponentially with increasing depth – a physical impossibility. Hence, in order that $c_1 = 0$, the net flux, F_0, and the surficial concentration, c_0, *must* satisfy the relation

$$F_0 = c_0(V/2 - \alpha D). \tag{6.32}$$

And then $c_2 = c_0$, so that the physical solution is

$$c(z) = c_0 \exp\{[(V/(2D) - (V^2/(4D^2) + 4AD)^{1/2}]z\}. \tag{6.33}$$

Note, for later use, that equation (6.33) provides $\partial c / \partial z < 0$ everywhere. Thus, any observed profile of radioactivity which, at the surface or elsewhere, has a reversal in concentration cannot be described by a constant diffusion model. A simpler form for equation (6.33) is in terms of the scale-length, L, where

$$L^{-1} = -V/2D + ((V/2D)^2 + 4A/D)^{1/2} > 0 \tag{6.34a}$$

when

$$c(z) = c_0 \exp(-z/L). \tag{6.34b}$$

Thus the two unknowns in this constant diffusivity model are c_0 and L – which can readily be determined by least squares fitting to a set of data.

However, note the requirement of an exponential decrease with depth in this case, which is not necessarily how a set of data will behave. There is no requirement that Nature slave itself to the dictates of any given such model. And observations indicate the lack of such conformity to exponential behavior with depth, as has been shown earlier.

For this reason, attempts have been made over the years, as referenced earlier, to allow both the velocity, V and the diffusion coefficient D to vary with depth in efforts to accommodate for the non-exponential with depth behavior of observed activity profiles. We consider this aspect next.

4.2. *Variable diffusion coefficient with depth*

There are general constraints on concentration gradient with depth for any diffusion coefficient dependence. For suppose equation (6.28) is integrated from $z = 0$ to a depth z, then

$$V(z)c(z) - D(z)\partial c/\partial z = F_0 - \Lambda \int_0^z c(z)\,dz. \tag{6.35}$$

At infinitely large depths, one must have $c \to 0$, $\partial c/\partial z \to 0$. Hence

$$F_0 = \Lambda \int_0^\infty c(z)\,dz > 0. \tag{6.36}$$

Therefore, at any shallower depth, it follows that the right-hand side of equation (6.35) is intrinsically positive. Thus at all depths there is the requirement $V(z)c(z) > D(z)\partial c/\partial z$.

Suppose $c(z)$ were to exhibit a maximum at some depth. Then at the maximum one requires $\partial c/\partial z = 0$ and $\partial^2 c/\partial z^2 < 0$. However, from equation (6.28), at any place where $\partial c/\partial z = 0$ one has

$$D(z)\partial^2 c/\partial z^2 = c(\Lambda + \partial V/\partial z) > 0. \tag{6.37}$$

Hence, there cannot be a maximum. Thus, because $c(z) \to 0$ as $z \to \infty$ it follows that $c(z)$ must be decreasing from the surficial value, c_0, at all depths, i.e., $\partial c/\partial z \leqslant 0$ everywhere. Thus any profile showing a reversal of $c(z)$ with depth cannot be described by a steady-state diffusion model no matter what spatial dependence is chosen for the diffusion coefficient.

These general considerations allow one to provide a simple numerical scheme for constructing $c(z)$ for any depth dependence choice of $D(z)$, as follows. Split the subsurface into a suite of N layers, as depicted in Fig. 6.1, with the diffusion coefficient and sedimentation velocity constant within each layer. For depths, z, greater than z_N, it is taken that the diffusion coefficient is constant at the value D_{N+1}. The diffusive solution for $z > z_{N+1}$ can then be written down using the solution given in the previous subsection:

$$c(z > z_{N+1}) = c_* \exp(-\beta_{N+1}z), \tag{6.38a}$$

where

$$\beta_{N+1} = \left(V_{N+1}/2D_{N+1} - \left[(V_{N+1}/2D_{N+1})^2 + 4\Lambda/D_{N+1}\right]^{1/2}\right). \tag{6.38b}$$

Here c_* is a constant that has to be related to the surficial concentration and/or net flux. The procedure for effecting this connection is now to work upwards. Thus in $z_N > z > z_{N+1}$, where $D = D_N$ and $V = V_N$ one can write the general diffusion solution as

$$c(z) = c_{N1} \exp(\beta_{N1}z) + c_{N2} \exp(\beta_{N2}z), \tag{6.39a}$$

where β_{N1}, β_{N2} are given by the two independent solutions

$$\beta = V_N/2D_N \pm \left[(V_N/2D_N)^2 + 4\Lambda/D_N\right]^{1/2}. \tag{6.39b}$$

Here c_{N1} and c_{N2} are two constants to be determined. Conservation of $c(z)$ across $z = z_{N+1}$, and of the net flux $Vc - D\partial c/\partial z$ yields the constants c_{N1}, c_{N2} in terms of c_* as

$$c_{N2} D_N (\beta_{N1} - \beta_{N2}) \exp(\beta_{N2} z_{N+1})$$
$$= (V_{N+1} - V_N) c_* \exp(-\beta_{N+1} z_{N+1})$$
$$+ c_* \exp(-\beta_{N+1} z_{N+1})(\beta_{N1} D_N + \beta_{N+1} D_{N+1}), \tag{6.40a}$$

$$c_{N1} D_N (\beta_{N1} - \beta_{N2}) \exp(\beta_{N1} z_{N+1})$$
$$= -(V_{N+1} - V_N) c_* \exp(-\beta_{N+1} z_{N+1})$$
$$+ c_* \exp(-\beta_{N+1} z_{N+1})(-\beta_{N1} D_{N+1} - \beta_{N2} D_N). \tag{6.40b}$$

Thus by repeating the match of the various piecewise solutions in each interval, one gradually and systematically iterates the solution, in terms of c_*, eventually arriving at the sedimentary surface where both the surficial concentration *and* the surficial flux will be expressed in terms of the *one* unknown constant c_*. Hence, the only technical point to address is, for a given functional dependence of $D(z)$, how one choose the "constant" values for D_n in each depth interval $z_{n+1} \geqslant z \geqslant z_n$. The easiest way is, of course, to write

$$D_n = (z_{n+1} - z_n)^{-1} \int_{z_n}^{z_{n+1}} D(z)\, dz. \tag{6.41}$$

Clearly, for numerical schemes it is the constant c_* and the parameters in the constant values of D in the N domains that have to be evaluated to provide minimum mismatch to any available data set of activity variation with depth.

Thus the spatially variable diffusion coefficient situation can be brought to the same level of computation as can the constant diffusion coefficient case – which was the main point of this section.

5. Near surficial rapid mixing

Prior to the biological mixing model introduced by Gardner et al. (1987), there was already available (Robbins, 1978) a simple near-surficial rapid mixing model that is still occasionally invoked (Edgington et al., 1991). In the rapid mixing model, it is assumed that there is some form of shallow sediment mixing (usually unspecified in detail), whose operation is to take the concentrated activity deposited to a depth z_{mix}, and arrange for that activity to be uniformly spread from the sediment surface to the depth z_{mix}. The purpose, apparently, is to handle situations in which there is slow sedimentation relative to biological overturning and homogenization of deposited material to depth z_{mix}. Thus one fits a constant concentration model to the near-surficial measurements to some depth z_{mix}. As z_{mix} is increased, the root mean square departure of the assumed constant concentration from the observations should first decrease and then, as the data included incorporate depths greater than the depth to the physical mixing depth, so the mismatch increases. Thus, in this way, one obtains an operational estimate of the mixing depth plus an estimate of the equivalent constant concentration (Fig. 6.3).

Often it is further assumed that this mixing depth scale then applies to all deeper measurements so that the intrinsically deposited concentration, $I(z)$, now at depth z, is modulated

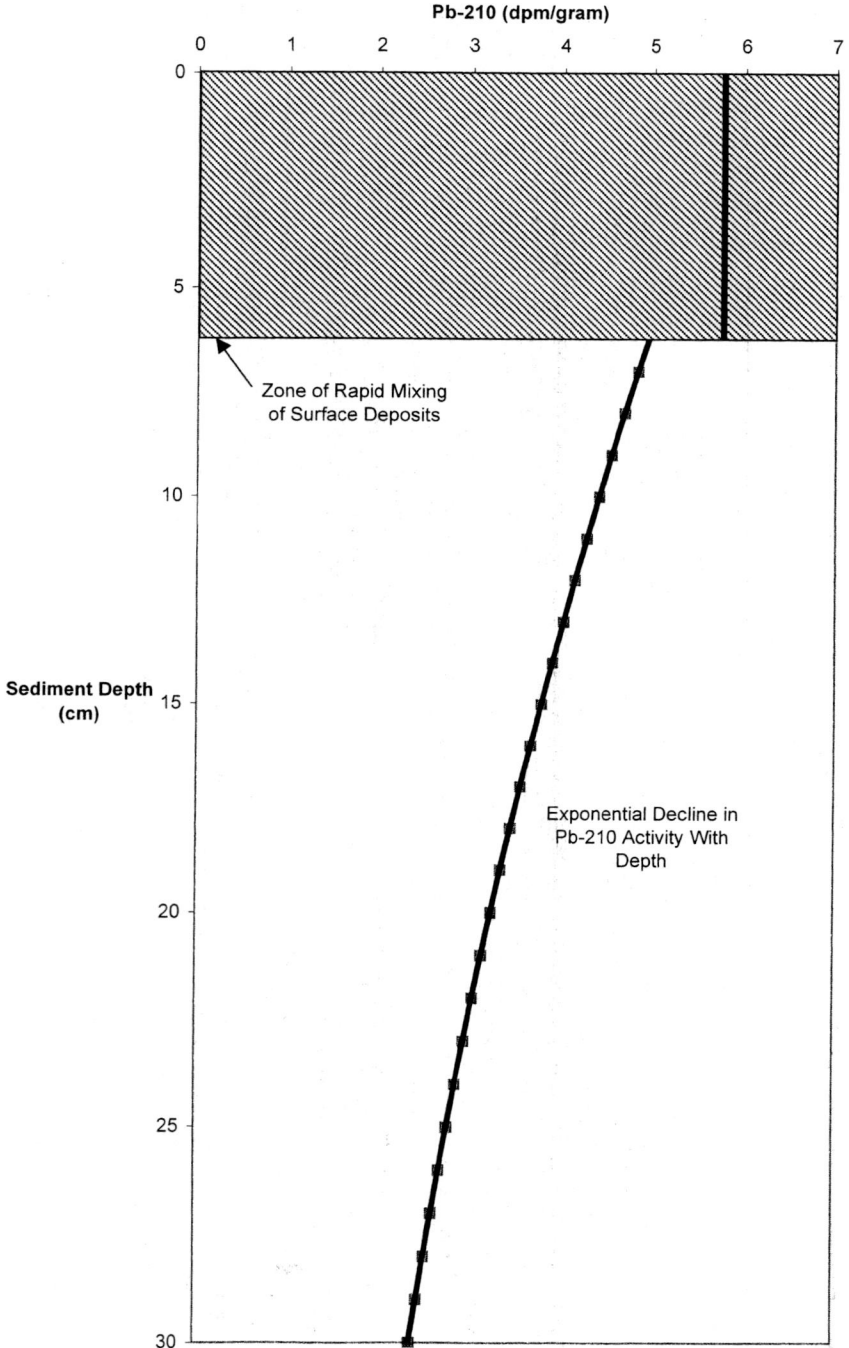

Fig. 6.3. Sketch of the model for a rapid mixing layer at shallow depth.

by a "square-box" filter of spatial width z_{mix}, at deposition. The observed concentration with depth, $A(z)$ (without any decay) would then be given through

$$A(z) = \frac{1}{z_{mix}} \int_z^{z+z_{mix}} I(z) \, dz. \tag{6.42}$$

From the observed concentration variation, $A(z)$ one then recovers the intrinsically deposited and mixed material, $A_{ob}(z)$, through

$$A_{ob}(z) = A(z) \exp(-\Lambda z/V) \tag{6.43}$$

under the constant sedimentation rate scenario.

Thus the intrinsically deposited non-mixed intensity, $I(z)$, is related through

$$z_{mix} A_{ob}(z) \exp(\Lambda z/V) = \int_z^{z+z_{mix}} I(z) \, dz. \tag{6.44}$$

This equation can be easily inverted analytically to obtain $I(z)$ with

$$I(z) = -(4\pi)^{-1} \int_0^\infty G(z') \, dz' \int_{-\infty}^\infty \frac{\sin[k(z - z' + z_{mix}/2)] \, dk}{\sin[k z_{mix}/2]}, \tag{6.45}$$

where

$$G(z) = \partial \left[z_{mix} A_{ob}(z) \exp(+\Lambda z/V) \right] / \partial z. \tag{6.46}$$

Equation (6.45) can be written in the simpler form

$$I(z) = -\sum_{n=0}^\infty G(z + n z_{mix}). \tag{6.47}$$

In terms of the observational values, the intrinsic value, $I(z)$, can be written

$$I(z) = -z_{mix} \sum_{n=0}^\infty \exp\left(-\Lambda V^{-1}(z + n z_{mix})\right)$$

$$\times \left[\Lambda V^{-1} A_{ob}(z + n z_{mix}) + \partial A_{obs}(z + n z_{mix})/\partial z \right]. \tag{6.48}$$

Under the rapid mixing scenario, one requires

$$\left| \partial A_{ob}(z)/\partial z \right| \geqslant \Lambda V^{-1} A_{obs}(z) \tag{6.49a}$$

and

$$\partial A_{ob}(z)/\partial z \leqslant 0 \tag{6.49b}$$

in order that $I(z)$ be positive. Thus only observed activities that systematically decrease with increasing sedimentary depth are permitted. If activity anywhere increases then, once again, the rapid mixing assumption cannot be sustained. Thus one has extracted the intrinsic depositional concentration behavior from the rapid mixing model under the assumption of steady, constant, sedimentation.

6. Partition coefficient effects

In the marine or estuarine environments, there is a problem of concentration control that is not present in dry continental climates. The problem stems from the fact that all radioactive components are partially soluble in water. Thus sediments introduced into a basin, perhaps by riverine influx, have two factors to contend with before their deposition. First, the intrinsic radioactive concentration on sedimentary particles in, say, a riverine input will depend on the solubility in the river. Secondly, after being introduced to the sedimentary basin, sedimentary particles may either add to the particular radioactive concentration by sorption from the standing waters, or may give up radioactivity to the water if the water is undersaturated. In either event, the sedimentary particles then have an altered radioactive concentration. The deposition rate of sedimentary particles versus the residence time of radioactivity in the waters then determines precisely the level of depositional activity. Fine-grain particles, which spend a long time in the water column compared to coarse-grained materials, presumably reach a closer equilibrium balance (partitioned activity) of radioactive exchange with the surrounding water milieu. Thus deposited sediments have variable activity per gm due to both intrinsic variations, modifications during burial influx, and modifications due to the relative partitioning of radionuclides in the water column versus on sedimentary particles.

Observations can occasionally be used to provide some idea of relatively long or short residence times. For instance, in the case of Lake Baikal, Edgington et al. (1991) have noted that two of the sites they investigated were completely devoid of ^{137}Cs; they, therefore, argued for a high partition coefficient in water of ^{137}Cs and a very long residence time compared to sedimentary deposition time through the water column. However, the remaining eight sites they investigated *did* show significant ^{137}Cs concentrations so that either the water partition coefficient Edgington et al. (1991) used is smaller than they estimated for the other two sites or, more likely, the sediment deposition time is not that long compared to water residence times – thereby accounting for the variability in the observations of ^{137}Cs seen in the ten different sedimentary profiles.

Clearly, what is needed to understand the process of potential sedimentary particle pick-up (or stripping) of radionuclide activity relative to that in the water column is an estimate of the water partition coefficient, P_w, for each radionuclide. In turn, this partition coefficient is tied to the water concentration, C, of each radionuclide and to a coefficient, K_d, which measures the relative affinity of sediment and water for reaching an equilibrium balance in a radionuclide when full solubility can be reached.

Two major problems arise in attempting to use P_w (or P_p) to assess the likely changes in sedimentary activity brought about prior to deposition (or at the sedimentary surface). First, the radionuclide concentration is highly variable in the water column so that there is no precise value that can be used to evaluate the water (or sediment) partition coefficient. Instead one must allow for the measured (or inferred) concentration variability. Second, even if the concentration were to be precisely known, there is still considerable uncertainty in determinations of the K_d values for particular radionuclides and, it would seem that K_d values are pressure, temperature, and water salinity-dependent (Dursma & Carroll, 1996) so that some range of values is always present. Thus the partition coefficient will also have a dynamical range. Part of the task of this section of the chapter has been to show, by illustration, how one goes about determining this range and how it can be tied to the cumulative probability for estimating the likelihood of activity at deposition in excess of a specified amount per gm. This factor is important because it provides an idea of the fraction of radioactive concentration that is likely to be original to the sedimentary particles, and also the fraction that is likely "allochthonous" due to a combination of sorption from the particles into riverine or standing waters in the basin and sorption by the sedimentary particles from the waters during their journey to final deposition.

Chapter 7

Marine contaminant studies

As has been noted earlier in this volume, the importance of radionuclide methods is not only in their ability to determine sedimentation rates and, simultaneously, source variations, but also to help understand the transport of contaminants in marine environments – ranging from toxic chemical pollutants to radioactive products from reactors, submarine and nuclear power plants. Thus, even if contaminant products are not themselves radioactive, their depositional histories can be traced with radionuclide markers. And, if contaminant products are long-lived radionuclides then, again, their modern history can be traced with short-lived radionuclides.

Clearly, multiple site studies in a contamination area are the sine qua non if one is to be able to sort out the input, and later transport, of contaminant products from naturally occurring "background" effects.

Accordingly, this chapter develops in some detail contaminant aspects of radionuclides and modern sedimentation. Four case histories are presented to illustrate applications of some of the procedures developed in earlier chapters. In the first case, we provide an example of how even the most basic information on the ages of sediment horizons can prove helpful in interpreting sources of polyaromatic hydrocarbons in a marine bay. The final three cases utilise the procedures developed at length in Chapter 2 to interpret sedimentation histories and to further apply these to other constituents attached to sediments. The reconstruction of historical inputs of radioactive contaminants is presented for a lake and marine system. The final example, from Lake Baikal, uses sediment burial histories to trace and interpret the movement of organic carbon through this system. Although not a contaminant itself, organic carbon provides a useful example because it serves as an important carrier phase for many hydrophobic contaminants. Together, these examples provide practical examples that illustrate the methods and utility of interpreting sedimentation as part of any investigation of contaminants in aquatic systems.

1. Sources of polyaromatic hydrocarbons in the White Sea

The group of chemical compounds known as polycyclic aromatic hydrocarbons (PAHs) represents an important class of petroleum hydrocarbons that contributes to the overall hydrocarbon load in the Arctic environment (Boehm et al., 1998; Robertson, 1998). PAHs are formed by (1) low to moderate temperature diagenesis of sediment organic matter to fossil fuels, (2) direct biosynthesis by bacteria, fungi, higher plants and insect pigments, and (3) high temperature pyrolysis of organic materials (Page et al., 1999). The pervasiveness and variety of

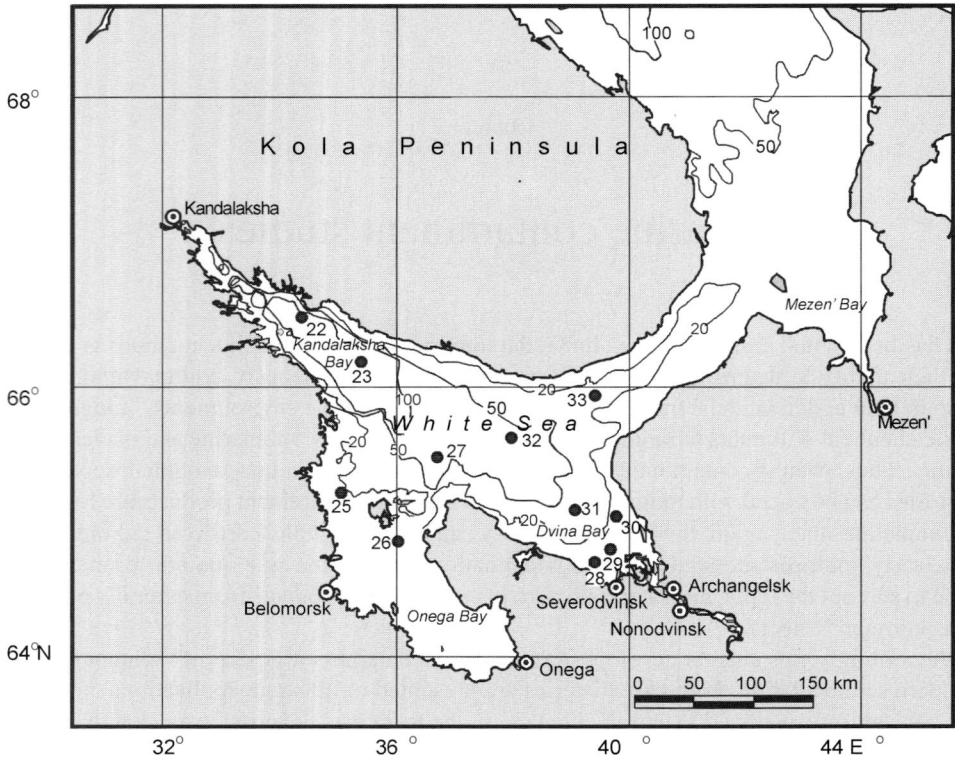

Fig. 7.1. Map of the White Sea with sampling locations indicated.

processes, both natural and anthropogenic, which are responsible for generating PAH compounds, make it particularly difficult to extract information on sources from environmental
data (Mitra et al., 1999). Yet information on sources is needed because of concerns over the
persistence of some PAH compounds in the environment and associated high bioaccumula-
tion potential in aquatic organisms (Robertson, 1998; McElroy et al., 1989). A number of
PAH compounds are known carcinogens (IARC, 1987).

The White Sea (Fig. 7.1) is a relatively small coastal sea with a surface area of only
90×10^3 km^2 and a volume of 6×10^3 km^3. It contains three main sections: *Voronka* (Fun-
nel), *Gorlo* (Throat) and the Basin. The Basin is comprised of three large bays: Kandalaksha,
Onega and Dvina. Average and maximum depths in the sea are 67 and 350 m, respectively
(Dobrovolsky & Zalogin, 1982). A sill extends across the entrance to the open sea with the
sill depth increasing from 10 meters near the Kola Peninsula to 50 meters on the opposite
shoreline (Scarlato, 1991). The combined freshwater discharge for rivers entering the White
Sea is 180 km^3/year. The largest rivers are Severnaya Dvina River (108 km^3/y), Mezen' River
(24 km^3/y) and Onega River (15 km^3/y) (Mikhailov, 1997).

In the White Sea, municipal and industrial activities in adjacent cities produce a variety of
waste products that are subsequently introduced into the marine environment by land-based

discharges, runoff, and atmospheric deposition. Aluminum smelter operations in Kandalaksha City are one of the main industrial pollutant sources (MREC, 1997). Several large pulp and paper mills that operate in Arkhangelsk and Novodvinsk are additional sources (AREC, 1998; Hansen et al., 1996). Previously, pollution levels were monitored in the White Sea by the Regional Hydrometeorological Service of the Former Soviet Union. However, few data are available concerning the status of oil contamination (Girkurov, 1993; Melnikov et al., 1993). Furthermore, the historical data on hydrocarbons do not contain information on the content and composition of PAHs in bottom sediments.

1.1. *Sediment analyses*

Sediment samples were classified as pelite (silt + clay) (< 0.063 mm in diameter of grains), sand (0.063–2 mm) and gravel (> 2 mm). Pelite predominated in bottom sediments collected at Stations 22, 23 (Kandalaksha Bay), 27, 31, 32 (Basin) and 30 (Dvina Bay). Sediments containing high C/N ratios (> 13.7) were found at Stations 26 (Onega Bay) 28, 29, 30 (Dvina Bay) and 33 (Gorlo), indicating a high relative content of terrigenous matter probably in association with inflows from the Rivers Onega and Severnaya Dvina. The C/N ratios determined at other stations varied from 7.6 to 8.6; a range that is normally associated with marine sediments.

^{210}Pb analyses were conducted on one sediment core collected from Station 32. The method of analysis is described in Pheiffer-Madsen & Sørensen (1979). Mass and linear sedimentation rates were determined for one sediment core collected at Station 32 based only on a linear regression of log excess ^{210}Pb activity against cumulative mass depth (Savinov et al., 2000). No evidence of mixing was observed in the upper section of the core.

1.2. *PAH distribution*

The composition and concentrations of aromatic hydrocarbons in bottom sediments of the White Sea indicate two primary station groupings. Pyrogenic PAHs predominate in seven of the eleven stations evaluated (Stations 22, 23 (Kandalaksha Bay), 25, 26 (Onega Bay), 27, 31 and 32 (Basin)). The highest pyrogenic PAH level (including $\sum$CPAH) are detected in bottom sediments of the Kandalaksha Bay. Atmospheric discharges from the Kandalaksha aluminum smelter are the most likely source of PAHs to the White Sea. At other stations, primarily located within the Dvina Bay, terrigenous and petrogenic PAHs predominate (Stations 28, 29, 30 (Dvina Bay) and 33 (Gorlo)). Bottom sediments in Dvina Bay contain higher percentages of petrogenic PAH and perylene compared with those from all other regions investigated. The main source for these compounds the Severnaya Dvina River is impacted by discharges from the pulp and paper mill industry.

1.3. *Historical levels of PAHs*

Constant sedimentation rates at Station 32 are low (390 g m^{-2} year^{-1} and 1.6 mm year^{-1}). The associated sediment ages of the individual depth intervals indicate that below 10–12 cm, sediments are older than 100 years (Table 7.1). Below this depth, ^{210}Pb geochronology cannot be used to determine ages for the sediment layers. However, the PAH concentrations

Table 7.1
Aromatic hydrocarbons (ng/g dw) and corresponding ages of sediment layers for Station 32

Compounds	0–1 cm	1–2 cm	2–4 cm	10–12 cm	28–30 cm
	1988–1994	1977–1989	1957–1978	< 1904	< 1904
$\sum$PAH	47	86	103	27	12
$\sum$CPAH	16	36	38	8.1	2

associated with these depths must correspond to pre-industrial times and therefore must represent background levels. These background concentrations ($\sum$PAH = 27.0 ng/g dw and $\sum$CPAH = 8.2 ng/g dw) are roughly 2–5 times lower than the concentrations found in sediment intervals deposited after 1957.

PAH concentrations also increased from the sediment surface (1994) to a depth of 2–4 cm (1957–1978). To explain a decrease in PAH concentrations with time, data were acquired on air discharges from non-ferrous metallurgy industries compiled by the State Committee of the Russian Federation on Statistics (Statistical book, 1998). These data consist of annual emissions estimates for Russia but only for the time period 1991–1997. Over the 3-year time period 1991–1994, emissions have decreased from 5089 to 3502 tons/yr. While one would prefer to have site-specific data extending to earlier years, this trend of decreasing emissions might also explain the changes observed in PAH concentrations at Station 32.

1.4. *Conclusions*

Based on sediment age dating results at one location, present-day $\sum$PAH and $\sum$CPAH concentrations are 2–5 higher in pre-industrial times. Since at least 1957, both $\sum$PAH and $\sum$CPAH concentrations have decreased from 103 and 38.0 ng/g dw, respectively in 1957–1978 to their present-day concentration of 46.9 and 15.9 ng/g dw, respectively (1994).

Comparing the White Sea data to data from adjacent areas, PAH levels are lower than levels detected in sediments from the Barents Sea. The PAH signature identified for the White Sea (not including alkyl-substituted homologs of parent PAHs or perylene) is similar to the signature reported for the SE Barents Sea. However, if one includes the alkyl-substituted homologs of parent PAHs and perylene, the PAH signatures identified for these two areas are not similar enough to suggest that atmospheric transport of particles emitted from the Kandalaksha smelter is the main anthropogenic source of PAH contaminants to both regions.

2. Burial history of radioactive waste in the Kara Sea

2.1. *History of radioactive dumping*

The Arctic has been impacted by several well-known sources of radioactive contamination: nuclear weapons testing, releases from nuclear installations, European reprocessing plants and the Chernobyl accident (Table 7.2). In 1992, Russian authorities first revealed a new source of radioactive contamination to several shallow Arctic seas (White Book 3, 1993). Beginning

Table 7.2
Sediment and water (surface and bottom) activities for selected radionuclides from stations in the open Kara Sea (compiled by Hamilton et al., 1994 and Strand et al., 1994). Only stations with measurable activities were used in the determinations

	Median	Maximum	Minimum	
Water				
^{137}Cs (Bq/m^3)	7.8	20.4	3.3	$n = 39$
^{238}Pu (mBq/m^3)	0.3	1.4	0.1	$n = 17$; n.d. $= 18$
^{239,240}Pu (mBq/m^3)	6.0	16.0	1.8	$n = 35$; n.d. $= 1$
^{90}Sr (Bq/m^3)	4.0	12.1	3.0	$n = 39$; n.d. $= 13$
^{241}Am (mBq/m^3)	0.85	0.2	2.6	$n = 16$
Sediment				
^{137}Cs (Bq/kg)	24.5	32.0	17.9	$n = 8$
^{239,240}Pu (Bq/kg)	0.78	1.25	0.44	$n = 8$
^{238}Pu (Bq/kg)	0.02	0.05	0.01	$n = 8$
^{241}Am (Bq/kg)	0.28	0.46	0.20	$n = 8$

n.d. = non detectable; n = number of measurements.

in the mid-sixties, several thousand containers with low- and intermediate level wastes (total activity $= 0.6$ PBq) and some liquid waste (1 PBq) were dumped into bays located along the margin of Novaya Zemlya and into the Novaya Zemlya Trough by the former Soviet Union. A total of 17 nuclear reactors from submarines and the icebreaker "Lenin" were also dumped (88 PBq) (White Book 3, 1993). The containers and nuclear reactors contain a wide variety of radionuclides (fission products, activation products, actinides), leading to concerns for the ecosystem and human health. The primary list of potential contaminants includes ^{60}Co, ^{90}Sr, ^{137}Cs, ^{239}Pu, ^{240}Pu and ^{241}Am (Fig. 7.2). As a result of these dumping activities, a wide variety of radionuclides now reside in the shallow Arctic marine environment of the Kara Sea (Cooper et al., 1998). The International Atomic Energy Agency investigated the risks posed by radioactive contamination to biological and human health and considered the feasibility of possible remedial actions (IASAP, 1998).

2.2. *Regional description*

The Kara Sea region includes the river deltas for the Ob and Yenisey Rivers, the Kara Sea (a shallow continental sea including the Novaya Zemlya Trough) and the nearshore bays of the island of Novaya Zemlya. These Arctic marine systems encompass a wide range of depositional environments that may be temporary or permanent sinks for radionuclide releases from the dumping grounds.

The annual cycle of winter ice formation and summer melting supplies large quantities of new sediment to the continental margin each year through the bays along the island of Novaya Zemlya where most of the waste containers are currently deposited. If significant waste release has previously occurred, the bays may act as temporary sinks for contaminated sediments and/or contaminated sediments may be transported into the open Kara Sea (Harms

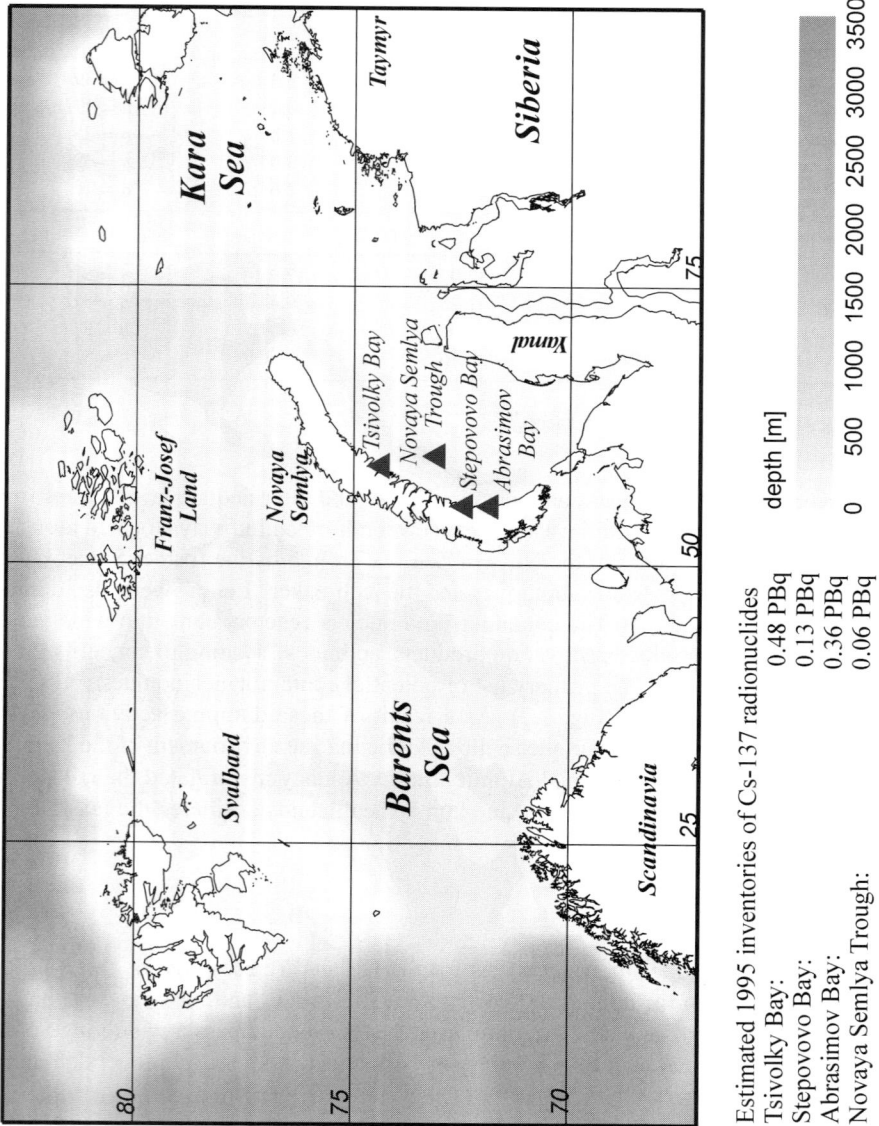

Estimated 1995 inventories of Cs-137 radionuclides

Tsivolky Bay:	0.48 PBq
Stepovovo Bay:	0.13 PBq
Abrasimov Bay:	0.36 PBq
Novaya Semlya Trough:	0.06 PBq

Fig. 7.2. Locations of dumped nuclear materials in the Kara Sea. Estimated inventories of radionuclides refer to 1995 and pertain to naval reactors disposed at major dumpsites based on working documents of the International Atomic Energy Agency's 'International Arctic Seas Assessment Programme (IASAP)'.

Cs-137 Inventory (kBq/m²)

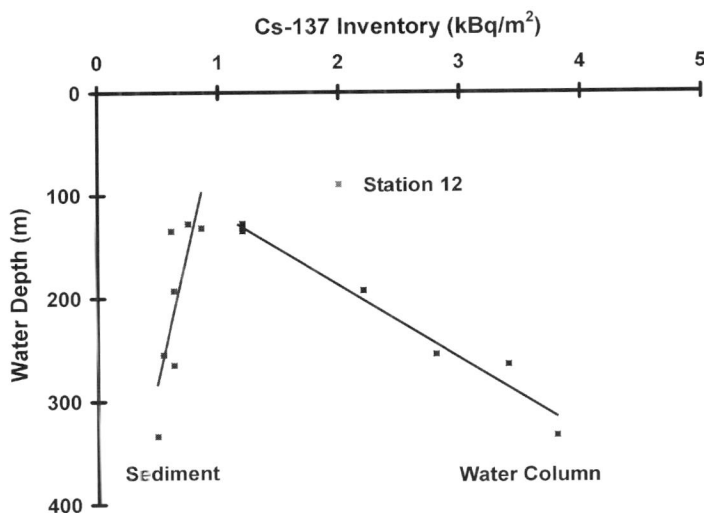

Fig. 7.3. ^{137}Cs inventories in sediment and seawater for the Kara and Barents Seas as a function of water depth (from Hamilton et al., 1994).

& Karcher, 1999; McClimas et al., 2000). Assessment of sediment accumulation rates in the open Kara Sea has shown that the rates are on the order of a few millimeters per year (Hamilton et al., 1994). Thus any signal of past radionuclide releases from the shallow dumping grounds should be observed within a few centimeters below the sediment–water interface of open Kara Sea bottom sediments. However, observations of the distribution of some of the contaminants (^{137}Cs, ^{238}Pu, ^{239,240}Pu) do not show appreciable increases in activity above the levels expected from other sources of contamination. Given that the sedimentary signal of contamination from the dumping grounds is weak, there is no clear present-day evidence of adverse environmental impacts associated, as of yet, with the dumped nuclear waste in the open Kara Sea. However, evidence of leakage does exist within some of the shallow bays bordering the island of Novaya Zemlya as we show next.

2.3. *Age dating and contaminant flux*

An investigation of sediments from one of the dumpsites in a shallow bay along the coastline of Novaya Zemlya (Stepovovo Bay) shows relatively high concentrations of both ^{137}Cs (30–290 Bq kg^{-1}) and trace amounts of ^{60}Co, that may indicate a local source of contamination. As verification, the radionuclide input histories were reconstructed from the ^{210}Pb age-to-depth relationship as interpreted by the Sediment Isotope Tomography (SIT) model described earlier. This SIT method was chosen because the ^{210}Pb activities below 5 cm depth in the sediment profile do *not* decrease uniformly (Fig. 7.4). The SIT method was able to disentangle the effects of sediment accumulation from flux variations to determine age-to-depth relationships (Fig. 7.5). The corresponding contours of surface activity for ^{210}Pb versus average sediment accumulation rate (depicted in Fig. 7.6) indicate an activity of 250 ± 20 Bq/kg and an accumulation rate of 22 ± 4 mm/yr. Based on the age-to-depth relationship at this lo-

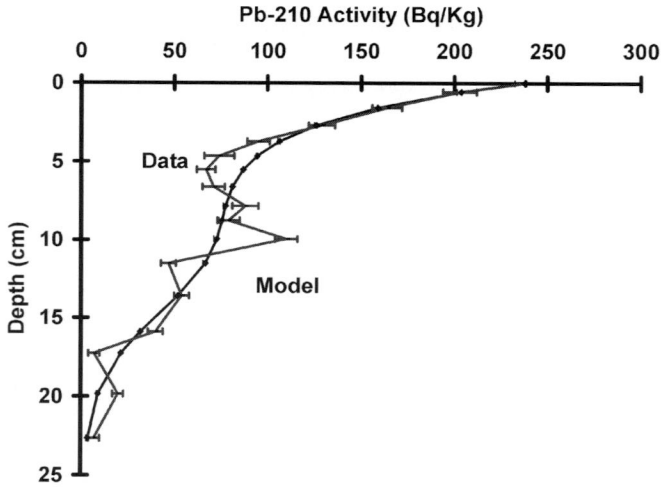

Fig. 7.4. Probability distribution of model-determined fits to the data ($\chi^2 < 10$).

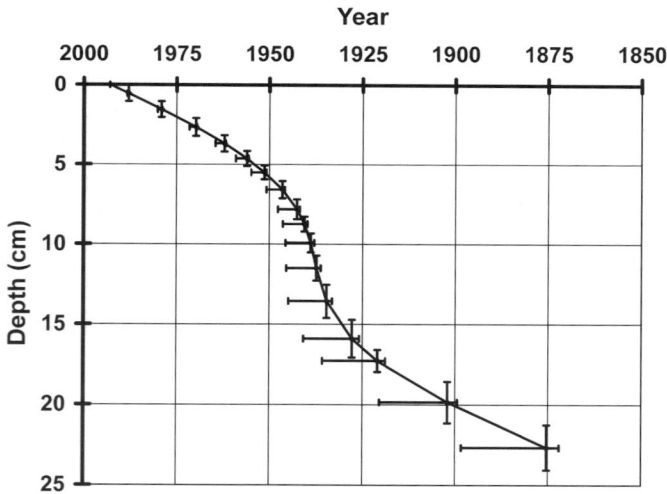

Fig. 7.5. Probability distribution of model-determined ages at each sediment depth for $\chi^2 < 10$. The uncertainties in sediment ages with depth for the nuclear era are less than 4 years.

cation, a local source of contamination is clearly responsible for the observed radionuclide distributions (Figs 7.7 and 7.8).

2.4. Conclusions

The marine systems of the Kara Sea encompass a wide range of depositional environments that may be temporary or permanent sinks for radionuclide releases from the dump-

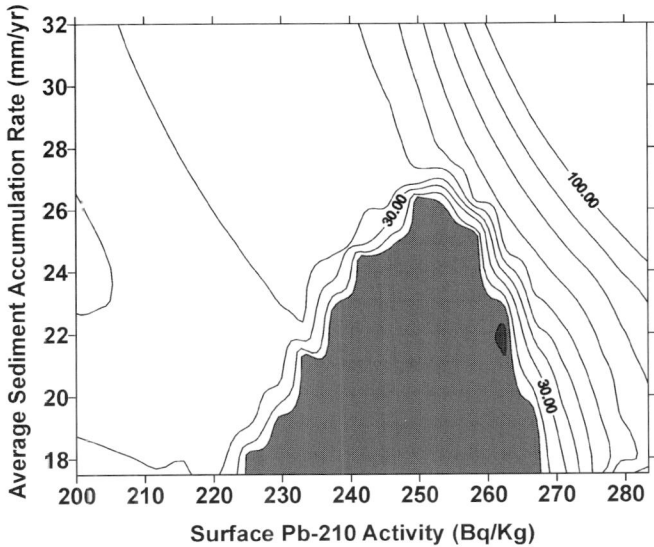

Fig. 7.6. Sediment accumulation rate versus surface ^{210}Pb activity values. Contours of χ^2 are lightly shaded.

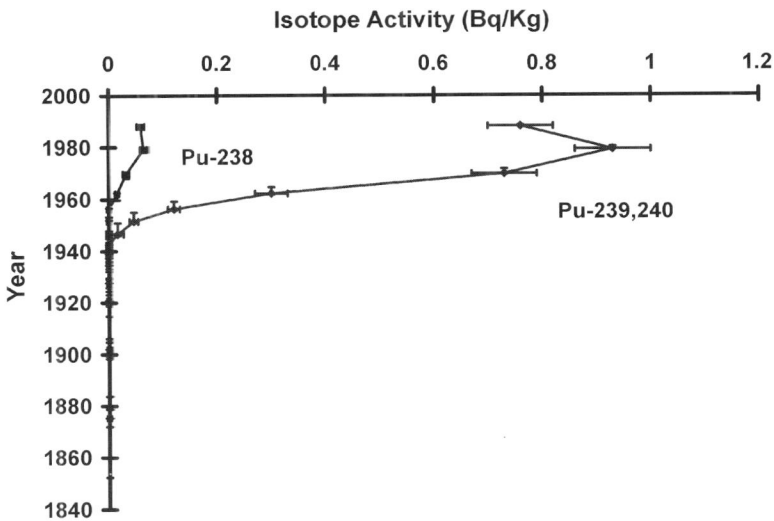

Fig. 7.7. ^{238}Pu and ^{239,240}Pu versus year. The maximum in ^{239,240}Pu corresponds to 1979 ± 1 year. This age is younger than would be expected for global fallout and indicates a local source of contamination.

ing grounds. Assessments of radionuclide inventories and accumulation rates for open Kara Sea sediments show that there is no discernible signal of past radionuclide releases from nuclear waste dumpsites located along the island of Novaya Zemlya. Sediment ages were reconstructed by the Sediment Isotope Tomography model for a complex ^{210}Pb profile collected

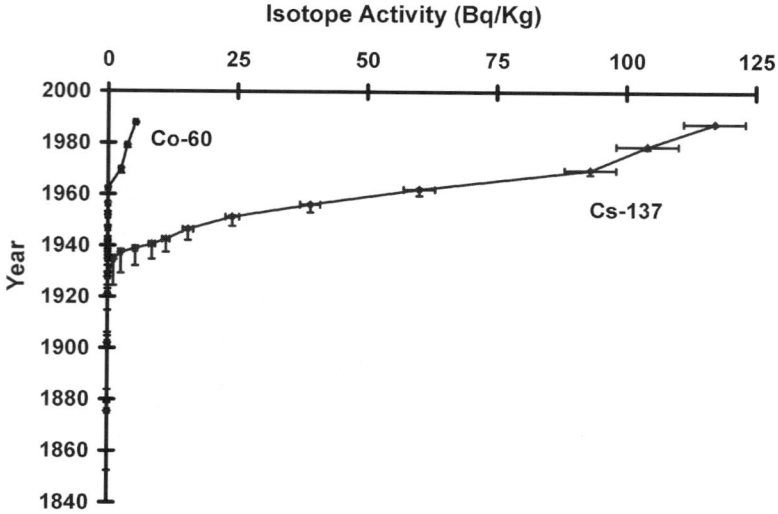

Fig. 7.8. [137]Cs and ^{60}Co versus year. Relatively high activities of ^{137}Cs and trace amounts of ^{60}Co also indicate a local source of contamination.

near a dumpsite in Stepovovo Bay. Based on the age to depth probability distribution determined by the SIT model, some leakage of radioactivity has already occurred. The extent of the uptake to sediments of the released radioactivity from the investigated site is thought to have resulted in only local-scale impacts.

3. Incorporation of uranium mill tailings into North Pond, Colorado

Concern by the U.S. Department of Energy over contamination of a small pond (0.02 km^2), led to an investigation at the U.S. Department of Energy Grand Junction Projects Office (GJPO) in Grand Junction, Colorado. The United States War Department purchased the site in August 1943 to stockpile uranium ore and extract uranium from the ore for the Manhattan Engineering District.

A sediment core was collected along a margin of the pond using a hand-held coring device capable of retrieving a sediment core 10 cm in diameter and 25 cm in length. The core was sectioned into 1 cm slices, weighed, dried, and reweighed to determine sediment porosities. Sediment sub-samples were subsequently analyzed for ^{137}Cs, ^{210}Pb, and ^{226}Ra.

^{137}Cs and ^{210}Pb profiles from North Pond display broad peaks in activity around 12 cm (Fig. 7.9 (a and d)). The SIT model was applied to the profiles utilizing two historical events as time markers. First, the age of the sediments can be no older than the 1920s, when the pond was excavated for gravel (UNC Geotech, 1989) and, because ^{137}Cs is still present at the base of the core, the base must be younger than approximately 1950. Second, the peak in ^{137}Cs at 12 cm is interpreted as the 1963 maximum in the atmospheric fall-out of ^{137}Cs (Ritchie & McHenry, 1990).

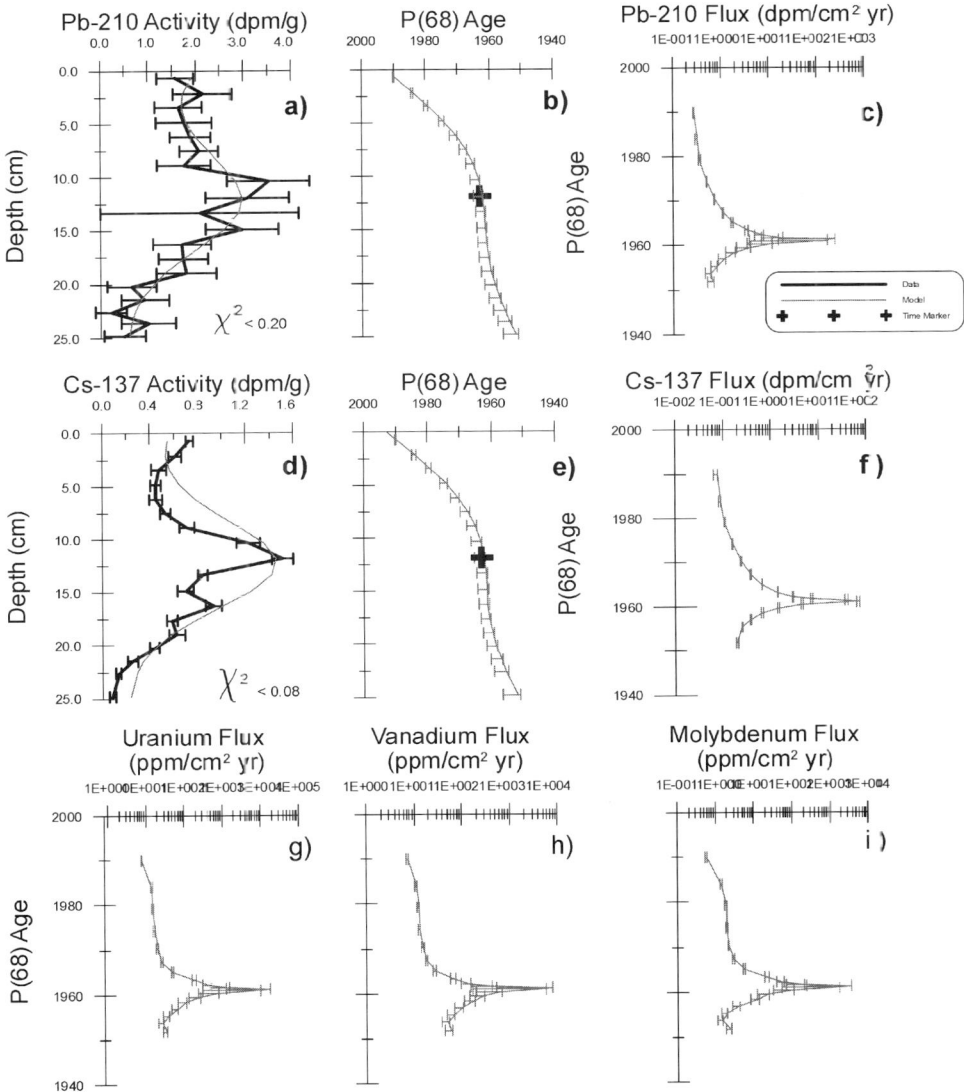

Fig. 7.9. North Pond: (a) Measured ^{210}Pb (dpm/g) versus depth (m) and associated counting errors. The smooth curve represents SIT model $P(68)$ ^{210}Pb activities versus depth ($\chi^2 < 0.2$). (b) Probability distribution of age-to-depth variations consistent with the smooth curve in Fig. 7.9a. The cross represents the ^{137}Cs age marker of 1963 ± 3 yrs. (c) Probability distribution of surficial flux (dpm cm^{-2} yr^{-1}) variation with time. (d) Measured ^{137}Cs (dpm/g) versus depth (m) and associated counting errors. The smooth curve represents SIT model $P(68)$ ^{137}Cs activities versus depth ($\chi^2 < 0.08$). (e) Probability distribution of age-to-depth variations consistent with the smooth curve in Fig. 7.9d. The cross represents the ^{137}Cs age marker of 1963 ± 3 yrs. (f) Probability distribution of surficial flux (dpm cm^{-2} yr^{-1}) variation with time for ^{137}Cs. (g) Probability distribution of surficial flux (ppm cm^{-2} yr^{-1}) variation with time for uranium. (h) Probability distribution of surficial flux (ppm cm^{-2} yr^{-1}) variation with time for vanadium. (g) Probability distribution of surficial flux (ppm cm^{-2} yr^{-1}) variation with time for molybdenum. Note the peaks in all of the flux curves are around 1960, representing the time of peak uranium processing at the facility.

Table 7.3
Physical depth to decompacted depth conversions

	Station 12	
Measured depth (cm)	Porosity	Decompacted depth (cm)
0.125	0.58	0.00
0.375	0.54	0.31
0.625	0.58	0.93
0.875	0.58	1.54
1.125	0.58	2.16
1.375	0.57	2.78
1.625	0.57	3.40
1.875	0.59	4.01
2.25	0.59	4.62
3.75	0.59	5.56
6.00	0.60	9.26
	Station 5A	
Measured depth (cm)	Porosity	Decompacted depth (cm)
0.25	0.22	0.00
0.75	0.24	0.46
1.25	0.25	2.31
1.75	0.23	4.17
2.5	0.28	6.48
3.5	0.31	9.26
5.0	0.32	12.03
6.5	0.33	14.81

The SIT model satisfactorily reproduced the ^{137}Cs and ^{210}Pb profiles (Table 7.3), yielding a most probable model fit to each data profile (Fig. 7.9 (a and d)). When the region of best fit for [210]Pb (with time restrictions) is graphically overlain with the region of best fit for the [137]Cs (with time restrictions), optimum average sediment accumulation rates are identified for the North Pond core (Fig. 7.9a). Indeed, both age-to-depth distributions agree within the probability ranges (Fig. 7.9b). The most probable age-to-depth profiles for ^{137}Cs and [210]Pb (Fig. 7.9 (b and e)) show striking similarities; sedimentation rates increase for both ^{137}Cs and ^{210}Pb from 1960 to 1965.

The distribution of most probable fluxes of ^{210}Pb displays a pronounced peak during the early 1960s (Fig. 7.9c), associated with the timing of the processing of uranium ore in the late 1950s and early 1960s. For ^{137}Cs, the peak is associated with the maximum in ^{137}Cs atmospheric fallout.

Visual inspection, as well as chemical analyses of the sediments, indicates the presence of uranium mill tailing residuals within the depth range around 12 cm (GJPOAR, 1993). Sediment sub-samples were analysed by the GJPO Analytical Chemistry Laboratory to determine the concentrations of constituents commonly associated with uranium mill tailings (Morrison & Cahn, 1991).

Total uranium concentrations were measured on an inductively-coupled-plasma-mass-spectrometer (ICP-MS); vanadium, and molybdenum concentrations were measured on an inductively-coupled-plasma-atomic-emission-spectrometer (ICP-AES) (GJPOAR, 1993). Based on the model-determined most probable sediment accumulation rates, the most probable fluxes of total uranium, vanadium, and molybdenum to the bottom of North Pond were determined (Fig. 7.9 (g, h, and i)). The most probable sedimentation rates in the region around 12 cm yield uncertainties in the magnitude of the fluxes. However, there is a clear increase in the flux of mill tailing residuals to the sediments of North Pond, coinciding with the peak period of uranium mill tailings extraction at the GJPO. Since the completion of this study, the margins of North Pond have been remediated to remove the mill tailings residuals in accordance with the U.S. Department of Energy Uranium Mill Tailings Remedial Action (UMTRA) Program.

4. Organic carbon preservation in Lake Baikal Sediments, Russia

Several factors influence organic carbon preservation in sediments (rate of burial, bioturbation, water depth, etc.) but two factors generally dominate: (1) primary production, and (2) oxygen content of water column and sediments (DeMaison & Moore, 1980; Calvert, 1987). Anoxic conditions are especially favorable for organic carbon preservation by decreasing or eliminating bioturbation (Demaison & Moore, 1980; Somayajulu et al., 1999). Lake Baikal is an oligotrophic lake with abundant dissolved oxygen throughout the water column including bottom waters (Demaison & Moore, 1980). In addition, oxygen is found in the surficial sediments with a depth of penetration measured at about 15 mm in the Selenga River area (Martin et al., 1993), and at about 50 mm in the northern basin (Martin & Grachev, 1994); these values correlate well with the thickness of the near surface brownish, oxidized layer. In Lake Baikal the preservation of TOC is dependent on several factors (i) low productivity because Lake Baikal is oligotrophic; (ii) the O_2 content throughout the water column which oxidizes TOC in water; (iii) the depth (900–1600 m) of the water column and the rate of sinkage of TOC through the water column; (iv) the times of low sedimentation rate when TOC stays in the oxygen zone (both in the water column and in the oxygenated near surface sediments) for considerable time; (v) the temporal variation of O_2 in bottom waters and in the sediments penetrated by O_2.

4.1. *Methods/materials*

After collection, box cores were promptly extruded in the ship laboratory for ^{210}Pb analysis using a water-driven piston mechanism. Based on studies by Edgington et al. (1991), showing that sedimentation rates vary with depth, the sampling intervals were varied at increments ranging from every 0.25–0.50 cm at the top of the cores, to every 2–3 cm at the lower portions of the cores. This sampling interval was chosen because, in the upper fractions of the core sediments, very fine laminations (of order 0.5–1.0 cm thick) were observed (some diagenetic but some not) which showed that the sediment regime was changing and the rate of sedimentation also presumably varying. The sampling at 0.25–0.5 cm can help to understand these changes, although in hindsight, based on ^{210}Pb fluctuations, a sampling regime of 0.1–0.2 cm would have been considerably more useful. Weighed, homogenized samples from two cores at Stations 5A and 12 (Fig. 7.10) (with water depths of 930 and 305 m, respectively) were

North Pond

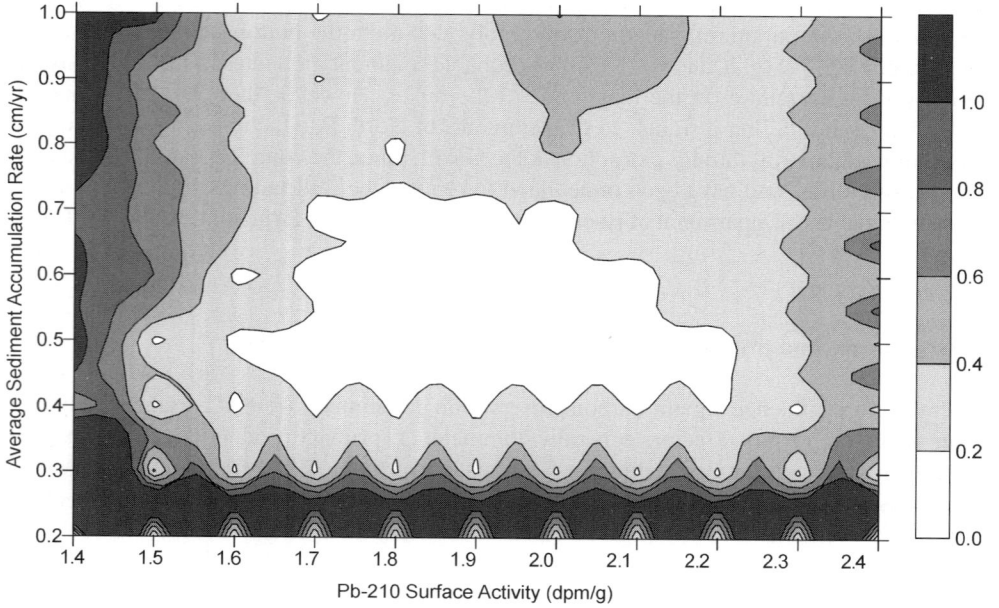

(a)

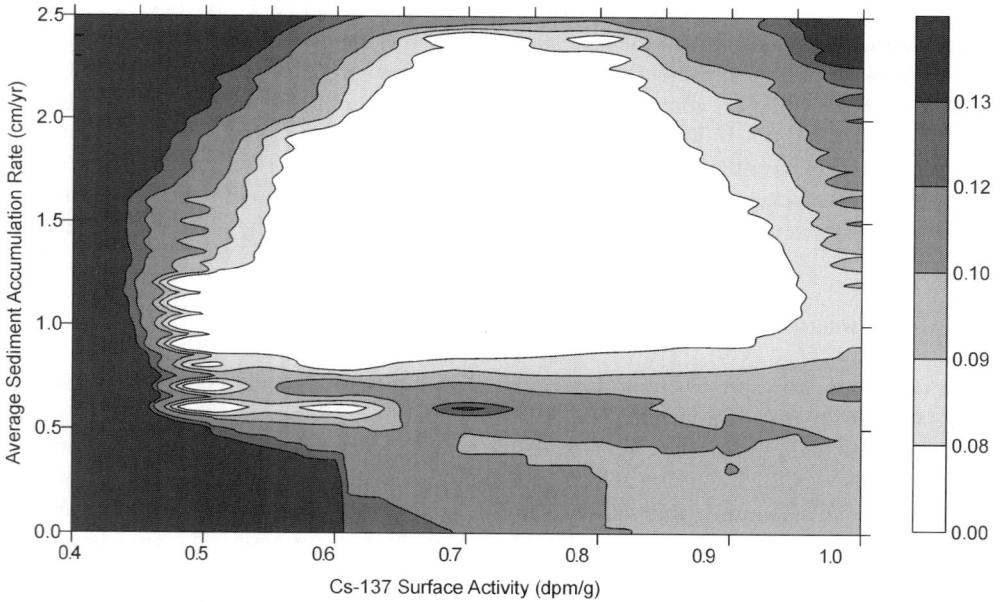

(b)

Fig. 7.10. Intersection of ^{137}Cs and ^{210}Pb average sediment accumulation rates (0.58–0.68 cm/yr) for North Pond, Grand Junction, Colorado versus surface values (a) for ^{210}Pb, (b) for ^{137}Cs.

measured for ^{210}Pb using high purity germanium detectors using procedures developed by Moore (1984). TOC content was determined by CHN analysis. Sediment porosity was measured in each of the sampling intervals and depths corrected to decompacted depths, so that any radionuclide disentanglement procedure will then produce directly sedimentation rates in terms of physical thickness deposited, rather than measured sediment thickness at present day. Unless otherwise stated, all results discussed in this section refer to the use of decompacted thickness as a basic reference frame. Table 7.3 provides measurement depth intervals, porosities, and decompacted depths for both Stations 12 and 5A. Accumulations of total organic carbon were then determined.

Sediment mixing can have a profound effect on the distribution of radionuclides, and thus the determination of sedimentation rates, especially in environments where deposition amounts are low relative to sediment mixing thicknesses (Chapter 5). Mixing disturbs the sedimentary record of accumulation, erasing the signature of time given by the decrease in radionuclide concentration with sediment depth. In Lake Baikal it is known from cores that, in general, mixing zones occur, typically from 1.5–2 cm thick and rarely as thick as 5 cm. There appears to be more bioturbated sediments in the Selenga River delta region than in the northern basin. There are no visible presences of bioturbation in either of the upper parts of the cores from Stations 12 and 5A. There is evidence of bioturbation below the oxidized zone in Station 5A at about 1.5–2 cm thickness. The sharp laminations (in the core from Station 5A), and the nominal rapid shifts in ^{210}Pb values in the upper parts of both cores on a 0.25–0.5 cm sampling interval, would argue against any significant bioturbation at these two sites, as would the variations in ^{137}Cs. Therefore no correction for bioturbation effects was required.

4.2. ^{210}Pb and ^{137}Cs geochronology results for Lake Baikal

4.2.1. Station 12
Station 12 is located on top of a small ridge on the general deltaic slope of the Selenga River in 305 m of water. A small valley divides the site of Station 12 from the prevailing delta slope, thereby preventing the transport of coarse-grained sediments to the site. Hemipelagic sediments in the core are fine-grained. Terrigenous, very siliceous, muds, with 20–30% of diatom frustuls, are seen in core but no turbidite sediments are seen. There is a brown, oxidized zone, 1.5 cm in thickness, in the surficial core sediments, with sediments being reduced and of gray to olive-gray in color deeper than the oxidized zone. An earlier, brown oxidized zone is also found in the 2.5–3 cm depth range. Plotted on Fig. 7.11 are the measured ^{210}Pb activities (dpm/g) with the superposed theoretical curve best fitting to the data with depth using the radionuclide disentanglement procedure of Carroll et al. (1995a). The uncertainty bars on the measured ^{210}Pb activities are taken to be at an average of ± 2 dpm/g, while the bars on measurement resolution over depth intervals are shown by the horizontal shading. To be noted is that, with the exception of the value of around 71 dpm/g, centered at 3.4 cm depth, and the next deepest value of 33 dpm/g, centered at 4 cm depth, the inverse method fits through the data to within resolution. Indeed, if the average of 52 dpm/g, centered at 3.7 cm depth, is used to replace the two exceptional neighboring points, then an almost perfect fit to the data is obtained.

Because the radionuclide disentanglement procedure separates sedimentation and source effects, age-to-depth determination is an automatic component of the method. Shown in

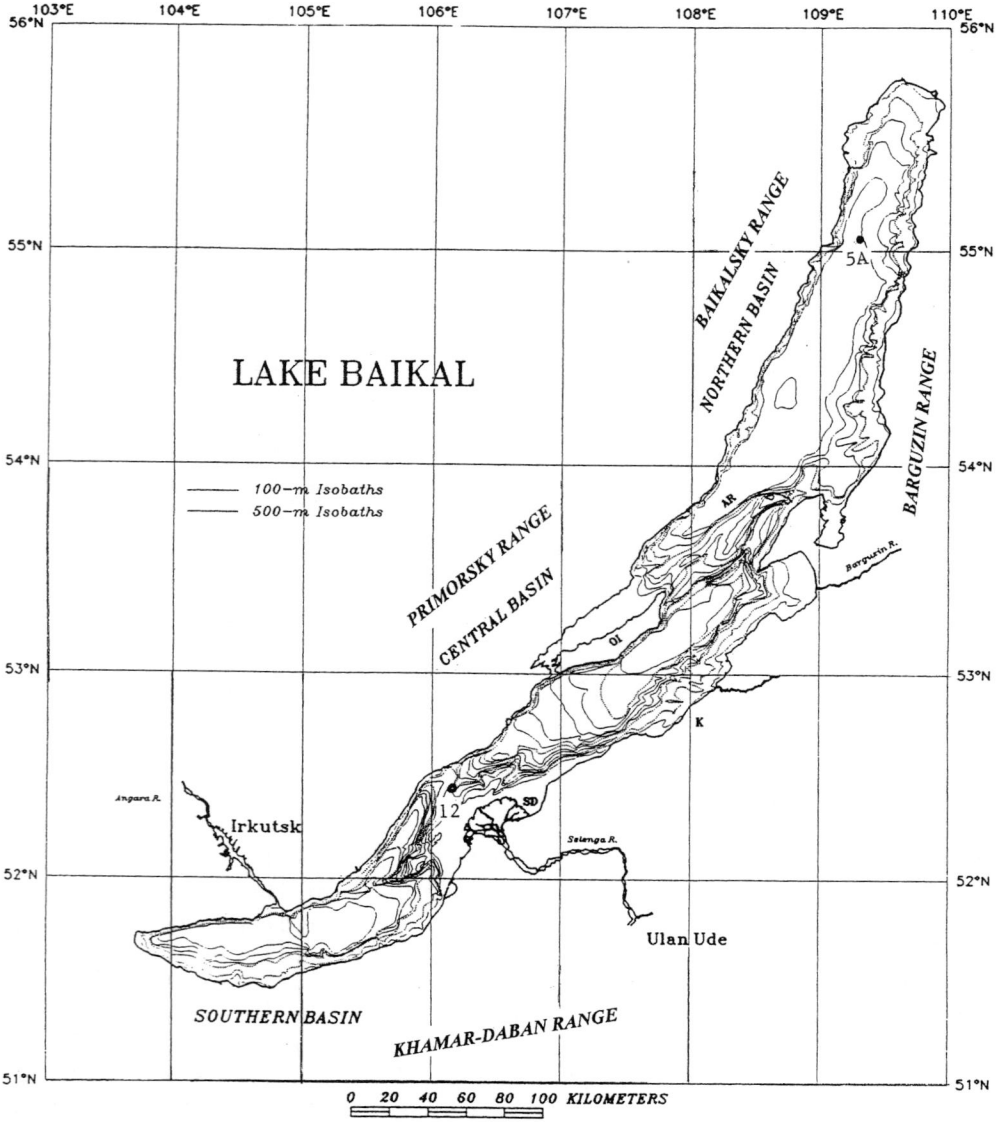

Fig. 7.11. Map of Lake Baikal showing locations of Stations 12 and 5A.

Fig. 7.12 is the corresponding age-to-depth determination with associated uncertainty error bars resulting from both the uncertainty in activities and, more importantly, the width range of the sampling interval in depth. Deeper than about 6–8 cm (corresponding to mean ages of years 1960–1930) there is very little resolution of the data, implying that lower reliance should be placed on interpretations deeper than about 6–8 cm. The corresponding sedimentation rate (cm/yr) versus depth is drawn in Fig. 7.13a, while the mean curve of age-to-depth conversion

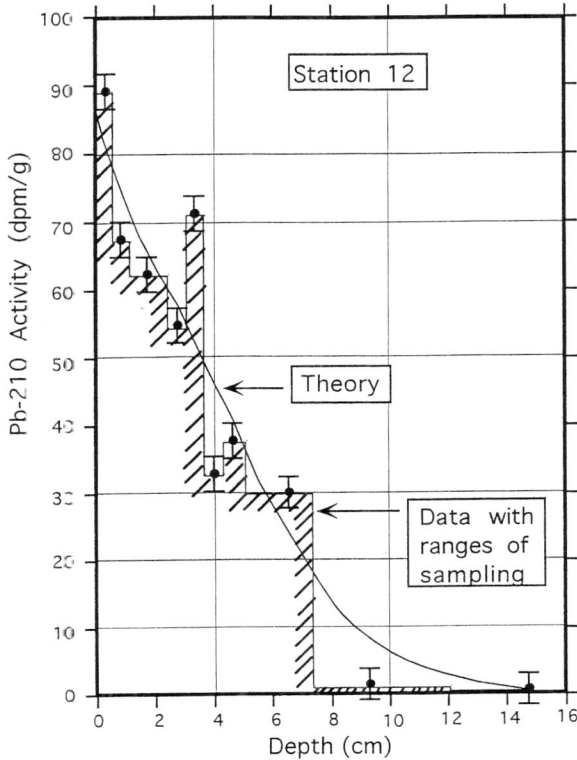

Fig. 7.12. Measured ^{210}Pb activities (dpm/g) for Station 12 versus decompacted depth (cm) with the superposed theoretical curve determined from the radionuclide disentanglement method given in text. The error bars on activity and depth determination for the measurements are indicated.

of Fig. 4.18 enables the sedimentation rate with time to be drawn as in Fig. 7.13b. Deeper than 6 cm (earlier than about year 1957) the uncertainties in the sampling width cause such large uncertainties on the sedimentation rate that it is not relevant to plot the values. Note that there has been a systematic decrease in the mean sedimentation rate from past to present. At 5 cm depth (year 1960) the sedimentation rate was about $0.38^{+0.03}_{-0.23}$ cm/yr, but by 2 cm depth (year 1980) the rate was nearly halved to about $0.22^{+0.02}_{-0.08}$ cm/yr, and has been at roughly 0.2 cm/yr since year 1980.

In addition to the sedimentation rate, the radionuclide disentanglement code also automatically provides an intrinsic source variability with depth (or time). The source variability is taken relative to a unit strength source at the date the core was taken (1994). Shown on Fig. 7.14a is the intrinsic source variability with depth including error bars of uncertainty inherited from the finite spatial sampling, while Fig. 7.14b shows the source variability with time using the mean age-to-depth curve of Fig. 7.12. Note that, despite the error uncertainty, there is a systematic trend with depth (age) indicating about a 40% higher source contribution at 5 cm (year 1960) than at core date of 1994.

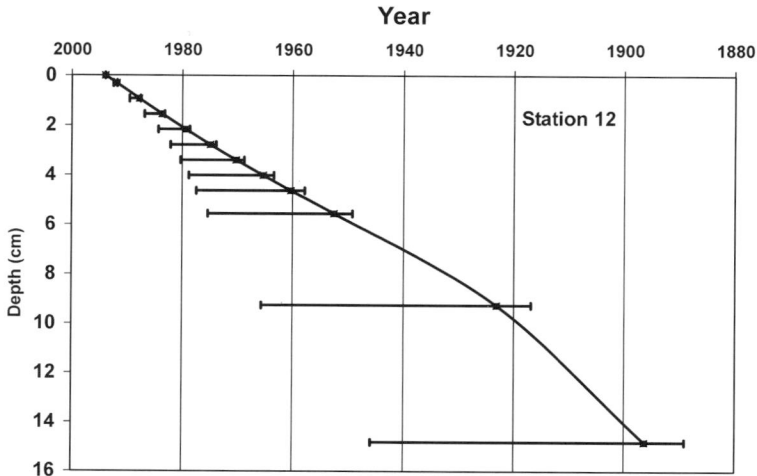

Fig. 7.13. Age-to-depth curve for the ^{210}Pb profile for Station 12 determined from the radionuclide disentanglement method given in text. The associated error bars on the curve are given and arise from both the uncertainty in activities and from the sampling interval in depth. Depth is for decompacted sediments.

In addition to the ^{210}Pb information, ^{137}Cs activities were also measured at Station 12. These activities were *not* used to constrain the ages of sediments as determined by ^{210}Pb; instead, after the ^{210}Pb age-to-depth conversion was independently determined, the ^{137}Cs data were then plotted against age as shown in Fig. 7.15. The rationale here is that if the ^{210}Pb age-to-depth conversion is adequate then the ^{137}Cs data should automatically be at ages younger than about 1945–1950, which is the onset of atmospheric atomic bomb testing. As can be seen from Fig. 7.15, this expectation is substantiated, providing corroborative support for the age-to-depth determination from ^{210}Pb information. It is also of interest to note that the ^{137}Cs activity at Station 12 has steadily increased in time from around 7 dpm/g at around year 1950 to about 37 dpm/g at year 1994. There is no indication of any slowing of the rate of ^{137}Cs accumulation at Station 12, with the steady increase arguably caused by the Chinese atmospheric bomb testing in the 1970s and/or the later atmospheric release of ^{137}Cs from the Chernobyl disaster in the 1980s.

4.2.2. *Station 5A*

Station 5A was taken from the basin floor of the northern basin in 935 m of water. The core site is located in the distal part of a large underwater fan arising from the Tompuda river. The core contains turbidite sediments, together with fine-grained hemipelagic deposits that are interbedded by coarse silt and turbidite sands. The hemipelagic sediments seem to be mainly diatomaceous ooze with 30–70% diatom frustuls. Two silt turbidites occur at 20–25.5 cm and 30–38.5 cm core depth, respectively, and contain much wood debris. A modern, dark-brown, oxidized zone is found at 0–5 cm depth, and an older, brown, oxidized zone (at 13–19 cm depth) is consolidated into a crust. Elsewhere sediments are reduced and are gray to gray-green in color.

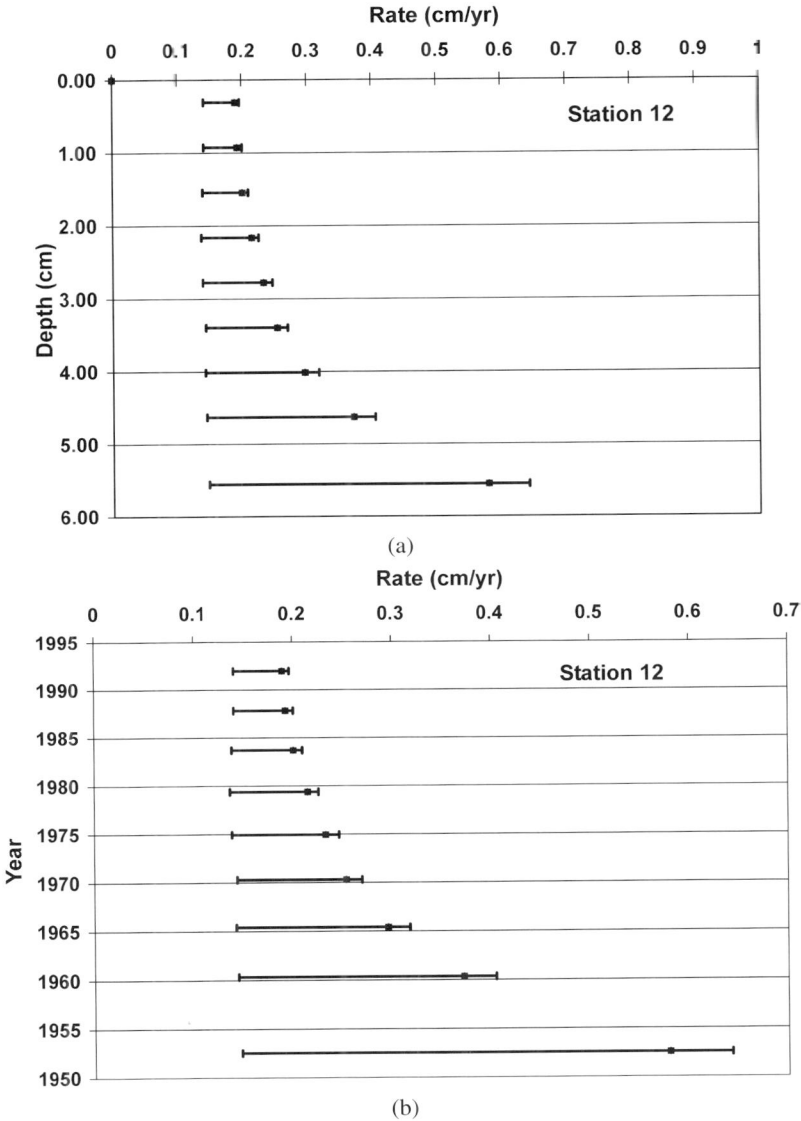

Fig. 7.14. (a) Sedimentation rate (cm/yr) versus decompacted depth, with associated error bars, for Station 12; (b) sedimentation rate (cm/yr) versus time, using the mean age-to-depth conversion of Fig. 7.13.

In a similar vein to Station 12, the SIT disentanglement code was used solely on ^{210}Pb activity data with decompacted depth in order to determine an age-to-depth curve, and then the ^{137}Cs data (3 points only) superposed to determine if the ages were acceptable in terms of the onset of ^{137}Cs at around year 1950. Figure 7.16 plots the observed ^{210}Pb activities (with ± 2 dpm/g error bars) versus decompacted depth but also including the sampling interval ranges of uncertainty. Superposed is the best theoretical fit from the radionuclide disen-

SOURCE

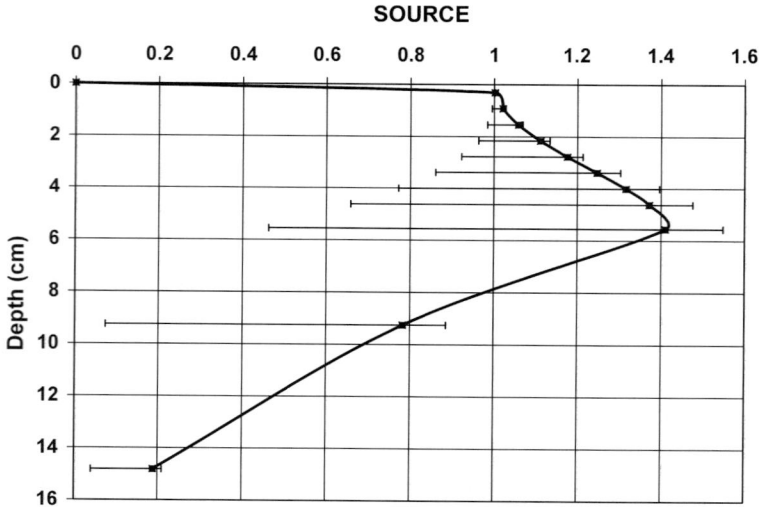

(a)

SOURCE

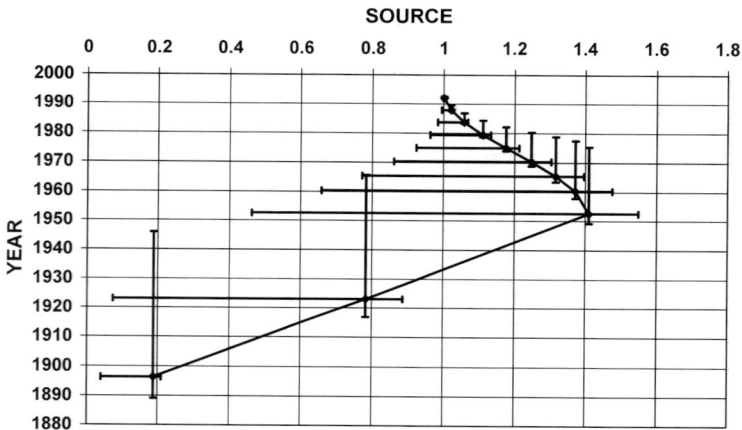

(b)

Fig. 7.15. (a) Intrinsic source variability (dimensionless) versus decompacted depth for ^{210}Pb profile of Station 12. Error assessments include uncertainty in measured activity values and sampling interval; (b) source variability versus time using the mean age-to-depth conversion of Fig. 7.13.

tanglement model. The only discrepant points of significant mismatch are the first and second activity values (centered at 1.25 and 1.4 cm mean depths, respectively) which are at 39 dpm/g and 70 dpm/gm, respectively, for an average of 55 dpm/g – precisely the same as the theoretical model predictions to within resolution, suggesting that a finer depth sampling interval would be needed to improve resolution. Deeper than about 6 cm depth there is no resolution because ^{210}Pb activities are extremely low, as shown in Fig. 7.17.

Using the resulting disentangled radionuclide information yields the age-to-depth curve shown in Fig. 7.18, with 6 cm depth corresponding to year 1905^{+20}_{-10}. The corresponding sed-

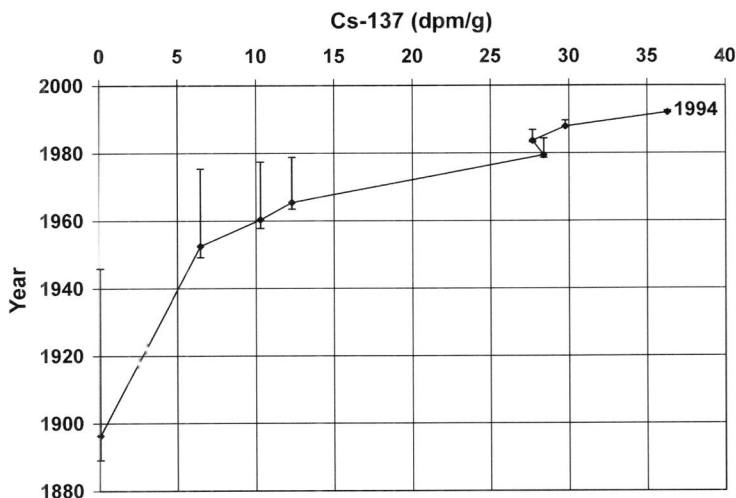

Fig. 7.16. Variations of measured ^{137}Cs activities (dpm/g) versus time for Station 12 using the mean age-to-depth conversion of Fig. 7.13.

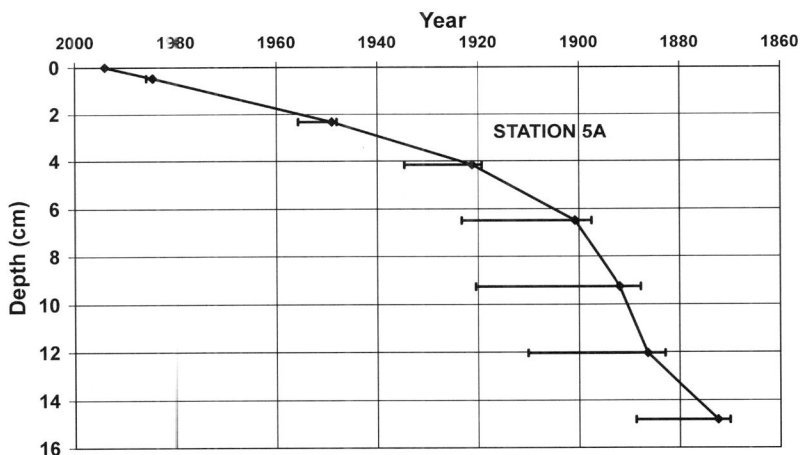

Fig. 7.17. Measured ^{210}Pb activities (dpm/g) for Station 5A versus decompacted depth (cm) with the superposed theoretical curve determined from the radionuclide disentanglement method given in text. The error bars on activity and depth determination for the measurements are indicated.

imentation rate with depth is shown in Fig. 7.19a and, using the mean curve of age-to-depth of Fig. 7.18, the sedimentation rate with time is shown in Fig. 7.19b.

There seems to be a roughly uniform rate of sedimentation at site 5A after year 1905 (i.e., shallower than 6 cm), at about 0.1 ± 0.05 cm/yr – a factor of between 2–4 less than at Station 12 since about year 1960.

The corresponding source variation with time is shown in Fig. 7.20, including error ranges from age-dating and from the finite depth ranges of the samples, indicating that, while having a

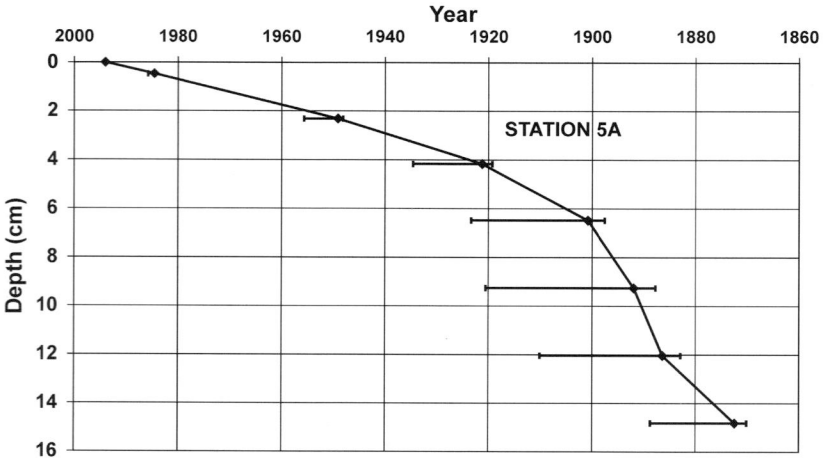

Fig. 7.18. Age-to-depth curve for the ^{210}Pb profile for Station 5A determined from the radionuclide disentanglement method given in text. The associated error bars on the curve are given and arise from both the uncertainty in activities and from the sampling interval in depth. Depth is for decompacted sediments.

significantly larger error than for Station 12, nevertheless there is an indication of a systematic source peak of about 40% higher than present day at around years 1950–1960. Uncertainty errors preclude unscrambling any significant source variation earlier than year 1950.

While only 3 activity values for ^{137}Cs are available for Station 5A, it is instructive to super-impose those activity values on an age curve determined from the mean age-to-depth curve of Fig. 7.18. As shown in Fig. 7.21, the oldest ^{137}Cs activity corresponds to around year 1950, the start of atomic testing, while the peak at 5 dpm/g corresponds to around year 1970, with the shallowest value at 3 dpm/g corresponding to year 1985. It is also important to note that both the magnitude and pattern of behavior of ^{137}Cs at the two stations are remarkably differ-ent. The magnitude of ^{137}Cs activity is around 3–5 dpm/g at Station 5A and is between 6–10 times higher at Station 12, with the pattern at Station 12 showing an increase towards the present day. It may be that this pattern and magnitude difference are tied to sediment sources containing ^{137}Cs or to sediment gathering regions by the Selenga River, or to dispersal pat-terns of sedimentation into Lake Baikal. It would be pointless to speculate further on the ^{137}Cs differences without more finely sampled data and/or more stations providing lateral variations of information.

4.3. *Sediment accumulation*

The prevailing changes in rates of sedimentation that occur at both Stations 12 and 5A are important for understanding the processes of sedimentation in Lake Baikal. For instance, the large (factor $\sim$ 100) decrease in sedimentation rate in the last 100 yrs at Station 5A implies very fast sediment rate changes can take place (in accord with turbidite depositions at Sta-tion 5A). The decrease could be tied to changes in precipitation in the lake basin or to changes in lake level that occur both naturally and, more recently, anthropogenically. The percentage of diatoms increases with core depth, in direct association with the earlier sedimentation event

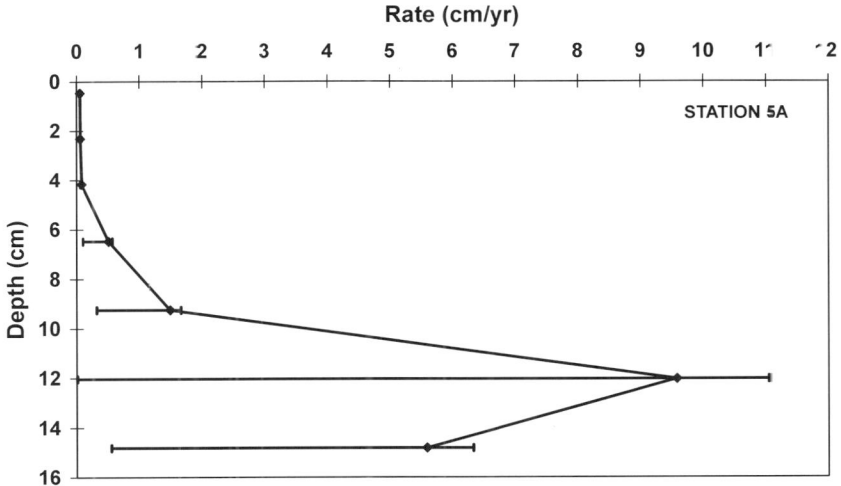

(a)

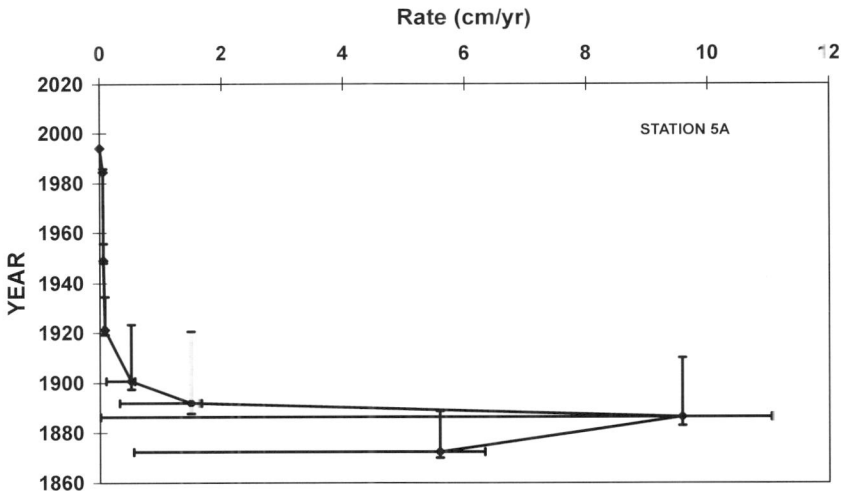

(b)

Fig. 7.19. (a) Sedimentation rate (cm/yr) versus decompacted depth, with associated error bars for Station 5A; (b) sedimentation rate (cm/yr) versus time, using the mean age-to-depth conversion of Fig. 7.18.

recorded around 1890 by ^{210}Pb. Clearly, further data at more sites are needed to more cleanly delineate such events without speculation or equivocation.

The presence of an identified event at 1955–1960 (Station 12), correlated with a high sedimentation rate (~ 0.6 cm/yr) at that time, has an immediate identifiable cause. After the dam across the Angara river was constructed in 1957, the lake level rose by 1 m with erosion of lake shores recorded after the lake level rise. In the Selenga delta, the rate of erosion was around 50 m/yr until 1961. Stabilization of erosion then occurred with an associated progra-

Source

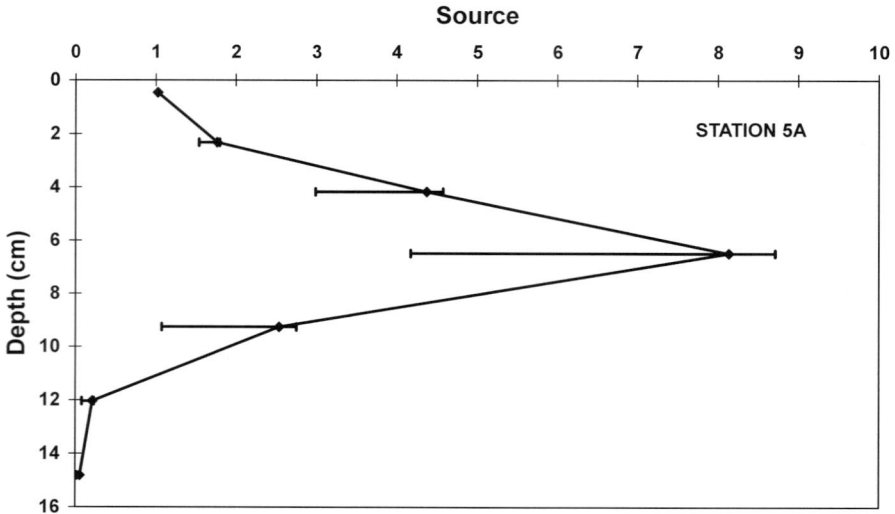

(a)

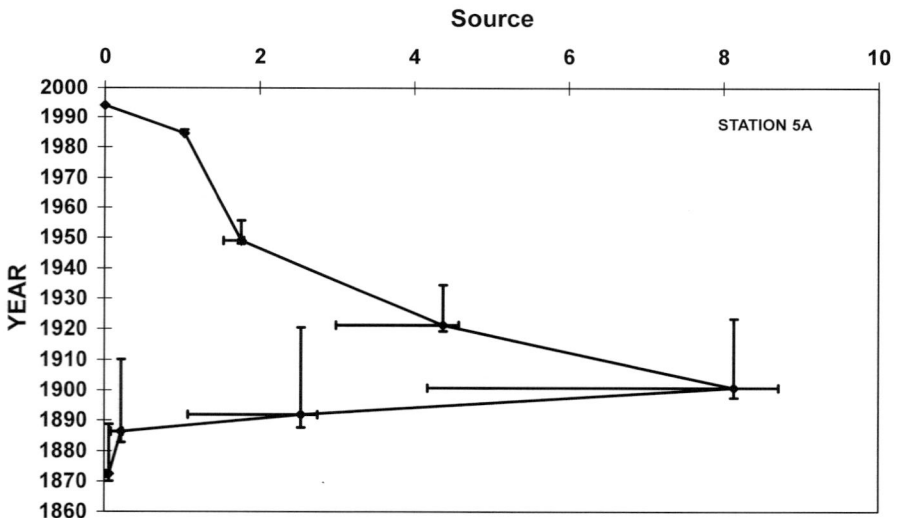

(b)

Fig. 7.20. (a) Intrinsic source variability (dimensionless) versus decompacted depth for ^{210}Pb profile of Station 5A. Error assessments include uncertainty in measured activity values and sampling interval; (b) source variability versus time using the mean age-to-depth conversion of Fig. 7.18.

dation commencement of the delta (Rogozin, 1993). It is highly likely that the change of sedimentation rate at Station 12, inferred from ^{210}Pb inversion, corresponds with the massive sedimentation during the lake level rise and erosion phase (1957–1961), and the slowing of sedimentation with later erosional stabilization.

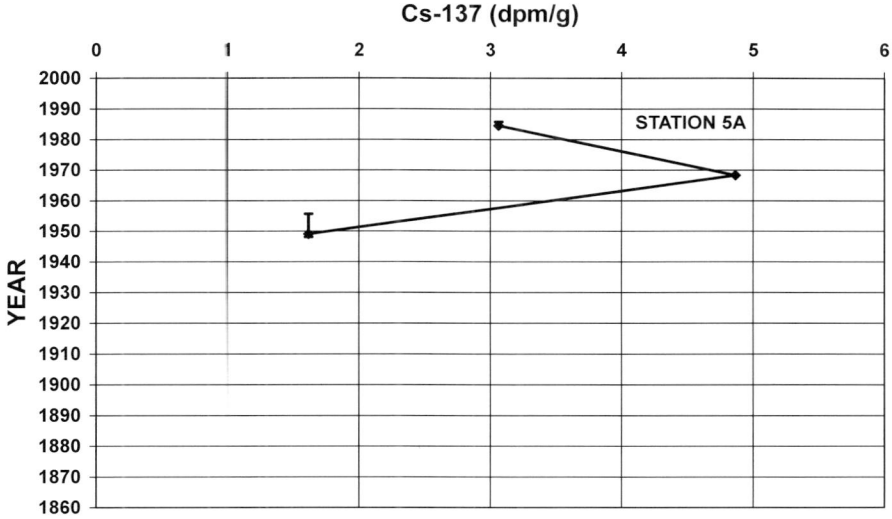

Fig. 7.21. Variations of measured ^{137}Cs activities (dpm/g) versus time for Station 5A using the mean age-to-depth conversion of Fig. 7.18.

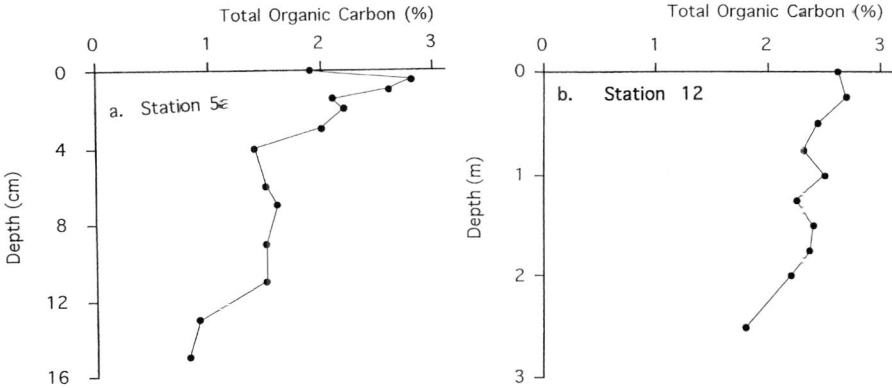

Fig. 7.22. Total organic carbon (TOC) content (%) versus measured depth (cm) for Stations 5A (Fig. 7.22a) and 12 (Fig. 7.22b).

4.4. *Organic carbon accumulation*

Total organic carbon content was generally low in core 5A as compared with core 12 (Fig. 7.22). Near the sediment surface the TOC content was approximately 1.9%, with a peak of 2.5% at a depth of 0.5 cm. Values decreased overall with depth with the lowest concentrations in the present-day depth interval of 13–17 cm within the core. By way of comparison, total organic carbon content at Station 12 remained fairly uniform with depth, with a surfi-

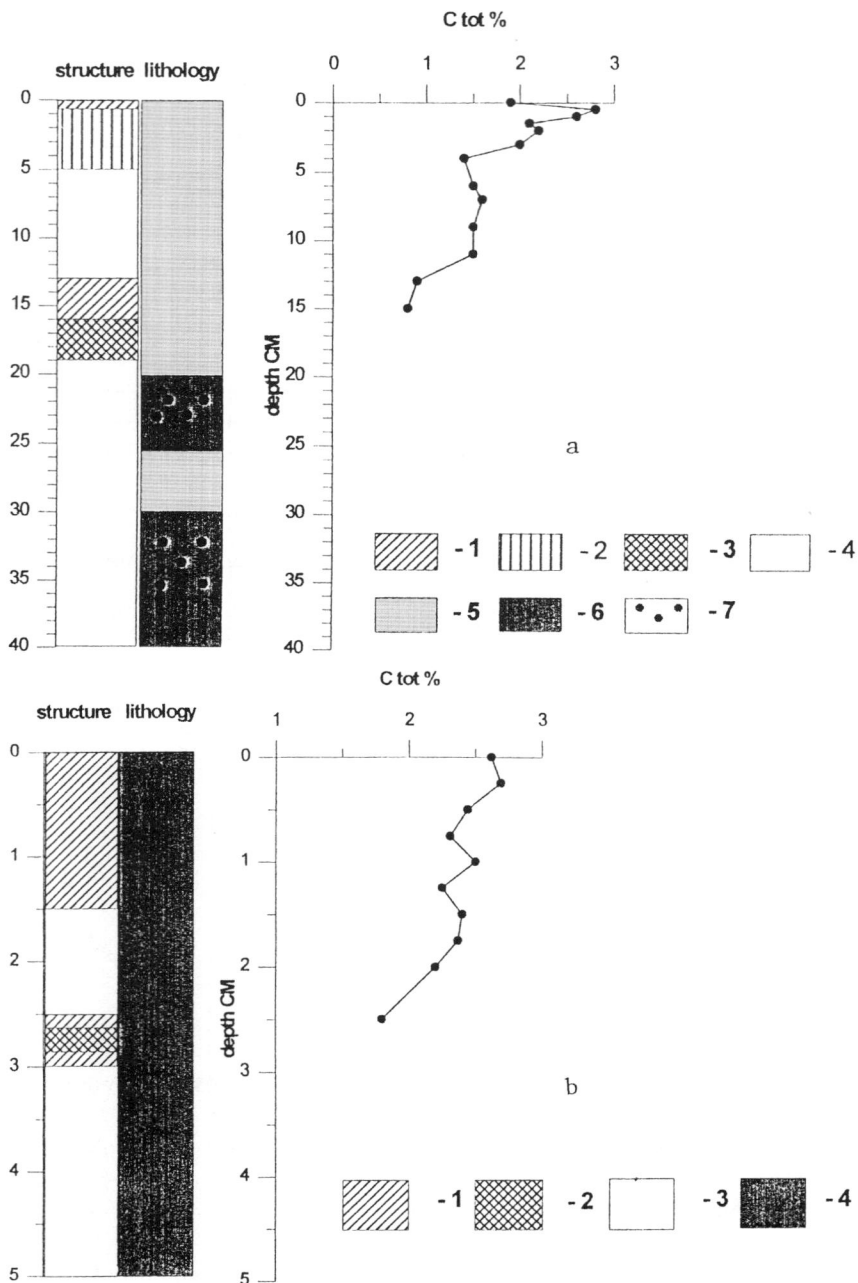

Fig. 7.23. (a) Lithology of the boxcore for Station 5A with measured depth drawn on the same scale as the TOC (%) content with depth. Legend: 1 = oxidized sediments; 2 = spotted reduced sediments; 3 = oxidized crust; 4 = reduced sediments; 5 = diatom ooze; 6 = turbidites; 7 = wood debris. (b) As for Fig. 7.23a but for Station 12. Legend: 1 = oxidized sediments; 2 = oxidized crust; 3 = reduced sediments; 4 = terrigenous mud with 20–30% diatoms.

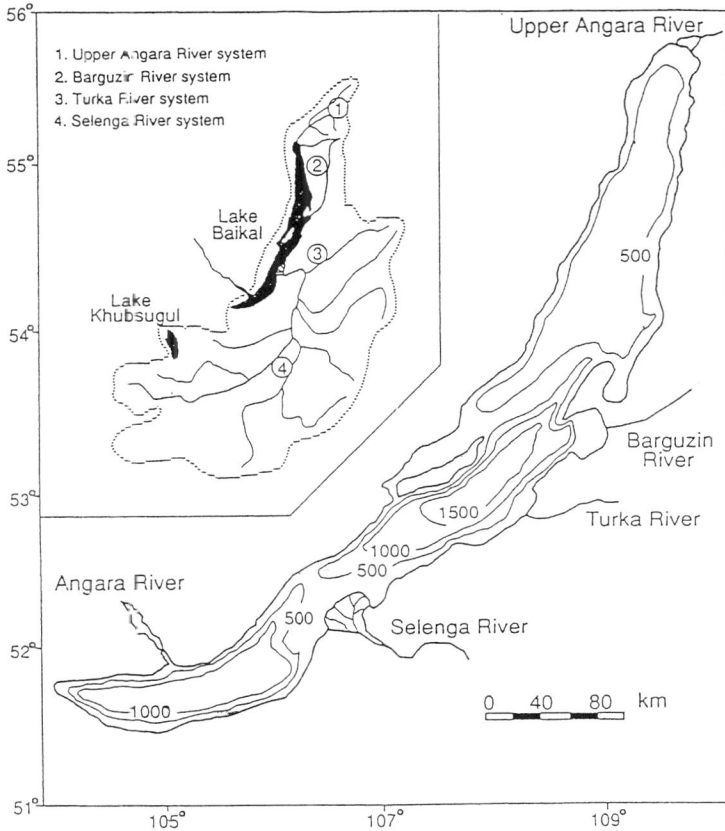

Fig. 7.24. Major drainage systems and bathymetric map (contours in meters) of Lake Baikal.

cial concentration at about 2.7% decreasing to a minimal concentration of 1.9% at a depth of 2.5 cm.

Accumulations of total organic carbon were higher (by roughly a factor two) in the southern basin core than in the northern basin core in accordance with a factor 2–4 higher sediment accumulation rate in the southern basin of Lake Baikal. Primary productivity in the southern region is considerably higher than that of the northern basin (Votinsev & Popovskaya, 1967), and may be attributed to two factors: first is the shorter duration of ice cover; the southern area is the last to be covered by a layer of ice in fall and the first to be exposed as the ice cover recedes north in late spring, resulting in a longer period of direct exposure to sunlight in the euphotic zone, and thus a greater degree of primary productivity. The second factor is the nutrient input of the Selenga River, which supplies approximately 50% of the nutrients utilized by the entire lake. The proximity of Station 12 to the Selenga River Delta allows for high concentrations of nutrients in the area and thus high primary production, which is fed by the nutrients. The resulting degree of primary production causes the bottom sediments in this region to be higher in organic carbon accumulation than the northern basin (Weiss et al.,

1991). However, evidence for marine sediments suggests that the majority of organic carbon is oxidized prior to burial in the ocean (Moore & Dymond, 1988); it is possible that the same is true for lake sediments, leading to an underappreciation of the true TOC accumulation. Indeed, maps showing differences in distributions of total carbon (Williams et al., 1993) in the northern and southern basins can be used, in conjunction with C/N ratios, to infer source supply for TOC. For example, in the Selenga river region the C/N ratio is larger than 10, implying that the sources of TOC are allochthonous rather than autochthonous. Old Russian references (1979–1980) show that the distribution of total carbon in surficial sediments is dependent on riverine input to Lake Baikal, with the highest percentages of total carbon in front of river mouths and with C/N ratios indicative of allochthonous organic origin.

Total accumulations were low in both regions ($\sim$ 3% of total available) and is likely due to one or more of the following: minimal overall primary production, low rates of sedimentation, oxic bottom waters, yearly ice cover, or depth of the water column. Votinsev & Popovskaya (1974) have also estimated that only about 2.8% of organic matter (on an annualized basis) is buried in the sediments of Lake Baikal, with all other TOC being deteriorated in the water column. The northern basin of the lake has the lower percentage of organic matter because: (i) the water column is larger at Station 5A than in the southern basin at Station 12, leading to a greater degree of TOC deterioration; (ii) Station 12 is located in front of the Selenga River mouth, which has a high input of material. In contrast, Station 5 in the northern basin is not near any rivers carrying organic matter. Figure 2 of Williams et al. (1993) shows the general prevalence of sediments with higher percentages of total carbon in association with nearness of the sampling stations to rivers; (iii) there is the highest primary production of phytoplankton in the southern and central basins, and lowest in the northern basin, in direct association with highest nutrient input in the southern basin and lowest in the northern basin. Indications are that reason (ii) is dominant in lowering the total carbon at Station 5A relative to Station 12.

^{210}Pb activities were higher overall in the southern basin core, which may also be attributed to the Selenga River input within the region. The river likely carries surface terrigenous sediments, which would almost certainly be younger than those of the northern basin that is accumulating sediment more slowly; thus the sediments would contain lower activities.

Chapter 8

Concerning error estimates in age dating
of stage boundaries

The purpose of this chapter is to provide quantitative methods and techniques that answer three questions concerning errors and their influence on age determinations.

1. Introduction

As has been noted in previous chapters throughout this volume, the determination of ages for geological horizons is perhaps the quintessential ingredient required if one is to evaluate the evolution of sedimentary basins in respect of the sediment and fluid dynamics, the paleothermal regime, and the timing of hydrocarbon generation, migration and accumulation in relation to the on-going structural and stratigraphic development in a basin. Without accurate measures of ages of stage boundaries, there are severe problems in determining the relative and absolute timing of the different processes that are of geologic, environmental and economic importance in a basin.

The only "absolute" age dates available for stage boundaries are those from radiometric measurements; all other age dates are assigned by interpolation or extrapolation using secondary controls such as biozonations, magnetic striping, palynology, and so on.

Perhaps some of this uncertainty in departures from known radiometric horizons is recorded in the variety of ages attached by different age-scale crafters to individual stage boundaries that do not have radiometric ages. Over the years this variation in age assignments has led to some remarkably interesting discussions in the scientific literature. An appreciation of methods for determining age-scales, and also of the long-running debate on age-scales, which has often waxed and waned furiously depending on the polemic vigor of proponents and opponents of individual ages, can be gleaned from the listing of age-scale constructors given in the references.

The discussion in this chapter is not intended to add yet one more contribution to fuel the flames of controversy, but rather to set out quantitative procedures and methods to use in answering some questions related to age-scale determination.

In the middle of 1993, a discussion was held in Norway at which Gradstein (1993) raised three questions concerning age-scales to which he wondered if answers were available at all. This chapter cannot do justice to the verbal style of delivery of the questions put forward by Gradstein, but notes supplied by Gradstein put the questions in particularly sharp form and focus of intent.

The three questions raised are:

1. Is there a good interpolation technique for stage boundary ages, for which no satisfactory age error is available, using the ages and their uncertainties for known stages immediately above and below the desired stage boundary, and can one define "good" quantitatively?

2. Radiometric age estimates of stage boundaries contain uncertainties, and there is also an uncertainty (error) in relative stratigraphic position of radiometric data. Is it possible to combine both of these errors into one value appropriate to a radiometrically dated stage boundary?

3. If the answers to questions 1 and 2 are affirmative, is it then possible to use *all* available data on ages and their uncertainties for stage boundaries to determine the age of a stage boundary, and its error, for which no satisfactory age error is directly available?

In posing these three questions Gradstein added the notes that the age-error estimates proposed by Harland et al. (1989) are somewhat subjective estimates of uncertainty, and he also noted that Gradstein et al. (1988) had calculated errors for the Triassic through Lower Cretaceous from a generalized stage level maximum likelihood, which errors are then carried through all further calculations and interpolations. Gradstein went on to note that, for the Upper Cretaceous, Gradstein et al. (1988) used the error from Ar/Ar age estimates to provide a zonal standard exactly at the Upper Cretaceous stage boundary, so that any error in Ar/Ar age dating is then "spread" to other stage boundaries around the zonal template standard.

It would, then, seem of some relevance both to age-dating methods in particular, and for the more general uses of age dates and their uncertainties to basin analysis calculations, to provide quantitative methods to answer the three questions.

2. Minimum error interpolation

Consider two stage boundaries 1 and 2 in a well as shown in Fig. 8.1, with nominal ages $\langle T_1 \rangle$ and $\langle T_2 \rangle$, respectively, uncertainties on the nominal ages of $\pm\sigma_1$, $\pm\sigma_2$, respectively, and separated by a sediment thickness H. A stage boundary B is identified as lying between stage boundaries 1 and 2, at a physical distance x above stage boundary 2 ($H - x$ below stage boundary 1). It is required to determine an age, T_B, for stage boundary B as well as an error $\pm\Delta T_B$, using only the age information of the known stage boundaries 1 and 2, and, possibly, the distances of stage boundary B from stage boundaries 1 and 2.

One of the more popular methods for assigning an age to stage boundary B is to arbitrarily use direct proportionality of the physical distance of stage boundary B from stage boundaries 1 and 2, and so to write

$$T_B = \langle T_2 \rangle (x/H) + \langle T_1 \rangle (1 - x/H) \tag{8.1}$$

so that $\langle T_1 \rangle \leqslant T_B \leqslant \langle T_2 \rangle$. But this identification does not use any of the information available on the uncertainties of the ages of stage boundaries 1 and 2, which is one of the reasons for Gradstein's first question.

A more consistent interpolation scheme for determination of the age of stage boundary B, and its uncertainty, is as follows.

The error, $\pm\sigma_1$, in the nominal age, $\langle T_1 \rangle$, of stage boundary 1 argues that one can choose a value T, which has a Gaussian probability of being realized, centered on the nominal (average)

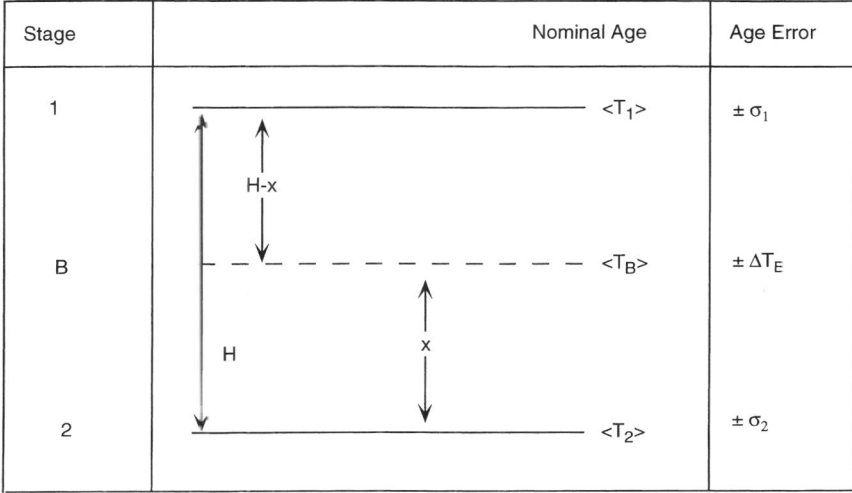

Fig. 8.1. Schematic diagram showing a stage boundary B lying between two other stage boundaries of nominal ages $\langle T_1 \rangle$ and $\langle T_2 \rangle$, respectively, with associated age errors $\pm\sigma_1$, $\pm\sigma_2$. The boundary B is at a physical depth $H - x$ deeper than stage boundary 1, and x shallower than stage boundary 2.

age $\langle T_1 \rangle$, and with a standard variance σ_1^2 around $\langle T_1 \rangle$; similarly for stage boundary 2. If a linear interpolation for stage boundary B is chosen between stage boundaries 1 and 2, then for any particular choices of T_1 and T_2, one writes

$$T_B = fT_1 + (1 - f)T_2, \tag{8.2}$$

where f is a positive fraction (less than unity). The expected value of T_B, $\langle T_B \rangle$, is then

$$\langle T_B \rangle = f\langle T_1 \rangle + (1 - f)\langle T_2 \rangle. \tag{8.3}$$

Equally, by squaring equation (8.2) and then taking the expected value one has

$$\langle T_B^2 \rangle = f^2 \langle T_1^2 \rangle + 2f(1 - f)\langle T_1 \rangle \langle T_2 \rangle + (1 - f)^2 \langle T_2^2 \rangle. \tag{8.4}$$

Then use the facts (Feller, 1966) that

$$\langle T_2^2 \rangle = \langle T_2 \rangle^2 + \sigma_2^2; \qquad \langle T_1^2 \rangle = \langle T_1 \rangle^2 + \sigma_1^2 \tag{8.5}$$

to rewrite equation (8.4) in the form

$$\langle T_B^2 \rangle = \left[f\langle T_1 \rangle + (1 - f)\langle T_2 \rangle \right]^2 + f^2 \sigma_1^2 + (1 - f)^2 \sigma_2^2$$
$$\equiv \langle T_B \rangle^2 + f^2 \sigma_1^2 + (1 - f)^2 \sigma_2^2. \tag{8.6}$$

The variance in the age estimate for stage boundary B is then

$$\Delta T_B^2 \equiv \langle T_B^2 \rangle - \langle T_B \rangle^2 = f^2 \sigma_1^2 + (1-f)^2 \sigma_2^2. \tag{8.7}$$

The variance ΔT_B^2 is minimized when the fraction f is chosen as

$$f_{min} = \sigma_2^2 / \left(\sigma_1^2 + \sigma_2^2 \right) \tag{8.8a}$$

when

$$\Delta T_B^2 (minimum) = \sigma_1^2 \sigma_2^2 / \left(\sigma_1^2 + \sigma_2^2 \right) \tag{8.8b}$$

and the corresponding estimate of $\langle T_B \rangle$ is

$$\langle T_B \rangle = \left(\sigma_1^2 + \sigma_2^2 \right)^{-1} \left(\sigma_2^2 \langle T_1 \rangle + \sigma_1^2 \langle T_2 \rangle \right) \tag{8.8c}$$

so the minimum error on the stage boundary B age estimate is $\pm \Delta T_B (minimum)$. Any other estimate of ΔT_B will be larger than the minimum.

For instance, if one chooses to force the fraction $(1-f)$ to take on the value x/H, corresponding to the popular choice of proportional distance, then

$$\Delta T_B^2 = \Delta T_B (minimum)^2 + \left(\sigma_1^2 + \sigma_2^2 \right)(1 - x/H - f_{min})^2$$

$$\geqslant \Delta T_B (minimum)^2. \tag{8.9}$$

Clearly then, without further evidence to force a choice of the fraction f, the answer to Gradstein's first question is that there is indeed a good choice for the age of stage boundary B (given by equation (8.8c)) which uses both the nominal ages and the uncertainties of the stage boundaries immediately above and below the unknown stage boundary, and which also provides a minimal error on the age determination (given by $\Delta T_B (minimum)$ of equation (8.8b)). Any other choice has a larger error so that in this sense a quantitative definition of "good" is also provided.

3. Combined absolute and stratigraphic radiometric errors

Consider a stage boundary of nominal age $\langle T \rangle$, with an absolute error in age of $\pm \sigma$ due to uncertainty in the radiometrics (or from other causes), together with a *fractional* age error $f = \pm (\delta T / \langle T \rangle)$ due to relative stratigraphic positioning, as per question 2 posed by Gradstein.

Two cases have to be distinguished: (1) statistical independence of the fractional and absolute errors; (2) partial dependence of the fractional error, f, on the absolute error, σ. Consider each in turn.

3.1. *Statistical independence*

If the fractional error, f, due to relative stratigraphic positioning and the absolute error, σ, due to radiometric uncertainty are independent of each other, then the fractional error contributes an amount

$$\delta T_{\text{strat}} = \pm f \langle T \rangle \tag{8.10}$$

to the age uncertainty of the stage, while the absolute error contributes

$$\delta T_{\text{absolute}} = \pm \sigma. \tag{8.11}$$

The total error, δT_{stage}, on the stage boundary age $\langle T \rangle$ is then just the square root of the sum of the squares of the two independent components (Feller, 1966), yielding

$$\delta T_{\text{stage}} = \pm \left(f^2 \langle T \rangle^2 + \sigma^2 \right)^{1/2}. \tag{8.12}$$

3.2. *Partial statistical dependence*

If the fractional error is partially dependent on the absolute error in a linear manner so that

$$f = f_0 + \lambda \sigma, \tag{8.13}$$

where λ is a known value (for instance, if the fractional error were 10% dependent on the absolute error then $\lambda = 0.1$), and f_0 is independent of σ, then

$$\delta T_{\text{strat}} = \pm \left\{ \langle T \rangle f_0 + \lambda \langle T \rangle \sigma \right\} \tag{8.14a}$$

and

$$\delta T_{\text{absolute}} = \pm \sigma. \tag{8.14b}$$

In this case, the total error on the nominal stage age, $\langle T \rangle$, is made up from the square root of the sums of the squares of the *independent* components (Feller, 1966), so that

$$\delta T_{\text{stage}} = \pm \left[\langle T \rangle^2 f_0^2 + \sigma^2 \left(1 + \lambda \langle T \rangle \right)^2 \right]^{1/2}. \tag{8.15}$$

Thus the second question asked by Gradstein is answered affirmatively.

4. General age/stage interpolations

Consider a numerical sequence $\{ i = 1, 2, \ldots, N \}$ of ordered stage boundaries B_i with increasing depth, with $B_1 < B_2 < B_3 < \cdots \leqslant B_N$. Let each stage boundary B_i have a nominal age $\langle T_i \rangle$, and an error $\pm \sigma_i$ on the nominal age. It is required to construct an age to sequence

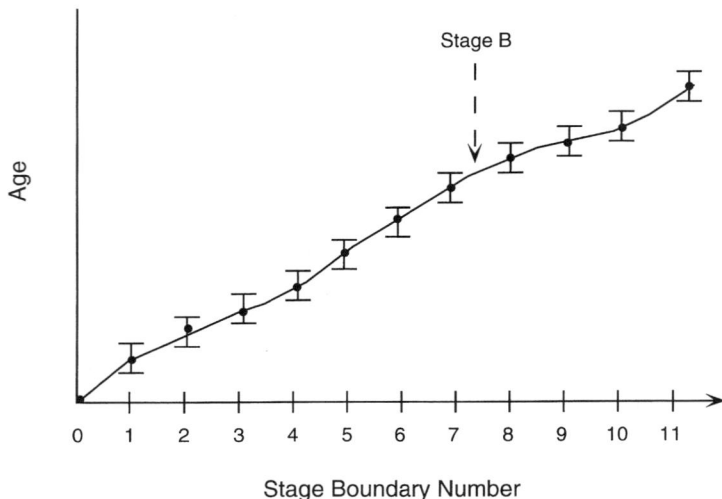

Fig. 8.2. Schematic of a stage boundary B in an age versus stage boundary plot with uncertainties attached to the ages of the known stage boundaries.

behavior which not only honors the sequence order, but also uses the nominal ages of the sequence stage boundaries and their uncertainties to construct a procedure for determining the stratigraphic age, and associated uncertainty, of any stage boundary B which lies between *any* pair of the known stage boundaries using *all* of the available data. The point here is that using a single pair of horizons (as done in Section 2) to bracket an intervening stage boundary may not be adequate if one stage boundary age has a large uncertainty. It may be preferable to use an horizon further from the unknown stage boundary, but with less uncertainty on the age of the further horizon. It may also be preferable to use more than one pair of horizons, anyway, to better constrain the age and uncertainty of a particular stage boundary.

This section of the chapter provides the general method for evaluating such concerns quantitatively. Consider the schematic representation of age versus stage boundary number shown in Fig. 8.2 together with errors on the known stage boundary ages.

For the known stage boundaries $i = 1, \ldots, N$ at depths $B_1, \ldots, B_N$, the sequence requirement is that $T_1 \leqslant T_2 \leqslant T_3 \leqslant \cdots \leqslant T_N$. For an unknown stage boundary B, lying between B_i and B_{i+1}, the requirement is to construct its age and age uncertainty, knowing that the age T_B must lie in $T_i \leqslant T_B \leqslant T_{i+1}$, i.e., the age to sequence stage number curve must be monotone increasing in B_1 to B_N.

To handle this general problem consider the function

$$Y(x) = \exp\left(ax + \sum_{m=1}^{M} \alpha_m \sin(m\pi x/B_N) \right) \tag{8.16}$$

in $0 \leqslant x \leqslant B_N$.

No matter the choice of a (positive or negative) or α_m (positive or negative) it follows that $Y(x) \geqslant 0$ in $0 \leqslant x \leqslant B_N$ with $Y(0) = 1$, $Y(B_N) = \exp(a B_N)$.

Then consider an age to sequence boundary relation of the form

$$T(B) = \xi_0 \int_0^B Y(x)\,dx \equiv \xi_0 Z(B),$$

(8.17)

where ξ_0 is a positive constant, which satisfies all of the monotonicity requirements of the stage to age criteria. What one would like to do is to use the functional forms (8.16) and (8.17) with the nominal age and uncertainty in age information of the known stages to determine the coefficients $\xi_0, a, \alpha_1, \ldots, \alpha_M$, as well as the number of terms M. Then, with the known depth B, one has an interpolation scheme for the age of stage B, together with its uncertainty. As is well known (Feller, 1966) a least-squares Gaussianly-distributed control function based on the nominal stage age sequence $\{T_i\}$ at depths $\{B_i\}$, together with the uncertainties $\{\sigma_i\}$ then requires a minimum for

$$X^2 \equiv N^{-1} \sum_{i=1}^{N} \{T_i - T(B_i)\}^2 \sigma_i^{-2}$$

(8.18)

which then allows a determination of the parameters $\xi_0, a, \{\alpha_m\}$ and M in equations (8.16) and (8.17).

The parameters enter equation (8.18) in two ways: quadratically for ξ_0, and non-polynomically for the remainder. We can capitalize on this difference as follows. For ξ_0, a minimum in X^2 with respect to ξ_0, with all other parameters held fixed, produces a linear equation for ξ_0 as

$$\frac{\partial X^2}{\partial \xi_0} = -2N^{-1} \sum_{i=1}^{N} \sigma_i^{-2} \{T_i - \xi_0 Z(B_i)\} Z(B_i) = 0$$

(8.19)

so that ξ_0 is given by

$$\xi_0 = \sum_{i=1}^{N} \sigma_i^{-2} T_i Z(B_i) \Big/ \sum_{i=1}^{N} \sigma_i^{-2} Z(B_i)^2 > 0.$$

(8.20)

Then, with the value (8.19) for ξ_0, equation (8.18) can be re-written

$$X^2 = \left[N^{-1} \sum_{i=1}^{N} \sigma_i^{-2} \left\{ T_i - Z(B_i) \sum_{j=1}^{N} \sigma_j^{-2} T_j Z(B_j) \right\}^2 \right] \left[\sum_{k=1}^{N} \sigma_k^{-2} Z(B_k)^2 \right]^{-1},$$

(8.21)

where $Z(B)$ (and so X^2) depends on the, as yet unknown, parameters $a, \{\alpha_m\}$, and M in non-polynomic form. These parameters can be determined from equation (8.20) as follows. For the jth parameter in the set $p = (a, M, \{\alpha_m\})$ let there be chosen values of minimum $\min_j$, and maximum $\max_j$, ranges.

Then set

$$b_j = (p_j - \min_j)/(\max_j - \min_j) \tag{8.22}$$

and regard b as a fundamental vector on which p depends, with $0 \leqslant b_j \leqslant 1$. Regard X^2 as depending directly on b. Then construct the iteration scheme that, at the nth iteration, returns the value

$$b_j(n+1) = \sin^2 \theta_j(n+1) \tag{8.23a}$$

with

$$\theta_j(n+1) = \theta_j(n) \exp[-\tanh\{\alpha_j(\partial X^2/\partial b_j(n))\delta_j(n)\}], \tag{8.23b}$$

where

$$\alpha_j = |\partial X^2/\partial b_j(0)|^{-1} \ln\{1 + (Rb_j(0))^{-1}\}, \tag{8.24a}$$

$$q_j(n) = \gamma^2 + |\theta_j(n) - \theta_j(n-1)|/|\theta_j(n-1)|, \tag{8.24b}$$

$$\delta_j(n) = q_j(n) / \frac{1}{J} \sum_j^J q_j(n) \tag{8.24c}$$

with R the number of times the non-linear loop is to be undertaken; $b_j(0)$ is the initial estimate of b_j $(0 \leqslant b_j(0) \leqslant 1)$; $\delta_j(n)$ is a sensitivity factor designed to reduce initially those components of b which are furthest from providing a minimum in X^2; and J is the number of terms being included. The factor γ^2 is the value in the numerical approximation scheme to the derivative

$$\frac{\partial X^2}{\partial b_j} \cong (\gamma b_j)^{-1}[X^2(b_1, \ldots, b_j + \gamma b_j, \ldots) - X^2(b_1, \ldots, b_j, \ldots)] \tag{8.24d}$$

and should be small (< 0.01) depending on the accuracy with which one chooses to calculate derivatives.

The iteration scheme given by equation (8.22b) guarantees to produce a minimum in X^2 for *all* parameters, guarantees to keep each and every parameter in the range $0 \leqslant b_j \leqslant 1$, and guarantees to fit the observed values of $\{T_i, B_i\}$ pairs with greater accuracy as more terms are included in the functional form (8.16).

Thus, based on the observed stage ages and their errors, all parameters are determined. The question of determining the age $T(B)$ of an unknown stage at a depth B, between known age-dated stages at depths B_i and B_{i+1}, is then resolved. The best estimate of the age is

$$T(B) = \xi_0 \int_0^B Y(x)\,dx \tag{8.25a}$$

with ξ_0 given by equation (8.19), and the remaining parameters a, $\{\alpha_m\}$ given through the minimum of equation (8 20). The variance, σ_B^2, around $T(B)$ is then given by

$$\sigma_B^{-2} = N^{-1} \sum_{i=1}^{N} \sigma_i^{-2}, \tag{8.25b}$$

so that the age of the stage at depth B in $B_i < B < B_{i+1}$, can be written $T(B) \pm \sigma_B$, including the uncertainty from the uncertainties on the N horizons used to determine the functional behavior.

It would seem from equation (8.24b) that it is best (in the sense of minimizing σ_B) to use horizons ordered with respect to the inverse of the square of each σ value.

For instance, if, arbitrarily, the a priori assumption were to be made that age was directly proportional to depth for the known horizons, then one would have

$$T(B) = B \sum_{i=1}^{N} \sigma_i^{-2} T_i B_i \Big/ \sum_{j=1}^{N} \sigma_j^{-2} B_j^2 \pm \sigma_B. \tag{8.26}$$

But then one can easily check if this arbitrary assumption is appropriate by inserting non-zero values of the parameter, a, in equation (8.20) and determining the value of a which produces a minimum in X^2. A value of $a \neq 0$ would then imply that it is better not to set up an a priori assumption of $a = 0$.

Note also that the best fit for $T(B)$ automatically produces the minimum error assessment of the age of the stage boundary. Thus Gradstein's third question is also addressed affirmatively.

5. A corollary

As was recognized by Gradstein (1993) in his presentation of the three seminal questions, there is also a corollary question related to correlations between pairs of wells. Gradstein noted that spline fitting of ages of stage boundaries versus depth in each well, and then production of error bars on correlation lines between wells, is somewhat primitive. Gradstein went on to note that the source of the problem from a practical viewpoint is that there are often not enough datum points in one of the wells for an *a posteriori* type of analysis of the error bars; he asked whether an *a priori* approach from generalized information for a basin may then be better.

Consider a set of sequential stages of ordered ages $T_1, T_2, \ldots, T_N$, which in the first well occur at the known depths $z_1, z_2, \ldots, z_N$, and in the second well at the known depths $y_1, y_2, \ldots, y_N$.

Suppose that a comparison is to be made of correlated behavior between the two wells. Two extreme options are possible: (i) correlate well 1 depths against well 2 depths as a template; (ii) correlate well 2 depths against well 1 depths as a template. Under the best of all conditions, the two cross-correlations should be identical. However, because the ages carry uncertainties, *which do not have to be the same in both wells*, the procedure which carries the minimum uncertainty on cross-correlating horizons of known ages between wells is not necessarily given by either of the extreme options above.

Fortunately, precisely this sort of problem has been addressed in great detail by Martinson, Menke & Stoffa (1982) for the case where lithologic correlations with depth are to be made between a pair of wells, and a variation of the procedure can be used to address the problem of converting chronostratigraphic records, measured with respect to depth, into records measured with respect to time. If the sedimentation rate and compaction velocity were constant from site to site, then measurements at depth Z_j would correspond precisely to a fixed time t_j for each and every location. However, if the sedimentation or compaction rates vary from location to location, then an identification of events at the same geologic time in the different stratigraphic columns becomes difficult. The problem is that the same depth on spatially distinct records corresponds to different times. To determine the best variation in velocity becomes a question of obtaining the best semblance measure from the recorded sequence of records.

To describe the sense of the argument, suppose a set of records of the same type have been measured with respect to sedimentary depth. To convert the measured depth records into records with respect to sedimentary age we must either independently date the formations or determine the rate of sedimentation and compaction with time. Assuming that we do not have very many radiometrically determined ages, we wish to use the record itself to determine a depth-to-time conversion for the formations between known age horizons. Suppose then that we have a set $x(L, Z_i)$ $(i = 1, 2, \ldots, N)$ of records at several neighboring spatial locations on the surface of the earth and at N equal depth intervals ΔZ, so that $Z_j = j \Delta Z$, where Z_j is the depth location of the jth sample at each site. The *relative* distortion of the different records can then be corrected. Let the rate of sedimentation with time at location L be $V(L, t)$, so that a current depth $Z(L)$ corresponds to a depositional time T given by

$$Z(L) = \int_0^T V(L, t) \, dt. \tag{8.27}$$

For a record at location L^1, with depositional rate $V(L^1, t)$, the depositional time T corresponds to the current depth $Z(L^1)$ given by

$$Z(L^1) = \int_0^T V(L^1, t) \, dt. \tag{8.28}$$

It follows that

$$Z(L^1) = Z(L) + \int_0^T \left[V(L^1, t) - V(L, t) \right] dt$$

$$= Z(L) + \int_0^T \Delta V(L, L^1, t) \, dt, \tag{8.29}$$

$$X[L, Z(L)] = \sum_{k=-N}^{N} x(L, k) \exp\left[\frac{i\pi Z(L)k}{N \Delta Z} \right], \tag{8.30a}$$

and

$$X[L^1, Z(L^1)] = \sum_{k=-N}^{N} x(L^1, k) \exp\left[\frac{i\pi Z(L^1)k}{N\Delta Z}\right],\tag{8.30b}$$

where $x(L, k)[x(L^1, k)x(L, Z)[x(L^1, Z)]$ and $N\Delta Z$ is the record length. Then, if we replace $Z(L^1)$ by equation (8.29) in the expression for $x(L^1, Z)$ we produce

$$X[L^1, Z(L^1)] = \sum_{k=-N}^{N} x(L^1, k) \exp\left[\frac{i\pi Z(L)k}{N\Delta Z}\right]$$

$$\times \exp\left[i\pi k(N\Delta Z)^{-1} \int_0^T \Delta V(L, L^1, t)\, dt\right].\tag{8.31}$$

In order to extract $\Delta V(L, L^1, t)$, we then proceed in a conventional manner.

To construct the semblance of $x(L, Z)$ and $x[L^1, Z(L^1)]$ for short times T, replace $\int_0^T M\Delta V\, dt$ by $\sum_{m=0}^{M} M\Delta V_m \Delta t$ where the record has been split into equal time intervals Δt, with $M\Delta t = T$. Then, for a fixed value of M, by systematically increasing ΔV_m, we can find the value that will maximize the semblance. This procedure is repeated for each and every measurement point (corresponding to different depths and so different times). In this manner we are able to construct a sequence of values of ΔV_m, for increasing M as the depth sequence of points is increased. Then, the relative velocity of sedimentation at any time is determined. The absolute velocity of sedimentation cannot be determined by this method for we use one record (or some average over all records) to supply the template for comparison against which the other records at different spatial locations are tested for distortion. Thus, we need to obtain some measure of absolute age dating in order to convert the template record (measured with respect to depth) into an equivalent template record in time.

However, this method will: (a) remove relative distortions between records; and (b) provide a means of determining the relative rate of sedimentation and compaction with time between spatial locations.

An alternative method is available. The problem of correlating two or more records when one has been compressed or expanded is also amenable to inverse methods of data analysis using mapping function techniques. Indeed, Martinson et al. (1982) have followed precisely this route to relate coherence of two signals measured at different locations with different sedimentation rates. They were able to determine the distortion of one record relative to another and so obtain relative sedimentation rates.

In order to understand the sense of their argument, consider two records R_1 and R_2 measured with respect to sedimentary depth below the sediment–water interface. Use $R_1(Z)$ as a reference signal, so that we are interested in obtaining the best correlation of R_2 with respect to R_1. Let $R_1(Z_1)$ be related to R_2 through a mapping function $Z_2(Z_1)$, so that $R_2[Z_2(Z_1)] = R_1(Z_1)$. The question is to obtain a method of finding the diffeomorphic map from Z_2 to Z_1, which maximizes the degree of agreement of the two records. Again the general underpinning logic of the method is least squares minimization. Martinson et al. (1982)

considered a mapping function made up of a linear trend plus a truncated Fourier series in the form

$$Z_2(Z_1) = a_0 Z_1 + \sum_{k=1}^{n-1} a_k \sin\left(\frac{\pi k Z_1}{Z_{max}}\right), \tag{8.32}$$

where Z_{max} is the length of the record R_1. The problem is reduced to finding the coefficients $a_0, a_1, \ldots, a_{n-1}$, which will minimize the difference between the reference signal R_1 and the signal R_2 distorted relative to R_1.

First, the two signals are formally normalized so that

$$r_1(Z_1) = R_1(Z_1)\left[\int_0^{Z_{max}} R_1(Z)^2 \, dZ\right]^{-1/2}, \tag{8.33a}$$

$$r_2(Z_2) = R_2(Z_2)\left[\int_0^{Z_{max}} R_2\big(Z_2(Z_1)\big)\right]^{-1/2}. \tag{8.33b}$$

Then, a least square error is defined by

$$E(a) = \int_0^{Z_{max}} \left\{r_1(Z_1) - r_2\big[Z_2(Z_1, a)\big]\right\}^2 dZ_1, \tag{8.34}$$

where $a = (a_0, a_1, \ldots, a_{n-1})$. This error is independent of magnitude differences between the records but is sensitive to the distortion of shape between the records. The clever observation made by Martinson et al. (1982) is that E is directly related to the coherence C between the two records by

$$E = 2(1 - C). \tag{8.35}$$

Thus, the shape error is minimized when the coherence is maximized with

$$C = \left\{\int_0^{Z_{max}} R_1(Z_1) R_2\big[Z_2(Z_1)\big] \, dZ_1\right\}$$
$$\times \left\{\left[\int_0^{Z_{max}} R_1(Z_1)\right]^2 dZ_1\right\}\left[\int_0^{Z_{max}} R_2\big(Z_2(Z_1)\big)\right]^{-1/2}. \tag{8.36}$$

The determination of the coefficients in the mapping function now proceeds. Pick an initial set of values a_i. Then, if a be perturbed from a_i by δa, we have $C(a_i + \delta a)$ replacing $C(a)$. Then, with ΔC as the change in coherence, we have

$$\Delta C \cong \delta a_i \cdot \nabla_a C, \tag{8.37}$$

where $\nabla_a C = (\partial C/\partial a_1, \partial C/\partial a_2, \ldots, \partial C/\partial a_n)$. But we want the increment in δa to be such that the coherence heads most rapidly to its maximum value. Thus, δa should be in the direction of $\nabla_a C$, which is the normal direction to contours of constant coherence. Then,

$$\delta a = \Delta C \nabla_a C / |\nabla_a C|^2. \tag{8.38}$$

But, as Martinson et al. (1982) noted, this form of δa is of limited use since the coefficients a can be changed iteratively to make $\nabla_a C$ tend to zero and so drive δa unstable. They overcome this problem by making the vector δa of fixed magnitude k. Then,

$$\Delta C = k^2 |\nabla_a C|^2 / [\nabla_a \tilde{C} \delta a]. \tag{8.39}$$

This process also aids convergence because iterations are continued as long as C increases and ΔC is decreasing. Resolution is maximal where ΔC begins to decrease. Martinson et al. (1982) also showed that the error in the mapping function $Z_2(Z_1)$ at any depth Z_i is given as follows. The mapping is linear in the a's in the sense that at any depth $Z_1 = Z_i$,

$$Z_2(Z_i) = a_j F_{ij}, \tag{8.40a}$$

where F_{ij} is the coefficient of a_j evaluated at depth $Z_1 = Z_i$ in the map. Then, the error σ_i in the map at depth Z_i is

$$\sigma_i = \left[(\text{cov } Z_2)_{ii}\right]^{1/2}, \tag{8.40b}$$

where $(\text{cov } Z_2)$ is the covariance matrix of Z_2 at $Z_1 = Z_i$. Numerous synthetic examples and case histories have been provided by Martinson et al. (1982) to illustrate the strength and accuracy of the method.

The problem of determining the best n number of coefficients $a_0, \ldots, a_{n-1}$ to use is also addressed in a conventional manner. As the number of coefficients is increased, there comes a point where the uncertainty in the determination of the nth coefficient is as large as the coefficient itself. At this point the noise limit has been reached and there is no reason to continue to add coefficients. In short, we have available a method of estimating a mapping function that will remove distortions of one record relative to another – which latter is regarded as a template. Absolute conversion of depth to geologic time then requires age dating of the template sedimentary column.

One point, however, should be mentioned. Because both the distortion and mapping function enter nonlinearly, there is no guarantee that the mapping function is unique. The resolution and precision of given choices of mapping functions can be, and have been, addressed using the procedure advocated by Martinson et al. (1982), but the question of uniqueness remains an outstanding concern.

Indeed, to put the point differently, Matthews (1984) has boldly stated, "To a first approximation, many deep sea cores can be regarded as having a constant sedimentation rate. To a first approximation this rate can be estimated from carbon-14 dating of the upper portion of the core." Extrapolation to lower parts of the core with the assumed constant sedimentation rate is then done to provide a depth-to-time conversion. But how do we know, and not assume, that we can regard the sedimentation rate as constant? What control do we have on the validity and accuracy of such extrapolations to lower parts of the core? Clearly, unless we have some form of age dating as a control at deeper depths, we have nothing except that which we believe to be true but cannot justify. Even when such age-dated depth horizons are available, we have no knowledge of variability of sedimentation rates on a scale finer than the sampled age-dated horizons, even though the records may be sampled on a much finer scale. What portion of the record we ascribe to intrinsic variations and what portion we ascribe to variations in sedimentation rate are then matters of emotion rather than, more properly, matters of quantitative science. It is all too easy to impose on the record, either consciously or subconsciously, a variation that we hope will agree with some preconceived notion, such as orbital cycle effects of the earth. One cannot resolve the signal from the noise (sedimentation variability) on a time scale finer than that of the age-dated horizon samples.

Chapter 9

Epilogue

The importance of obtaining ages for sediments is the quintessential problem throughout the whole of the geological record. Without accurate ages there is no way to reconstruct the history of sedimentary deposition in a quantitative sense. For modern sediments, this problem is all the more acute because of the problems of contaminant production, transport and later incorporation in sediments. A major need is a procedure which can recall the components of the contaminant from the sediments and provide a method of sorting out intrinsic source variations at a location from components that could otherwise mask a true age effect. One could otherwise obtain not only an incorrect age but also an incorrect estimate of sediment accumulation rate and contaminant accumulation. This latter problem could then be severe in attempts to suggest appropriate courses for remediation of contaminant from the sediments.

The purpose of this monograph has been to investigate several methods and incorporated processes for obtaining ages of sediments based on radioactive decay of various unstable nuclides. For modern sediments, the currently best understood such workhorses are the cesium and lead radionuclides because their half-lives fall into the 100-year time scale of massive anthropogenic influences on the environment. The problem in any investigation is that the data one obtains contain some mix of intrinsic flux variations with time and also the effects of the temporal decay of the unstable nuclide. In order to unscramble the two, some assumptions must be made. These assumptions may be relatively benign or may be sufficiently constraining that only under exceptional circumstances can one even hope to extract any information. And the most unclear aspect for a highly constrained procedure is how one would go about identifying when the conditions for that procedure would unequivocally obtain. More general procedures, capable of disentangling flux and sediment variations with time in a wide variety of geological settings, are to be preferred. Such procedures must also contain the capability of being generalized easily to include processes that are but poorly known (or poorly quantified) now or are not yet recognized as relevant at all. In addition, any procedure must have the capability for handling extreme variations of data because it is often the understanding of such extreme cases that sheds enormous light on processes in general.

From this perspective, it is clear that the historical procedures of radiometric age dating of modern sediments have developed apace over the last thirty years since the days of Goldberg's (1963) demonstration of how to use ^{210}Pb as an age dating device. More and more complex situations are now capable of being disentangled, and more and more chemical, physical and biological processes are capable of being included in modern methods of determining both contaminant flux and sediment accumulation rates. Perhaps the major point of this monograph has been to show how such things are done and to show the assumptions and limitations of various facets of various model developments applied to such problems. The individual case studies represent a cross-section of the capabilities currently available and also show when different processes are dominant players so that one can more easily portray how disentanglement methods operate under varying and various conditions.

One factor stands out from the investigations reported here. The variations of contaminant, of natural radionuclide flux, and of sediment accumulation with time must all be handled simultaneously if one is to understand better the temporal behavior of such systems. This factor is paramount in just about all the case histories. Of course, there is no preclusion of situations where either sediment accumulation or flux variations dominate a system, but one would need to prove (and not surmise), ahead of any investigation based on such a premise, that such truly was the case based on other compelling evidence.

We would be remiss, however, if the impression is given to the reader that we have reached some sort of completion to techniques available for investigations of modern sediments based on radiometric fluctuations measured with depth in core. For instance, all disentanglement methods to date operate on the principle that radionuclides are bound to the sediments once the sediments fall through the water column. The sole exception is the idea of subsurface diffusion and/or advection of contaminant or nuclide by connate water flow, bioturbation or physical turbation. Such intrinsically time-dependent processes call for a more general procedure of investigation than currently exists. Any such procedure must also include surficial flux variations of contaminant and radionuclides at different sites and must also provide lateral and depth transport mechanisms which can also be disentangled by such procedures. Generation of such methods would truly enable one to link dynamically the data from site to site at multiple locations across an area of interest and so determine lateral sediment accumulation and denudation variations, and thus tie together the various concentrations of contaminants and radionuclides measured at the various sites into a comprehensive picture of dynamical evolution of such a system.

Perhaps, however, we will have been successful with this monograph if others, more capable than ourselves, can extend the methods and procedures to a higher level than we have done or, indeed, than we are capable of doing, and so bring forth a sharper understanding of the intertwined nature of sediment accumulation and contaminant deposition in evolving sedimentary systems.

Appendix A
Sediment Isotope Tomography software, general requirements and operation

Introduction

The Sediment Isotope Tomography (SIT) model was developed as an MS-DOS-based computer software program to analyze complex depth profiles of radionuclide activity measured in marine and freshwater sediment cores. The model is designed to reconstruct the history of unmixed sediment deposition and to recover past events, such as human impacts and natural chemical alterations that are preserved in buried sediments. The approach, based on the SIT model, applies inverse numerical analysis techniques to disentangle components of variations in radionuclide activity with sediment depth caused by variations in sediment accumulation rate and radionuclide flux. This appendix provides a user's guide for the SIT model that includes simulations using sample data sets, and the input and output listings from these simulations. The source code for the SIT model is presented as well.

The SIT model software was developed using the following system requirements:

- Computer: IBM-compatible PC
- CPU: 80386 minimum; 80486 recommended
- Coprocessor: Recommended
- Hard Drive: 2 to 5 Megabytes (M) recommended
- Floppy Disk Drive: 1.2 or 1.44 M
- Monitor: VGA recommended
- Printer/Plotter: Not supported
- Mouse: Not supported
- MS-DOS: Version 5.0 or higher

1. Applications and general limitations

The SIT model, like all models, has its limitations; this model cannot be used to reconstruct sediment accumulation histories in all cases. The most important limitation of this model is that, at present, it may not be used for areas where sediments are mixed, physically or biologically, over short time scales relative to the rates of sediment accumulation, although the numerical basis for such an improvement was presented in Chapter 5. Mixing causes the signal of radioactive decay to be erased through homogenization of the sediment profile; no inverse model is capable of recovering a signal that is not present. Another approach that is currently being tested is to interpret sediment cores where mixing predominates by combining a forward model of sediment mixing with the inverse SIT model. However, at present, the use of the SIT model for mixed sediment profiles must be conducted with great care.

Many sediment profiles exhibit fluctuations in deposition rates at frequencies that are higher than the practical resolution limit of the core-sectioning process (about 0.5 cm) or of the radionuclide itself. A lack of depth resolution in the core sections themselves will lead to difficulties in analyzing and interpreting model results. This difficulty may occur when ^{210}Pb is used to interpret depositional histories in environments where excess ^{210}Pb is only observed a few centimeters below the core surface. Lack of depth resolution may also occur when ^{210}Pb is used in environments where the sediment accumulation rates are so high that the depth of ^{210}Pb/^{226}Ra equilibrium is never observed.

To use the SIT model successfully, consideration must always be given to the characteristics of the environment under investigation, the sampling strategy to be used, and the handling of the sediment cores. As a guideline, careful description of the sedimentary structure of the core should be completed to aid in the interpretation of results. The analyses should include sedimentary structure, size distributions, bioturbation, and noticeable biogeochemical colorations or artifacts. High-depth resolution sampling of the cores for nuclide analysis increases the model's confidence level and will result in a better reconstruction of the ^{210}Pb profile by the SIT model. Density and porosity measurements made through the length of the core are essential to properly normalize the data for compaction and for calculating radionuclide flux. Possible further chemical analysis may be needed to identify "key beds" or time markers within the sediments of the core.

Even with all reasonable precautions, cases will still exist where current methods of interpretation, such as CFCS, CSA, and SIT, are unable to satisfactorily reconstruct sediment burial histories.

2. Model design

2.1. *General requirements*

The SIT model was configured to run on an IBM-compatible computer in base memory. The SIT model consists of three main parts: (1) input files, (2) the SIT.EXE program, and (3) the PROB.EXE program (Fig. A.1). The executable programs were coded and compiled by using Borland Turbo C++.[1] Six file types are used or created by the SIT model: (1) *.DAT, (2) SIT.CFG, (3) SIT.PPP, (4) *.XYZ, (5) *.OUT, and (6) *.PRO files; these files are in ASCII format and can be viewed and edited in a DOS-based text editor (e.g., EDIT in MS-DOS version 5). The *.DAT, SIT.CFG, and SIT.PPP files are used by the SIT.EXE program to produce *.XYZ and *.OUT files; *.PRO files are created by the PROB.EXE program. The "*" symbol is a wildcard character that can be used to delineate files with the same extension in a directory. For example, the input file that contains the data from Lake Hart might be called HART.DAT.

---

[1] This is not an endorsement of the Borland compiler; this information was supplied to help the user understand the program.

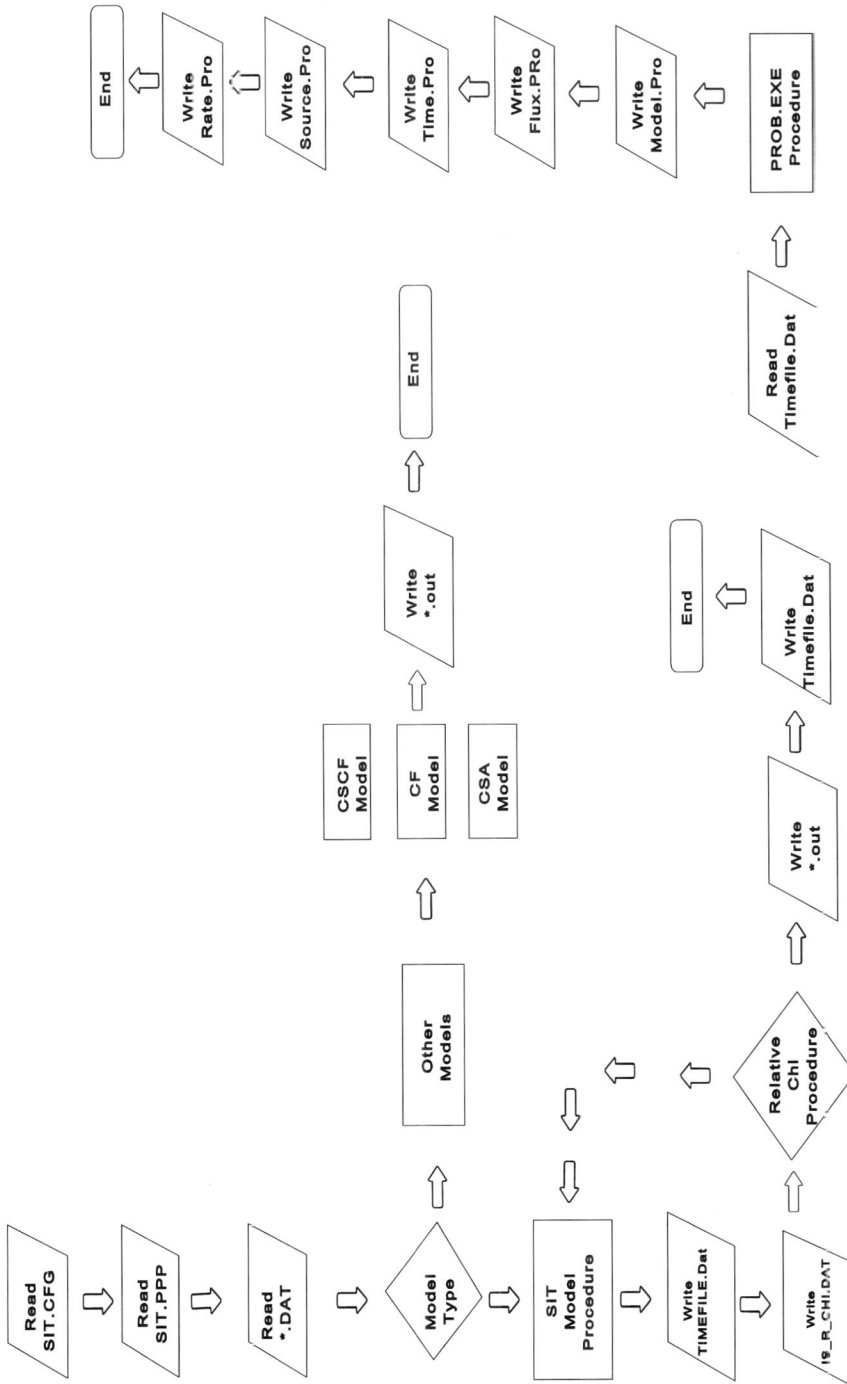

Fig. A.1. Generalized flow of the SIT simulation.

Data Name

Depth (cm)	Data (dpm/g)	DBD (cm³/g)	Error (dpm/g)
0.73	1.58	0.544	0.38
2.20	2.15	0.550	0.61
3.42	1.65	0.524	0.49
4.85	1.76	0.532	0.58
6.22	1.89	0.511	0.43
7.55	2.07	0.494	0.41
8.86	1.75	0.489	0.56
10.37	3.52	0.567	0.87
11.89	3.08	0.565	0.87
13.34	2.09	0.542	2.08
14.85	2.97	0.561	0.76
16.25	1.71	0.524	0.60
17.63	1.73	0.515	0.52
18.97	1.80	0.499	0.63
20.21	0.66	0.465	0.52

Fig. A.2. Example of *. DAT file.

Output files generated on subsequent model executions would be called XHART.OUT, where $X = a, b, c, \ldots, 8, 9$.

2.2. Input files

The purpose of the input files is to supply the model with the initial information needed to successfully execute the SIT.EXE program. The input files consist of three types: (1) data files (*.DAT), (2) a parameter file (SIT.PPP), and (3) a configuration file (SIT.CFG).

2.2.1. Data files
A *.DAT file contains the depth profile of radionuclide activity (e.g., ^{210}Pb, ^{230}Th, ^{137}Cs) for a sediment core. These data are preceded by two lines of user-defined comments. The first comment line contains the title of the data and may be up to 30 characters in length. The second comment line is a user-friendly aid for reading the *.DAT file. This comment line identifies the columns in the data file in the following order: depth, isotope activity, dry bulk density (DBD), and the standard errors in the activities of the radionuclide measurements. Data units are not specifically defined in the SIT code. In the example *.DAT file provided in Fig. A.2, the authors elected to use the units of centimeters, disintegrations per minute, and grams. The user may elect to use other units; however, the same units must be used consistently in all program modules.

2.2.2. SIT.PPP parameter file
The SIT.PPP file contains all the information required to run the SIT model. This parameter file serves two basic purposes: (1) the file contains information to direct the analysis of the data, and (2) the file identifies the type of output files to be generated during model execution. The SIT.PPP file is the most complicated portion of the SIT model, and the user must

have a complete understanding of the parameter file to successfully execute the SIT model. A SIT.PPP file, with modified labels for each entry, is presented in Table A.1.

The following is a detailed explanation of each line in the SIT.PPP file:[2]

Line 1

Entry 1 (number of iterations): The number of iterations used to adjust the coefficients (a_n, b_n) for each set of parameter values (P_0, V).

Entry 2 (radionuclide [Am241, Be7, Be10, Cs137, Pa231, Pb210, Po210, Pu239, Pu240, Ra226, Th230, and U234]): The radionuclide that is to be modeled. If any of the previous symbols are input by the user, the model will automatically choose the appropriate decay coefficient in units of yr^{-1}. If the user would prefer to use a decay coefficient based on a different unit of time or to use an alternate radionuclide, that value can be input in line 1, entry 3, of the SIT.PPP file.

Entry 3 (user t-1/2): The user-specified radionuclide half-life.

Entry 4 (time units): The user-specified time units (e.g., yr, sec, min, and months).

Entry 5 (activity units): The user-specified units of activity (e.g., dpm/g).

Line 2

Entry 1 (operational mode [0–6]): The first entry is the operational switch: $0 =$ Full-core SIT model [select full-core SIT with 0 terms (line 3, entry 1) to execute a CFCS model]; $3 =$ Core sectioning with SIT model; $4 =$ Full-core SIT model with probability data generated; $5 =$ Constant Flux model; and $6 =$ Constant Specific Activity model.

Entry 2 (graphics [on/off]): Graphics switch.

Entry 3 (spline [nospline/spline]): Spline smoothing of data.

Entry 4 (tension factor [1–100]): Spline tension factor.

Line 3

Entry 1 (number of terms [1–9]): Number of terms used in SIT model. Enter "10" when core cutting is to be tested using one, two, and three terms.

Entry 2 (contour file [Yes/No]): Option to create an *.XYZ file. The "no" switch will turn off all other restrictions to provide a complete file over the ranges of the parameters.

Entry 3 (no. of data points): Number of data points in the file.

Entry 4 (start depth switch [0/1]): Start location: 0 starts at the water interface; 1 starts at user-specified depth.

Entry 5 (end depth/start depth): If entry 4 was "0", last data depth to be modeled. If entry 4 was "1", the start location.

Entry 6 (year core was collected): The year the core was collected.

Line 4: Surface activity parameter

Entry 1 (surface [V/F]): Variable or fixed surface value.

---

[2]Values may be entered in float, integer, or exponential format.

Entry 2 (fixed): If entry 1 was "F", the fixed surface value.
Entry 3 (start): Minimum surface value.
Entry 4 (end): Maximum surface value.
Entry 5 (step): Step size for surface value.

Line 5: Average sediment accumulation rate parameter

Entry 1 (rate [V/F]): Variable or fixed rate value.
Entry 2 (fixed): If entry 1 was "F", the fixed rate value.
Entry 3 (start): Minimum rate value.
Entry 4 (end): Maximum rate value.
Entry 5 (step): Step size for rate value.

Line 6: Weight parameter (− 0.999 to 0.999 only)

Entry 1 (weight [V/F]): Variable or fixed weight value.
Entry 2 (fixed): If entry 1 was " F", the fixed weight value.
Entry 3 (start): Minimum weight value.
Entry 4 (end): Maximum weight value.
Entry 5 (step): Step size for weight value.

Line 7

Entry 1 (max χ^2): Maximum allowable χ^2 value (user specified).
Entries 2–5 (switches for writing probability data to TIMEFILE.DAT [NP/P][3]): "NP" will not write data to TIMEFILE.DAT; "P" will write data to TIMEFILE.DAT. The switches are placed in the following order: time, rate, source, flux, and model.

Line 8

Entry 1 (fit criteria [0–3]): Fit criteria switch: 0 = model must fit within the error bars for each data point; 2 = model must fit within the average error for the total data profile; 3 = no restriction of fit other than max χ^2.
Entry 2 (% error + data [1–99]): This entry allows the user to add an additional percentage of error to the data when restricted fit criteria are used (line 8, entry: 0 or 2).
Entry 3 (% error + time markers [1–99]): This entry allows the user to add additional error to the time markers.
Entry 4 (cut step [1–10]): This is the number of data points the model will progress through during core cutting. For example, if the user would like to model every second data point with core cutting, a "2" would be entered here.
Entry 5 (data for PROB.EXE): The file name for the probability data to be written by PROB.EXE (normally left at the default, TIMEFILE.DAT).

---

[3]Probability data can be written only if the operational model selected in line 2, entry 1, is the full-core SIT model with probability data (mode 4).

Line 9: Core age restrictions

Entry 1 (minimum age restriction [Yes/No]): A restriction on the minimum age of the bottom of the core.

Entry 2 (min age): If entry 1 was "yes", then user-specified minimum age.

Entry 3 (maximum age restriction [Yes/No]): A restriction on the maximum age of the core.

Entry 4 (max age): If entry 3 was "yes", enter the user-specified maximum age of the core; if entry 3 was "no", the SIT model will default to six times the half-life of the model isotope.

Line 10

Entry 1 (no. of time markers): The number (integer) of time markers in the core.

Line 11 through last line in file[4]

Entry 1 (depth): Depth of time marker.
Entry 2 (age): Age of time marker.
Entry 3 (error): Error in age of time marker.

2.2.3. *SIT.CFG configuration file*

The purpose of the configuration file is to inform the SIT model of the last data file name that was modeled and the number of the current iteration. The SIT.CFG file is created when the SIT.EXE program is executed for the first time. SIT.EXE has the ability to save up to 36 different modeled outputs of the same data. Model outputs are differentiated by an alpha-numeric character $(a, b, c, \ldots, 8, 9, 0)$ appended to the beginning of the *.OUT and *.XYZ files. For example, the Hart.dat input file can produce output files designated aHART.OUT, bHART.OUT, cHART.OUT, dHART.OUT, . . . ,8HART.OUT, 9HART.OUT, and 0HART.OUT. If the user runs the SIT.EXE program more than 36 times without changing the data file, the program will begin to overwrite the earlier model outputs. The SIT.EXE program will ask if you would like to change the data file (Y/N). If "Y" is chosen, the new data file name is entered (remember to change the SIT.PPP if there are any differences in the data, otherwise the model will crash). If "N" is selected, the model will continue from the last saved value in SIT.CFG. The user may wish to start the modeling sequence over by simply choosing the yes option and retyping the same file name, this will restart the SIT.CFG back at the initial "a*.*" position.

2.3. *SIT.EXE program*

The SIT.EXE program controls the execution of the SIT model. This program retrieves initial information from the input files, performs model calculations, and stores the results in the appropriate output file. Once all program files have been copied to a computer subdirectory, a model run is initiated by typing "SIT" followed by the ENTER key. If the graphics switch line 2, entry 2, is set to "on", the best χ^2 model up to that point in program execution is displayed in graphic form on the screen; if the graphics switch is set to "off" only the best χ^2 value is displayed.

---

[4]Use a separate line for each time marker.

```
cls
call sit
if "%1" == "P" call prob
```

Fig. A.3. Example of Batch file.

```
I   J Parameters Chi (Surface counter) (Rate counter) (Surface parameter)
(Rate parameter) (Chi²)
3   5   19.0  2.0    0.0012
a(n) = −1.6    2.1    −0.23          (aₙ coefficients)
b(n) =   1.2    0.3    −0.012         (bₙ coefficients)
```

Fig. A.4. Example of linear search output.

2.3.1. *SIT.EXE execution*

The SIT program is executed from the directory that contains the SIT.PPP and SIT.CFG files, the SIT.EXE program, and the respective input data file. For easy setup and execution, a separate directory should be established under the root directory (C:\SIT). The *.OUT, *.XYZ, IS_R_CHI.DAT, and TIMEFILE.DAT files are created during execution of the SIT program. The user may use a batch file to run the SIT.EXE and PROB.EXE programs. An example of a batch file is presented in Fig. A.3.

2.3.2. *Output files*

The output files are created in ASCII format during program execution. The SIT model creates the following four output files when the model is executed: *.OUT files, *.XYZ files, IS_R_CHI.DAT file, and TIMEFILE.DAT file. These files are designed to allow the user to interpret the results generated by the model. These files can be viewed using a standard DOS-based text editor.

2.3.2.1. **.OUT output files.*   The *.OUT files contain information on the input parameters, modeled isotope, abbreviated results from each successful iteration, and the best results based on the relative χ^2 procedure. As *.OUT files are generated, up to 36 runs can be saved without overwriting the files. An alpha or numeric character is appended to the front of the file that is being generated. The *.OUT files are organized in a manner that allows the user to easily retrieve information from the files. The first line in an *.OUT file identifies the data title that is specified in the first line of the corresponding *.DAT file. The second line identifies the date and time of the model run. The third line identifies the name and location of the corresponding *.DAT file. The next section of the file displays the parameters that were used to create the model (line 1 through line 9 of the SIT.PPP file), followed by the linear search outputs. This section presents each of the model solutions in an abbreviated format. Figure A.4 illustrates the linear search output.

The last segment of an *.OUT file displays the best answer calculated from the relative χ^2 procedure. The a_n, b_n, χ^2, and best parameters are displayed followed by the results for the best model. The information is presented in column format. Items that appear in the output are, from left to right, depth, age, time, rate, source, flux, data activity, and model activity.

2.3.2.2. **.XYZ output files.* The *.XYZ files store the best model result (contained at the end of the *.OUT files) in a format that can be imported into other computer applications to generate two-dimensional graphical displays of program results.

2.3.2.3. *IS_R_CHI.DAT output file.* The IS_R_CHI.DAT file contains four columns of data that can be used to create three-dimensional or contour plots of the SIT model linear search results. This file is in ASCII format and can easily be imported into several plotting programs.

2.3.2.4. *TIMEFILE.DAT output file.* The TIMEFILE.DAT file is created by the SIT model and is the input file for the PROB.EXE program. This file should not be edited before execution of the PROB.EXE program.

2.4. *PROB.EXE program*

The PROB.EXE program computes statistical parameters on the data produced by the SIT model. Simpson's rule is used to calculate an expected average $E_1(T)$ and the variance $\sigma(T)^2$ around the average. Accordingly,

$$E_1(T) = (T_{\min} + T_{\max} + T_{\text{midpoint}})/3$$

and

$$\sigma(T)^2 = \left[E_2(T) - E_1(T)^2\right]$$

may be calculated as

$$\sigma(T)^2 = \left\{\left[T_{\text{midpoint}} - (T_{\max} + T_{\min})/2\right]^2 + 3(T_{\max} + T_{\min})^2/4\right\}$$

and

$$\mu = \left[\ln\left(E_2(T)/E_1(T)^2\right)\right]^{1/2}.$$

For a distribution that is cumulatively log-normally distributed or approximately so (Feller, 1957), the cumulative probability of 68-percent $[P(68)]$ occurs at about $E_1(T)$, the 16-percent $[P(16)]$ at $E_1(T)\exp(-\mu - \mu^2/2) < E_1(T)$, and the 84-percent $[P(84)]$ at $E_1(T)\exp(\mu - \mu^2/2) > E_1(T)$. A log-probability graph can then be made and the 10-percent $[P(10)]$ and the 90-percent $[P(90)]$ values read. The distribution at each depth is then taken to be represented by the expected average value at $P(68)$, with uncertainty around the $P(68)$ value described through the $P(10)$ and $P(90)$ values. These values then can be used in plotting the error uncertainty of the time-to-depth, rate-to-depth, source-to-depth, isotopic flux-to-depth, and model activity-to-depth profiles. The output results are found in the TIME.PRO, RATE.PRO, SOURCE.PRO, FLUX.PRO, and MODEL.PRO files.

2.4.1. *PROB.EXE execution*

Before executing the PROB.EXE program, a TIMEFILE.DAT file must be created by the SIT.EXE program. The TIMEFILE.DAT file is the input file for the PROB.EXE program. On the basis of the number of options that were selected for probability data (line 7, entries 2 through 5, in the SIT.PPP file), the PROB.EXE program will create the appropriate *.PRO files.

2.4.2. **.PRO output files*

The PROB.EXE program creates files with a "PRO" extension. The PROB.EXE program is capable of creating *.PRO files for time, rate, source, flux, and model activity. The *.PRO files include values for depth, min, max, $P(68)$, $P(90)$, $P(10)$, $P(90)-P(68)$, $P(68)-P(10)$, volatility, risk, and relative risk. If insufficient data are available to complete a probability density function (i.e., less then 10 models), the PROB.EXE program will display simple descriptive statistics including the mean and standard deviation for time, rate, source, flux, and model activity.

3. SIT model usage

3.1. *Introduction and use of the SIT code*

The SIT model uses two basic steps to solve a problem. First, the problem must be formulated in terms that the model can use, such as isotope activity, density corrected depth, and historical time markers. Second, the computer code is run and the results processed for analysis. The computer run is completed through input of data collected from a core and through model operations that have been mapped out in the SIT.PPP parameter file. After a successful simulation has been completed, the user may choose to execute the PROB.EXE program to calculate probability density functions (see Section 2.4) from the results of the SIT.EXE program execution.

3.2. *Problem formulation*

Application of the SIT model to a specific problem requires the problem to be formulated in terms that the SIT model can interpret. Problem formulation entails (1) identifying spatial and temporal domains of influence for the location of core collection, (2) analyzing the sedimentary structure of the core, (3) analyzing the isotope activity of the core, (4) and identifying time markers that can be used to constrain the model results. Once the necessary data have been acquired and a general understanding of the area under investigation has been achieved, the appropriate program control options must be specified within the SIT.PPP parameter file.

3.3. *Data format*

All the files used by the SIT model are in ASCII format. The specific isotopic depth distributions and the sediment DBD are entered in the *.DAT file in the format described in Section 2.2.1. The information on the time markers and sedimentation rates as well as the statistical error allowed for the model must be entered in the SIT.PPP parameter file to run a simulation. Error factors and restraining environmental factors must also be identified in the

SIT.PPP parameter file. The SIT model will arrive at a solution quicker if the sedimentation rate and the presence of bioturbation and sediment disturbances of the area are approximately known.

3.4. *Model operation*

The SIT model is designed to run on a IBM-compatible computer with an MS-DOS version 5.0 or comparable operating system. Only a text editor is necessary for editing the files that control the models and for viewing the output files. The procedures discussed in the following subsections assume the use of an IBM-compatible 486/50-megahertz (MHz) computer running MS-DOS, version 5.0 or higher, with at least 8 megabytes of available RAM.

3.4.1. *Full-core linear search*
The most commonly used capability of the SIT model is the full-core linear search. The procedure steps through the parameters of surface activity and sedimentation rate with a user-specified range and step size. All data points are used to obtain a model of the radionuclide activity, and every data point is included in the fit criteria for the model acceptance.

The first step for performing a full-core linear search is to create a data file in the format explained earlier. The second step is to edit the SIT.PPP file to run the created data file. To run a full-core model, the operational mode (line 2, entry 1, of the SIT.PPP file) must be "0". The remaining lines and entries in the SIT.PPP file are set up as needed for specific uses. After saving the *.DAT file and the SIT.PPP file, the model is executed by typing "SIT" and by pressing the RETURN key. The program will prompt for a change in the data file name. The user selects "Y" and then types in the desired file name and presses the ENTER key. The program displays the solution on the screen. At the end of the simulation, the best answer based on the χ^2 fit and the time markers is printed to the *.OUT file. The *.XYZ file will contain an ASCII file that may be imported into a graphics program for further data analysis. By striking any key, the user can stop the program execution at any point during the simulation. The user can then go back to the SIT.PPP file and make any changes that are necessary to improve the solution and run the model again. If the user declines to change the data file name (by typing "N" at the prompt), the program will store the new output file using the same file name but with a new alpha or numeric character prefix to avoid overwriting the previous solution.

3.4.2. *Core cutting*
The SIT model is capable of producing models based on cutting the core to a desired length, starting at a depth below the surface, or stopping the simulation before reaching the bottom of the core. The procedure for creating the data file is the same procedure that is used in the full core linear search; however, the following changes should be implemented in the SIT.PPP parameter file.

Line 1: Set up as needed depending on the core.
Line 2, Entry 1: Set to "3" for a core-sectioning model.
Line 3, Entry 1: Identify the number of terms to be used during the core cutting or enter "10". An entry of "10" will run the core-cutting procedure over one, two, and three terms at the user-specified cut interval (line 8, entry 4).

Line 3, Entry 2: Enter "No"; this option should be always set to "No" when running the core-cutting procedure.

Line 3, Entry 3: Set to the number of data points in the file.

Line 3, Entry 4: Specify the start position of the core.

Line 3, Entry 5: Specify the bottom depth of the core.

The remainder of the file is set up based on the specific core.

Note: To stop the model during the multiple-term core cutting procedure, use the Esc key only.

3.4.3. *Probability density output*

To generate the probability density output of a model core, the operational mode (line 2, entry 1) must be set to "4" and the time, rate, source, flux, and model switches (line 7, entries 2 through 5) must be set either to "NP" (no probability density) or P (probability density output). The other parameters in the SIT.PPP file are set up as needed for the specific core. After the SIT model run is completed, the PROB.EXE program may be run to calculate the probability density functions of the output. Type "PROB" to execute the program. The PROB.EXE program will prompt for an input file name; type the file name that contains the data, which is usually the TIMEFILE.DAT file (line 8, entry 5).

3.4.4. *Alternative models*

The SIT model offers three alternative models for the calculation of the time-to-depth distribution of a sediment core: the CF model, CFCS model, and the CSA model. These models are executed from the SIT.PPP parameter file with the proper input in line 2, entry 1.

4. Example simulations

Three examples are provided in this section to illustrate how various SIT options work. These examples illustrate the use of the SIT model to characterize (1) a constant sedimentation accumulation rate/constant isotopic flux synthetic profile using ^{210}Pb, (2) a variable sediment accumulation/variable flux profile using ^{210}Pb, and (3) a preliminary demonstration of the core-cutting procedure. The example simulations discussed in the following subsections were performed using an IBM-compatible 486/50-MHz computer running MS-DOS version 5.0, with 8 megabytes of available RAM.

4.1. *Example 1: Constant sediment accumulation rate/constant isotopic flux*

One method of verifying that the model is operating properly is to examine a synthetic profile that does not exhibit any complex variations in sedimentation rates or isotopic flux. The profile is created by picking an arbitrary surface activity, then picking any sedimentation rate for one of the standard radionuclides, and then calculating an activity-to-depth profile using equations (1) and (2).

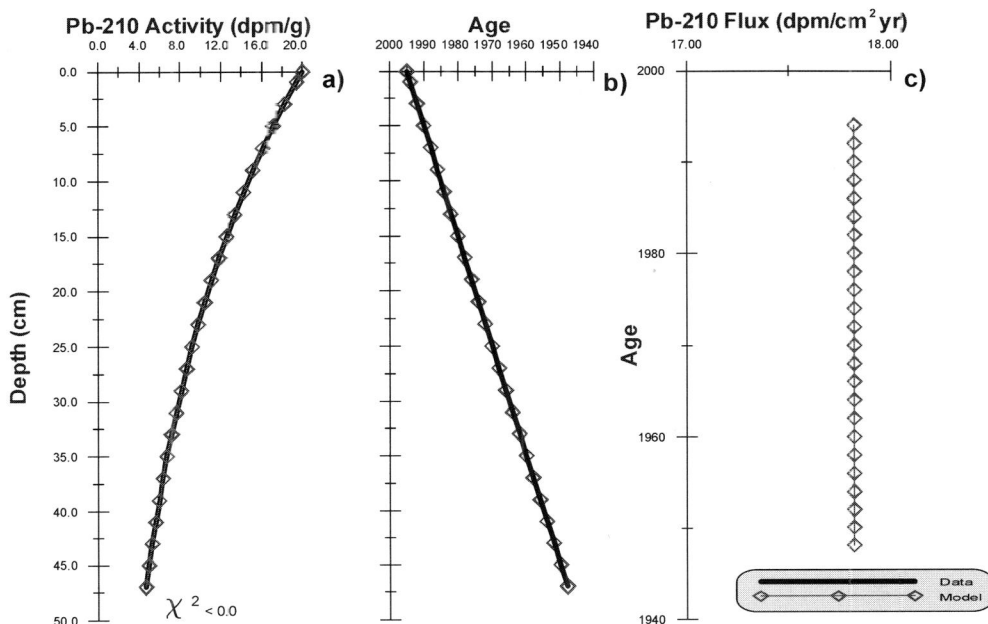

Fig. A.5. Example 1: (a) Synthetic ^{210}Pb (dpm/g) to depth (cm) and SIT code CFCS model output for ^{210}Pb to depth; (b) Synthetic age to depth (cm) and SIT code CFCS model output for age to depth; (c) SIT code CFCS model output for ^{210}Pb flux (dpm/cm^2 yr) to age.

4.1.1. *Execution of Example 1*

For this example, the surface activity is 20.00 dpm/g of ^{210}Pb (half-life $= 22.26$ yr) and the sediment accumulation rate is 1.00 cm/yr. A constant DBD of 0.89 g/cm^3 is assumed for this example. This profile may be modeled using the CFCS model. The SIT.PPP file is set up to run the CFCS model by placing a zero in line 3, entry 1. The surface-activity search range is entered in the SIT.PPP file as required to cover the characteristics of the profile. When using a CFCS model, the model will solve for the slope of the decay curve and will calculate a sedimentation rate based on the isotope modeled. When the model execution is completed, the results may be viewed in EX1.OUT and EX1.XYZ files. The EX1.XYZ file is then plotted in a graphics program to view the results of the model (see Fig. A.5). When using the CFCS model, no probability data are produced from the SIT.EXE program.

4.1.2. *Results of Example 1*

Figure A.5 illustrates the results obtained with the SIT model for the CFCS simulation. The output from this simulation may be found in Appendix A along with the input files used to generate the results. This simulation required 0.02 hour of processing time.

4.2. *Example 2: Variable sediment accumulation rate/variable flux rate*

In the natural environment, complex isotopic profiles are common; the SIT model was designed for these types of profiles. The profile used in this demonstration is from a sediment

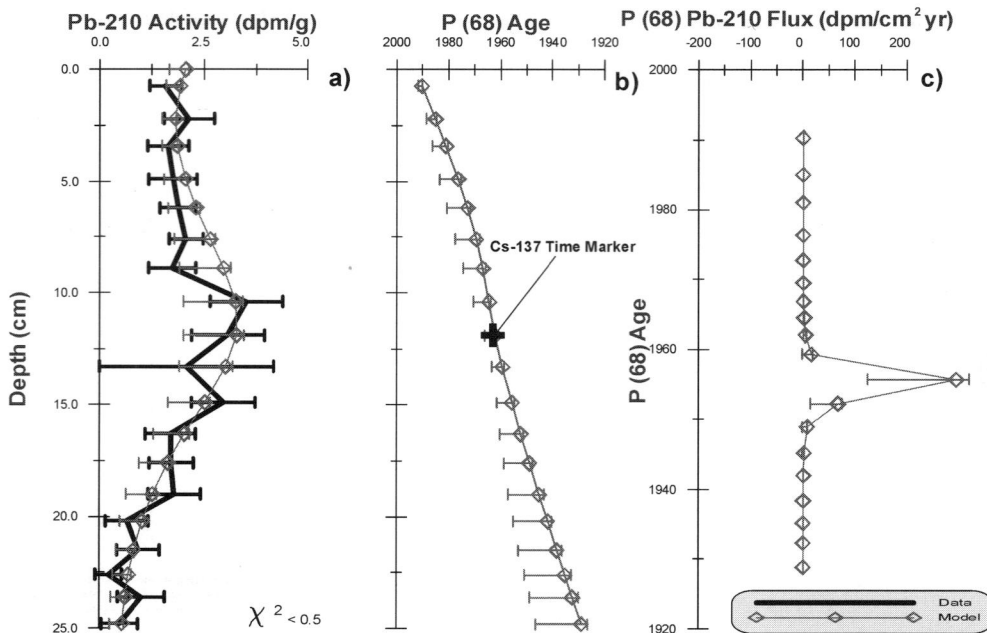

Fig. A.6. Example 2: (a) Measured ^{210}Pb (dpm/g) to depth (cm) and SIT model $P(68)$ activities for ^{210}Pb to depth; (b) SIT model $P(68)$ age to depth (cm) and the measured ^{137}Cs peak related to 1963 fallout to depth; (c) SIT model $P(68)$ ^{210}Pb flux (dpm/cm^2 yr) to age.

core collected in a pond. The profile demonstrates an elevated area of ^{210}Pb activity at a depth of 12.5 cm (see Fig. A.6). A two-term model simulates this type of profile best. Two time markers are available to restrict the model: (1) a peak in ^{137}Cs at 11.9 cm, which is related to the 1963 fallout from nuclear weapons testing (Ritchie & McHenry, 1990); and (2) the pond was created in 1922 therefore the sediments may not be older than 1922.

4.2.1. *Execution of Example 2*
The most important step in this simulation is the proper set up of the SIT.PPP parameter file, which is summarized as follows:

- Line 1 is set up for the use of ^{210}Pb in units of disintegrations per minute per gram (dpm/g).
- Line 2, entry 1, is set to "4"; this operational mode will generate the probability density data needed by the PROB.EXE program to complete the statistical analysis.
- Line 3, entry 1, is set to "2" for two terms; the remainder of line 3 is set up as dictated by the characteristics of the core.
- Lines 4 through 6 are set up for the desired search ranges and step sizes.
- Line 7, entry 1, is the maximum allowable χ^2 entry; for this example, all solutions to the profile will be accepted if χ^2 is below $(0.5 \text{ dpm/g})^2$. The remainder of line 7 activates the specific probability density outputs (see Section 5.2.2). For this example, all of the probability density data are saved to the data file TIMEFILE.DAT.

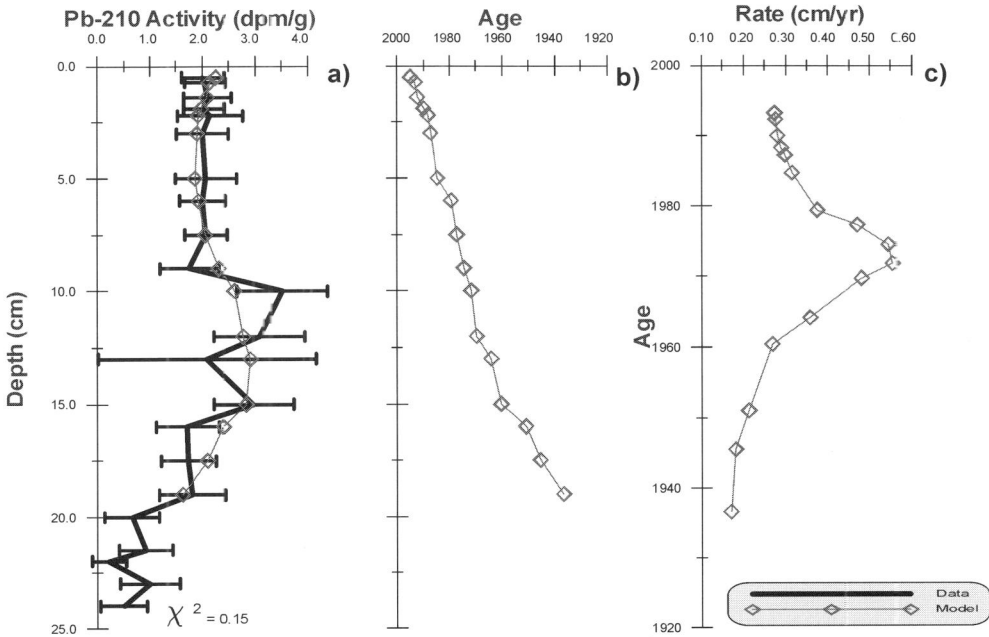

Fig. A.7. Example 3: (a) ^{210}Pb (dpm/g) to depth(cm) and the SIT model output for ^{210}Pb to depth; (b) Age to depth (cm) and the SIT model output for age to depth; (c) SIT model output for sediment accumulation rate (cm/yr) to age.

- Line 8 is set up to accept all answers having an error term with no added percent error required. The file where the probability density data are to be stored is the TIMEFILE.DAT file.
- Line 9, entry 3, is set to "yes", and entry 4 is set to "71.0" to limit the bottom age of the sediments.
- Line 10, entry 1, indicates there is one time marker that will be used in the simulation.
- Line 11 is set up with the depth and time plus error for the ^{137}Cs time marker.

The example SIT.PPP parameter file for Example 2 run is presented in Attachment A.

To run the simulation, execute the SIT.EXE program by typing "SIT" at the DOS prompt. At the conclusion of the program, execute PROB.EXE by typing "PROB" at the DOS prompt. The PROB.EXE program will calculate the probability density distributions for each of the parameters in Example 2 and will write the *.PRO files containing the ASCII solutions that can be edited in a DOS-based text editor.

4.2.2. *Results of Example 2*

The results of Example 2 simulation are listed in Attachment A. These listings include AEX2.OUT, TIME.PRO, RATE.PRO, SOURCE.PRO, FLUX.PRO, and MODEL.PRO. Seventeen models for the core were accepted for use in the probability density calculations on the basis of the fit criteria and maximum χ^2. Probability distributions were calculated for each depth for each parameter (i.e., the time distribution at a depth of 11.9 cm will be composed of

Table A.1
Example SIT.PPP file

	No. of iterations	Radionuclide	User t-1/2	Time unit[a]	Activity units[a]	
Line 1	10	Pb210	26.22	yr	dpm/g	
	Operational. mode	Graphics	Spline	Tension factor		
Line 2	4	On	Nospline	10		
	# of terms	Contour file	# of data points	Start depth/switch	End depth/ start depth	Year core was collected
Line 3	2	No	16	0.0	35.0	1985
	Surface (*/*)	Fixed	Start	End	Step	
Line 4	V	14.0	13.0	17.0	0.5	
	Rate (*/yr)	Fixed	Start	End	Step	
Line 5	V	0.9	0.6	0.8	0.2	
	Weight	Fixed	Start	End	Step	
Line 6	F	0.0	−0.9	0.9	0.1	
	Max χ^2	Time Prob	Rate Prob	Source Prob	Flux Prob	Model Prob
Line 7	0.5	P	P	P	P	P
	Fit criteria	% Error + data	% Error + time markers	Cut step	Data for PROB.EXE	
Line 8	3	0	0	1	TIMEFILE.DAT	
	Min age restriction	Min age (yr)	Max age restriction	Max age (yr)		
Line 9	No	16.0	No	37.0		
	No. of time markers					
Line 10	1					
	Depth[a]	Time[a]	Error[a]			
Line 11	11.0	19.0	2.0			

[a]The user must maintain consistency in the units.

17 values). The results of the simulation are presented in Fig. A.6. The original activity data plus the analytical error are plotted against the SIT-generated $P(68)$ activity values at each depth, and the $P(10)$ and $P(90)$ statistical limits are shown. Example 2 simulation required 0.04 hour of processing time.

4.3. Example 3: Introductory use of core cutting

Example 3 demonstrates the core-cutting capabilities of the SIT model. The SIT model allows the user to start the simulations at a minimum depth and then to add points based on the cut interval defined in line 8, entry 4, of the SIT.PPP parameter file. The SIT model will run a simulation at each of the cut-interval depths and will search the depth range based on the surface activity parameters and the average sediment accumulation rate parameters. The best answer is based on the best relative χ^2 if a time marker is used or is based solely on the χ^2 fit if there are no time markers. The best models for each of the cut intervals are compared to produce a final best model for the simulation.

4.3.1. Execution of Example 3
For core cutting, the SIT.PPP parameter file is set up in the same manner as in Examples 1 and 2. The only changes in the file setup are in the operational switches (line 2, entry 1),

and the cut interval (line 8, entry 4). No probability density data can be generated during the core cutting procedure; therefore, the switches in line 7 need to be set to "NP" to indicate no probability. Attachment A presents the SIT.PPP parameter file that was used in Example 3 simulation.

4.3.2. *Results of Example 3*

The SIT results of Example 3 are listed in Attachment A and include the AEX3.OUT output file. Taking into account the time marker, the SIT model chose the best answer with the core cut at the 17.5-cm depth. The model fit to the data is within the parameters set up in the SIT.PPP parameter file (see Attachment A). The SIT model generates an age-to-depth profile as well as a rate-to-age profile for the first 17.5 cm of the core (see Fig. A.7). Example 3 simulation required 0.03 hour of processing time.

Attachment A
Input and output listings from example simulations

Input and output listings for Example 1: Constant sediment accumulation rate/constant isotopic flux example simulation

Example 1: Data file EX1.DAT

Example one

Depth (cm)	Data (dpm/g)	DBD (g/cm^3)	Error (dpm/g)
1	19.39	0.89	0.01
3	18.22	0.89	0.01
5	17.12	0.89	0.01
7	16.09	0.89	0.01
9	15.12	0.89	0.01
11	14.21	0.89	0.01
13	13.35	0.89	0.01
15	12.54	0.89	0.01
17	11.79	0.89	0.01
19	11.08	0.89	0.01
21	10.41	0.89	0.01
23	9.78	0.89	0.01
25	9.19	0.89	0.01
27	8.64	0.89	0.01
29	8.12	0.89	0.01
31	7.63	0.89	0.01
33	7.17	0.89	0.01
35	6.73	0.89	0.01
37	6.33	0.89	0.01
39	5.95	0.89	0.01
41	5.59	0.89	0.01
43	5.25	0.89	0.01
45	4.93	0.89	0.01
47	4.64	0.89	0.01

Example 1: SIT.PPP file used in simulation

```
10  Pb210  26.22 yr  dpm/g          <==LINE 1
 0  on noSPLINE   5                 <==LINE 2
```

```
0 no 24  0  47.0 1995              <==LINE 3
V 20.0  15.0  25.0 .5              <==LINE 4
V 1.0 .5  1.5    .25               <==LINE 5
F 0.0 -0.9 0.90  0.10              <==LINE 6
500.0 nP nP nP nP nP               <==LINE 7
3 0.0 0.0  1 TIMEFILE.DAT          <==LINE 8
no 16.0 no 32.0                    <==LINE 9
0                                  <==LINE 10
```

Example 1: Output file A EX1.OUT

DATA TITLE = Example one.

PROCESSED ON 4/24/1995 AT 15:30.
THE DATA FILE NAME AND LOCATION: C:\PBMODELS\LEAD\EX1.DAT

```
10  pb210  26.22 yr dpm/g          <==LINE 1
0  on noSPLINE    5                <==LINE 2
0  no 24  0  47.0 1995             <==LINE 3
V  20.0  15.0  25.0 .5             <==LINE 4
V  1.0 .5  1.5    .25              <==LINE 5
F  0.0 -0.9 0.90  0.10             <==LINE 6
500.0 nP nP nP nP nP               <==LINE 7
3 0.0 0.0  1 TIMEFILE.DAT          <==LINE 8
no 16.0 ro 32.0                    <==LINE 9
```

PARAMETER FILE INPUT:

MTOM SET TO 0.
MAXIMUM CHI SQUARE VALUE IS 500.000.
THERE ARE 24 DATA POINTS IN THIS CORE.
THIS CORE WAS PROCESSED FROM THE SURFACE.
CONTOURING FILE (IS_R_CHI.DAT) IS NOT COMPLETE.
NUMBER OF ITERATIONS FOR ADJUSTING COEF IS 10.
DECAY COEFFICIENT FOR THE ISOTOPE PB210 IS: 3.113869E-02/yr.
ACTIVITY IS IN UNITS OF: dpm/g.
THE YEAR THE CORE WAS TAKEN: 1995.
NO DATA SPLINING.

THIS IS A LINEAR SEARCH TASK WITH THE FOLLOWING PARAMETERS:

```
V  20.0000  15.0000  25.0000  0.5000
V   1.0000   0.5000   1.5000  0.2500
F   0.0000  -0.9000   0.9000  0.1000
```

THERE ARE NO KNOWN TIME TO DEPTH OR VARIANCE VALUES INDICATED.

*****************************ZERO TERM*****************************
STEP 1
SURFACE =   15.0000 MSR =   6.8620.

STEP 2
SURFACE = 15.5000   MSR =   5.5496.

//

STEP 10
SURFACE = 19.5000   MSR =   0.0683.

STEP 11
SURFACE = 20.0000   MSR =   0.0000.

STEP 12
SURFACE = 20.5000   MSR =   0.0668.

//

STEP 20
SURFACE = 24.5000   MSR =   5.4000.

STEP 21
SURFACE = 25.0000   MSR =   6.6596.

THE PARAMETER(S) ADJUSTED: SURFACE RATE

THE BEST PARAMETERS ARE:

CHI-SQUARED =   0.0000 WHICH WAS OBTAINED AT 11TH ITERATION

PBMAX   =   20.0000
SLOPE   =   -0.0311
WEIGHT =   0.0000

DEPTH	AGE	TIME	RATE	SRCS	FLUX	DATA	MODEL
0.00	1995	0.00	0.0000	1.00	0.00	20.000	20.000
1.00	1994	1.00	1.0013	1.00	17.82	19.390	19.388
3.00	1992	3.00	1.0013	1.00	17.82	18.220	18.218
5.00	1990	4.99	1.0013	1.00	17.82	17.120	17.120
7.00	1988	6.99	1.0013	1.00	17.82	16.090	16.087
9.00	1986	8.99	1.0013	1.00	17.82	15.120	15.117
11.00	1984	10.99	1.0013	1.00	17.82	14.210	14.206
13.00	1982	12.98	1.0013	1.00	17.82	13.350	13.349
15.00	1980	14.98	1.0013	1.00	17.82	12.540	12.544
17.00	1978	16.98	1.0013	1.00	17.82	11.790	11.787
19.00	1976	18.98	1.0013	1.00	17.82	11.080	11.077
21.00	1974	20.97	1.0013	1.00	17.82	10.410	10.409
23.00	1972	22.97	1.0013	1.00	17.82	9.780	9.781
25.00	1970	24.97	1.0013	1.00	17.82	9.190	9.191
27.00	1968	26.97	1.0013	1.00	17.82	8.640	8.637
29.00	1966	28.96	1.0013	1.00	17.82	8.120	8.116
31.00	1964	30.96	1.0013	1.00	17.82	7.630	7.627
33.00	1962	32.96	1.0013	1.00	17.82	7.170	7.167

35.00	1960	34.96	1.0013	1.00	17.82	6.730	6.734
37.00	1958	36.95	1.0013	1.00	17.82	6.330	6.328
39.00	1956	38.95	1.0013	1.00	17.82	5.950	5.947
41.00	1954	40.95	1.0013	1.00	17.82	5.590	5.588
43.00	1952	42.95	1.0013	1.00	17.82	5.250	5.251
45.00	1950	44.94	1.0013	1.00	17.82	4.930	4.934
47.00	1948	46.94	1.0013	1.00	17.82	4.640	4.637

Input and output listings for Example 2: Variable sediment accumulation rate/variable isotopic flux example simulation

Example 2: Data file EX2.DAT

Example Two

Depth	Pb210	DBD	Error
0.73	1.58	0.544	0.38
2.2	2.15	0.55	0.61
3.4	1.65	0.524	0.49
4.9	1.76	0.532	0.58
6.2	1.89	0.511	0.43
7.6	2.07	0.494	0.41
8.9	1.75	0.489	0.56
10.4	3.52	0.567	0.87
11.9	3.08	0.565	0.87
13.3	2.09	0.542	2.03
14.9	2.97	0.561	0.76
16.3	1.71	0.524	0.6
17.6	1.73	0.515	0.52
19.0	1.8	0.499	0.63
20.2	0.66	0.465	0.52
21.5	0.93	0.462	0.51
22.6	0.22	0.426	0.33
23.6	1.01	0.373	0.57
24.8	0.5	0.455	0.44

Example 2: SIT.PPP file used in simulation

```
10  pb210   26.22 yr dpm/g         <==LINE 1
4  on  noSPLINE     5              <==LINE 2
2  no  19 0  24.8   1993           <==LINE 3
V  2.0   1.5  2.5  0 1             <==LINE 4
V  0.5   0.3  0.7   .05            <==LINE 5
F  0.0  -0.9  0.90  0.10           <==LINE 6
0.50 P P P P P                     <==LINE 7
3  0.0 0.0  1  TIMEFILE.DAT        <==LINE 8
no 16.0  yes 71.0                  <==LINE 9
1                                  <==LINE 10
11.9 30.0 2.0                      <==LINE 11
```

Example 2: Output file AEX2.OUT

DATA TITLE = Example two.

PROCESSED ON 4/24/1995 AT 16:14.
THE DATA FILE NAME AND LOCATION: C:\PBMODELS\LEAD\EX2.DAT

10 pb210 26.22 yr dpm/g	<==LINE 1
4 on noSPLINE 5	<==LINE 2
2 no 19 0 24.8 1993	<==LINE 3
V 2.0 1.5 2.5 .1	<==LINE 4
V 0.5 .3 0.7 .05	<==LINE 5
F 0.0 -0.9 0.90 0.10	<==LINE 6
0.50 P P P P P	<==LINE 7
3 0.0 0.0 1 TIMEFILE.DAT	<==LINE 8
no 16.0 yes 71.0	<==LINE 9

PARAMETER FILE INPUT:

MTOM SET TO 4.
MAXIMUM CHI SQUARE VALUE IS 0.500.
THERE ARE 19 DATA POINTS IN THIS CORE.
THIS CORE WAS PROCESSED FROM THE SURFACE.
CONTOURING FILE (IS_R_CHI.DAT) IS NOT COMPLETE.
NUMBER OF ITERATIONS FOR ADJUSTING COEF IS 10.
DECAY COEFFICIENT FOR THE ISOTOPE PB210 IS: 3.113869E-02./yr
ACTIVITY IS IN UNITS OF: dpm/g
THE YEAR THE CORE WAS TAKEN: 1993.
NO DATA SPLINING.
THIS IS A LINEAR SEARCH TASK WITH THE FOLLOWING PARAMETERS:

V	2.0000	1.5000	2.5000	0.1000
V	5.0000	0.3000	0.7000	0.0500
F	0.0000	-0.9000	0.9000	0.1000

THESE ARE THE KNOWN TIME TO DEPTH VALUES WITH A VARIANCE(S).
11.90 cm   30.00 years $\sigma^2 = 2.000$

LINEAR SEARCH TOMOGRAPHY OUTPUT

```
I J   PARAMETERS   M.S.R
2 6       1.6000    0.4500    0.4483
A(n) =  -1.66446   0.67986
B(n) =   0.72158   2.14273
RELATIVE CHI = 1.000000

//

7 1       2.1000    0.7000    0.3026
A(n) =   0.92932   0.86683
B(n) =  -0.18119   4.65701
RELATIVE CHI = 1.055037
```

```
7 2        2.1000    0.6500    0.1613
A(n) =     1.14577   1.03605
B(n) =    -0.09755   4.18738
RELATIVE CHI = 0.218556

7 3        2.1000    0.6000    0.1614
A(n) =     1.12724   0.98054
B(n) =     0.05288   4.17472
RELATIVE CHI = 1.760352

7 4        2.1000    0.5500    0.2817
A(n) =     1.10533   0.91516
B(n) =     0.23065   4.15916
RELATIVE CHI = 0.371897

8 1        2.2000    0.7000    0.4383
A(n) =     1.39208   0.87691
B(n) =    -0.18384   5.07938
RELATIVE CHI = 1.437696

8 2        2.2000    0.6500    0.4689
A(n) =     1.37911   0.83891
B(n) =    -0.08070   5.04452
RELATIVE CHI = 0.980847
```

THE BEST FIT IS GIVEN BY:

```
  A(n)       B(n)
1.14577   -0.09755
1.03605    4.18788
```

THE CHI-SQUARED FOR DATA TO MODEL IS 0.1613.

The Parameters adjusted: SURFACE and RATE.

```
I J   PARAMETERS   CHI   RELATIVE   CHI
7 2   2.1000  0.6500   0.1613   0.110439
```

DEPTH	AGE	TIME	RATE	SRCS	FLUX	DATA	MODEL
0.00	1993	0.00	0.0000	1.00	0.00	2.100	2.100
0.73	1989	3.18	0.2297	1.01	0.27	1.580	1.924
2.20	1983	9.41	0.2360	1.10	0.30	2.150	1.731
3.40	1978	14.14	0.2536	1.26	0.35	1.650	1.701
4.90	1973	19.36	0.2872	1.56	0.50	1.760	1.794
6.20	1969	23.12	0.3463	1.93	0.72	1.890	1.973
7.60	1966	26.27	0.4437	2.42	1.11	2.070	2.245
8.90	1964	28.38	0.5176	2.92	1.85	1.750	2.530
10.40	1963	29.91	0.9790	3.41	3.97	3.520	2.820
11.90	1962	30.67	1.9809	3.66	8.61	3.080	2.961
13.30	1962	30.90	5.9508	3.60	24.39	2.090	2.889
14.90	1962	30.92	89.334	3.20	336.7	2.970	2.566
16.30	1962	30.97	28.872	2.68	85.29	1.710	2.149
17.60	1961	31.21	5.3027	2.19	12.55	1.730	1.739

19.00	1961	31.81	2.3260	1.72	4.20	1.800	1.345
20.20	1960	32.66	1.4259	1.41	1.97	0.660	1.074
21.50	1959	33.90	1.0429	1.17	1.19	0.930	0.858
22.60	1957	35.19	0.8542	1.04	0.80	0.220	0.731
23.60	1956	36.50	0.7658	0.97	0.58	1.010	0.653
24.80	1954	38.15	0.7238	0.94	0.65	0.500	0.602

Probability density file information from Example 2

TIME.PRO
Number of Time Values: 17

DEPTH	MIN	MAX	P68	P90	P10	(P90–P68)	(P68–P10)	VOL.	RISK	RELATIVE RISK
0.000	0.00	0.000	0.000	0.000	0.000	0.000	0.000	0.000	0.000	0.000
0.730	1.603	3.202	2.683	2.865	1.436	0.182	1.246	0.533	0.808	0.202
2.200	4.708	9.488	7.955	8.487	4.322	0.531	3.633	0.523	2.436	0.610
3.400	7.127	14.28	12.026	12.805	6.701	0.779	5.325	0.508	3.798	0.951
4.900	10.453	19.61	16.679	17.704	9.665	1.026	7.014	0.482	5.546	1.389
6.200	13.565	23.494	20.248	21.412	12.288	1.164	7.960	0.451	7.202	1.804
7.600	17.147	26.825	23.548	24.750	15.334	1.201	8.214	0.400	9.441	2.364
8.900	20.622	29.127	26.133	27.245	18.526	1.112	7.607	0.334	12.555	3.144
10.400	24.438	30.921	28.550	29.404	22.710	0.854	5.840	0.234	19.518	4.888
11.900	28.118	31.966	30.946	31.548	26.831	0.602	4.115	0.152	32.542	8.150
13.300	30.842	35.665	33.700	34.316	29.488	0.616	4.213	0.143	37.701	9.442
14.900	30.919	39.842	37.319	38.162	31.555	0.843	5.764	0.177	33.787	8.462
16.300	30.967	43.347	40.750	41.893	32.934	1.143	7.816	0.220	29.709	7.440
17.600	31.212	46.736	44.054	45.475	34.333	1.421	9.721	0.253	27.919	6.992
19.000	31.814	50.547	47.755	49.461	36.083	1.707	11.672	0.280	27.323	6.843
20.200	32.656	53.920	51.030	52.961	37.827	1.931	13.203	0.297	27.581	6.907
21.500	33.902	57.705	54.677	56.826	39.982	2.149	14.695	0.308	28.449	7.125
22.600	35.190	61.039	57.802	60.106	42.045	2.304	15.757	0.312	29.652	7.426
23.600	36.496	64.081	60.676	63.114	44.010	2.437	16.667	0.315	30.891	7.736
24.800	38.154	67.736	64.152	66.746	46.409	2.594	17.742	0.317	32.437	8.124

RATE.PRO
Number of Time Values: 17

DEPTH	MIN	MAX	P68	P90	P10	(P90–P68)	(P68–P10)	VOL.	RISK	RELATIVE RISK
0.000	0.000	0.000	0.000	0.000	0.000	0.000	0.000	0.000	0.000	0.000
0.730	0.228	0.455	0.400	0.424	0.231	0.025	0.168	0.483	0.053	0.520
2.200	0.234	0.473	0.405	0.430	0.237	0.025	0.169	0.477	0.054	0.533
3.400	0.250	0.496	0.412	0.437	0.248	0.024	0.165	0.458	0.058	0.566
4.900	0.281	0.451	0.414	0.433	0.280	0.020	0.134	0.371	0.071	0.701
6.200	0.335	0.492	0.429	0.446	0.312	0.017	0.117	0.313	0.088	0.860
7.600	0.366	0.548	0.456	0.474	0.338	0.017	0.118	0.297	0.098	0.964
8.900	0.362	0.618	0.515	0.541	0.338	0.026	0.177	0.393	0.084	0.823
10.400	0.361	0.979	0.710	0.777	0.250	0.067	0.459	0.742	0.061	0.601

11.900	0.364	1.981	1.267	1.462	-0.068	0.195	1.334	1.207	0.067	0.659
13.300	0.370	5.951	3.997	4.677	-0.652	0.680	4.649	1.333	0.192	1.883
14.900	0.383	89.335	78.089	84.898	31.52	6.809	46.566	0.684	7.322	71.778
16.300	0.370	28.873	23.217	25.975	4.351	2.759	18.865	0.931	1.598	15.661
17.600	0.356	5.303	3.647	4.288	-0.732	0.640	4.379	1.376	0.170	1.665
19.000	0.343	2.326	1.569	1.857	-0.400	0.288	1.969	1.438	0.070	0.686
20.200	0.337	1.426	0.978	1.142	-0.141	0.164	1.119	1.311	0.048	0.469
21.500	0.328	1.043	0.735	0.842	0.009	0.106	0.727	1.133	0.042	0.408
22.600	0.310	0.854	0.618	0.694	0.094	0.077	0.524	0.972	0.041	0.399
23.600	0.299	0.766	0.563	0.626	0.136	0.062	0.427	0.870	0.042	0.407
24.800	0.293	0.724	0.538	0.594	0.157	0.056	0.382	0.813	0.042	0.416

SOURCE.PRO

Number of Time Values: 17

DEPTH	MIN	MAX	P68	P90	P10	(P90–P68)	(P68–P10)	VOL.	RISK	RELATIVE RISK
0.00	0.00	0.00	0.00	0.00	0.00	0.00	0.00	0.00	0.00	0.00
0.73	1.01	1.01	1.01	1.01	1.01	0.00	0.00	0.01	22.96	58.91
2.20	1.07	1.13	1.11	1.12	1.07	0.01	0.04	0.04	2.90	7.44
3.40	1.16	1.31	1.27	1.28	1.17	0.02	0.10	0.09	1.49	3.82
4.90	1.35	1.69	1.59	1.63	1.35	0.04	0.24	0.17	1.02	2.61
6.20	1.56	2.17	1.99	2.06	1.58	0.06	0.42	0.24	0.92	2.35
7.60	1.83	2.85	2.55	2.65	1.88	0.10	0.67	0.30	0.93	2.38
8.90	2.10	3.55	3.12	3.26	2.19	0.14	0.94	0.34	1.00	2.57
10.40	2.38	4.27	3.73	3.91	2.53	0.18	1.20	0.37	1.12	2.86
11.90	2.56	4.67	4.11	4.30	2.78	0.19	1.33	0.37	1.22	3.14
13.30	2.61	4.62	4.15	4.34	2.83	0.19	1.32	0.37	1.25	3.21
14.90	2.51	4.76	3.88	4.07	2.56	0.19	1.31	0.39	1.10	2.83
16.30	2.32	4.62	3.51	3.73	2.03	0.22	1.48	0.49	0.80	2.05
17.60	2.09	4.34	3.16	3.42	1.40	0.26	1.76	0.64	0.55	1.40
19.00	1.73	3.96	2.80	3.08	0.86	0.28	1.94	0.79	0.39	1.00
20.20	1.41	3.63	2.49	2.76	0.65	0.27	1.84	0.85	0.33	0.83
21.50	1.17	3.34	2.24	2.49	0.52	0.25	1.72	0.88	0.28	0.72
22.60	1.04	3.15	2.09	2.33	0.45	0.24	1.64	0.90	0.26	0.66
23.60	0.97	3.05	2.00	2.24	0.41	0.23	1.60	0.91	0.24	0.62
24.80	0.94	3.00	1.97	2.20	0.39	0.23	1.58	0.92	0.24	0.61

FLUX.PRO

Number of Time Values: 17

DEPTH	MIN	MAX	P68	P90	P10	(P90–P68)	(P68–P10)	VOL.	RISK	RELATIVE RISK
0.00	0.00	0.00	0.00	0.00	0.00	0.00	0.00	0.00	0.00	0.00
0.73	0.26	0.45	0.40	0.42	0.27	0.02	0.13	0.36	0.08	0.19
2.20	0.30	0.51	0.45	0.47	0.31	0.02	0.14	0.35	0.09	0.22
3.40	0.35	0.57	0.49	0.51	0.35	0.02	0.14	0.33	0.11	0.26
4.90	0.48	0.69	0.63	0.66	0.46	0.03	0.17	0.31	0.15	0.36
6.20	0.53	0.92	0.81	0.85	0.57	0.04	0.25	0.35	0.17	0.41
7.60	0.60	1.34	1.13	1.20	0.67	0.07	0.46	0.46	0.18	0.43

8.90	0.69	1.85	1.60	1.72	0.79	0.12	0.81	0.58	0.20	0.48
10.40	0.94	3.97	3.05	3.35	1.05	0.29	2.00	0.75	0.30	0.71
11.90	1.03	8.61	5.43	6.12	0.67	0.70	4.76	1.01	0.39	0.94
13.30	1.03	24.39	15.98	18.36	-0.25	2.37	16.23	1.16	1.00	2.40
14.90	1.06	336.75	294.01	318.69	125.19	24.69	168.82	0.66	32.49	78.12
16.30	0.91	85.29	67.88	75.74	14.14	7.86	53.74	0.91	5.44	13.08
17.60	0.79	12.56	8.46	9.88	-1.22	1.42	9.69	1.31	0.47	1.13
19.00	0.65	4.20	2.81	3.27	-0.33	0.46	3.14	1.28	0.16	0.38
20.20	0.52	1.97	1.38	1.56	0.12	0.19	1.26	1.05	0.10	0.23
21.50	0.41	1.19	0.91	1.00	0.26	0.10	0.65	0.82	0.08	0.19
22.60	0.33	0.82	0.67	0.73	0.25	0.06	0.41	0.71	0.07	0.16
23.60	0.26	0.69	0.52	0.57	0.21	0.05	0.31	0.68	0.06	0.14
24.80	0.31	0.83	0.61	0.66	0.25	0.05	0.35	0.67	0.07	0.16

MODEL.PRO

Number of Time Values: 17

DEPTH	MIN	MAX	P68	P90	P10	(P90–P68)	(P68–P10)	VOL.	RISK	RELATIVE RISK
0.000	1.600	2.200	2.070	2.129	1.666	0.059	0.404	0.224	1.085	8.418
0.730	1.527	2.058	1.936	1.985	1.595	0.050	0.340	0.201	1.128	8.748
2.200	1.450	1.952	1.830	1.877	1.508	0.047	0.321	0.201	1.067	8.273
3.400	1.445	2.011	1.871	1.926	1.494	0.055	0.377	0.231	0.952	7.382
4.900	1.495	2.247	2.053	2.126	1.554	0.073	0.499	0.279	0.865	6.705
6.200	1.572	2.576	2.311	2.407	1.654	0.096	0.657	0.326	0.833	6.457
7.600	1.672	3.026	2.659	2.786	1.788	0.127	0.871	0.375	0.832	6.449
8.900	1.758	3.464	2.983	3.140	1.911	0.157	1.072	0.412	0.850	6.595
10.400	1.819	3.878	3.267	3.451	2.005	0.185	1.262	0.443	0.866	6.717
11.900	1.803	3.861	3.289	3.474	2.022	0.185	1.266	0.441	0.875	6.785
13.300	1.659	3.432	3.035	3.198	1.919	0.163	1.115	0.421	0.846	6.560
14.900	1.425	2.641	2.532	2.661	1.645	0.130	0.886	0.401	0.741	5.745
16.300	1.192	2.178	2.047	2.156	1.308	0.108	0.740	0.414	0.580	4.501
17.600	0.980	1.822	1.645	1.745	0.962	0.100	0.683	0.476	0.406	3.146
19.000	0.780	1.475	1.286	1.380	0.648	0.093	0.638	0.568	0.266	2.060
20.200	0.627	1.220	1.035	1.116	0.481	0.081	0.554	0.614	0.198	1.536
21.500	0.446	0.997	0.824	0.889	0.379	0.065	0.445	0.619	0.156	1.212
22.600	0.349	0.852	0.696	0.751	0.319	0.055	0.377	0.621	0.132	1.020
23.600	0.292	0.750	0.612	0.661	0.280	0.049	0.333	0.623	0.115	0.896
24.800	0.251	0.661	0.547	0.591	0.248	0.044	0.299	0.627	0.103	0.795

Input and output listings for Example 3: Introductory core cutting

Example 3: Data file EX3.DAT

Example Three

Depth (cm)	Data (dpm/g)	DBD(g/cm^3)	Error (dpm/g)
0.5	2.0	0.544	0.4
0.73	2.05	0.544	0.38
1.4	2.1	0.544	0.45

1.9	2.03	0.5	0.39
2.2	2.15	0.55	0.61
3.0	2.0	0.524	0.49
5.0	2.07	0.532	0.58
6.0	2.01	0.511	0.43
7.5	2.07	0.494	0.41
9.0	1.75	0.489	0.56
10.0	3.52	0.567	0.87
12.0	3.08	0.565	0.87
13.0	2.09	0.542	2.08
15.0	2.97	0.561	0.76
16.0	1.71	0.524	0.6
17.5	1.73	0.515	0.52
19.0	1.8	0.499	0.63
20.0	0.66	0.465	0.52
21.5	0.93	0.462	0.51
22.0	0.22	0.426	0.33
23.0	1.01	0.373	0.57
24.0	0.5	0.455	0.44

Example 3: SIT.PPP file used in simulation

```
10  pb210   26.22 yr dpm/g          <==LINE 1
3  on noSPLINE    5                 <==LINE 2
2  no 19 0 24.8  1995               <==LINE 3
V  2.0   1.5  3.5 0 25              <==LINE 4
V  0.2  0.1  1.0    0.1             <==LINE 5
F  0.0  -0.9  0.90  0.10            <==LINE 6
20.0 nP nP nP nP nP                 <==LINE 7
3  0.0 0.0 1 TIMEFILE.DAT           <==LINE 8
no  16.0  no  70.0                  <==LINE 9
1                                   <==LINE 10
12.0  32.0  2.0                     <==LINE 11
```

Example 3: Output file AEX3.OUT

DATA TITLE = Example Three.
PROCESSED ON 4/26/1995 AT 10:46.
THE DATA FILE NAME AND LOCATION: C:\PBMODELS\LEAD\EX3.DAT

```
20  pb210  26.22 yr dpm/g           <==LINE 1
3  on noSPLINE    5                 <==LINE 2
2  yes 22 0 7.5 1995                <==LINE 3
V  2.0  1.5  3.5 .25                <==LINE 4
V  0.2 .10  1.0   .1                <==LINE 5
F  0.0 -0.9  0.90  0.10             <==LINE 6
20.0 nP nP nP nP nP                 <==LINE 7
3  0.0 0.0 1 TIMEFILE.DAT           <==LINE 8
no  16.0  no 300.0                  <==LINE 10
```

PARAMETER FILE INPUT:

MTOM SET TO 3.
MAXIMUM CHI SQUARE VALUE IS 20.000.
THERE ARE 22 DATA POINTS IN THIS CORE.
THIS CORE WAS PROCESSED FROM THE SURFACE.
THIS CORE WAS PROCESSED FOR A CONTOUR (ACCEPT ALL ANSWERS).
NUMBER OF ITERATIONS FOR ADJUSTING COEF IS 20.
DECAY COEFFICIENT FOR THE ISOTOPE PB210 IS: 3.113869E-02.
THE YEAR THE CORE WAS TAKEN: 1995.
NO DATA SPLINING.
THIS DATA HAS BEEN BROKEN INTO PARTS WITH 2 TERMS.

THIS IS A LINEAR SEARCH TASK WITH THE FOLLOWING PARAMETERS:

V	2.0000	1.5000	3.5000	0.2500
V	0.2000	0.1000	1.0000	0.1000
F	0.0000	-0.9000	0.9000	0.1000

THESE ARE THE KNOWN TIME TO DEPTH VALUES WITH A VARIANCE (σ^2).
12.00 cm   32.00 years $\sigma^2 = 2.000$

THE BEST FIT IS GIVEN BY:

A(n)	B(n)
-0.62786	2.34426
0.80183	0.28516

THE CHI-SQUARED FOR DATA TO MODEL IS 0.1507.

The Parameters adjusted: SURFACE and RATE.

SURFACE = 2.2500 AND RATE = 0.3000 WITH AN M.S.R. = 0.1507

DEPTH	AGE	TIME	RATE	SRCS	DATA	MODEL
0.00	1995	0.00	0.0000	1.00	2.25	2.25
0.50	1993	1.82	0.2743	1.00	2.01	2.13
0.73	1992	2.66	0.2760	1.01	2.05	2.09
1.40	1990	5.04	0.2805	1.03	2.10	1.98
1.90	1988	6.77	0.2898	1.05	2.03	1.92
2.20	1987	7.77	0.2986	1.07	2.15	1.90
3.00	1985	10.32	0.3147	1.14	2.00	1.86
5.00	1979	15.64	0.3754	1.40	2.07	1.94
6.00	1977	17.77	0.4708	1.59	2.01	2.06
7.50	1974	20.52	0.5450	1.95	2.07	2.31
9.00	1972	23.22	0.5556	2.39	1.75	2.61
10.00	1970	25.30	0.4815	2.71	3.52	2.78
12.00	1964	30.88	0.3586	3.39	3.08	2.92
13.00	1960	34.60	0.2685	3.70	2.09	2.84
15.00	1951	44.01	0.2125	4.20	2.97	2.40
16.00	1945	49.52	0.1814	4.36	1.71	2.10
17.50	1937	58.33	0.1703	4.45	1.73	1.63

Sediment Isotope Tomography Code

```c
/*
******************************************************************************
******************************************************************************
*****              Sediment Isotope Tomography Program              *****
*****                                                               *****
******************************************************************************
********************        NOTES:        ***************************
*****    ISOTOPE VALUES WITHIN THIS PROGRAM ARE FROM THE FOLLOWING:    *****
*****    CRC HANDBOOK OF CHEMISTRY AND PHYSICS 71ST EDITION,           *****
*****                                                               *****
*/

//***************************************************************************
//***Included C header Files************************************************
//***************************************************************************
#include <alloc.h>
#include <stdio.h>
#include <conio.h>
#include <stdlib.h>
#include <string.h>
#include <dos.h>
#include <dir.h>
#include <io.h>
#include <math.h>
#include <process.h>
#include <time.h>
#include <graphics.h>

extern unsigned _stklen = 20000U;

//***************************************************************************
//***Predefined Data Values and Strings*************************************
//***************************************************************************
#define NNTOM 3
#define  PI M_PI
#define PARAFILE "sit.ppp"
#define CFG_FILE "sit.cfg"
#define XYZ "xyz"
#define XYZ2 ".xyz"
#define OUT "out"
#define OUT2 ".out"
```

```
#define MAX_NUM_LINES 20
#define MAX_NUM_PTS 20
#define MIN_NUM_1_TERMS 12
#define MIN_NUM_2_TERMS 14
#define MIN_NUM_3_TERMS 16
#define FINC 0.2
#define BETA_0 0.01


//*********************************************************************
//***Parameter List: Files, long , floats, char, and pointers*****************
//*********************************************************************
FILE *in1, *in2, *out1, *CFG, *out_tmp, *out_time;


char tempstr[80], FNAME[50], DATFL[32],  OUTFL[32], NAMET[4][10], linear[32],
         timefile[32], sorttime[32], min_time[5], max_time[5],
         probfile[32], prob_ans[5], CONTOUR[5];
char *ptr, point = '.', ISOTOPE[6], VAR_ANS[2][5], GRAPHICS[5], SPLINE[10];
char XYZFL[32], PARA_NAME[2][20], BREAK_CORE[15], TERM[10], LTOM[4],
         PROB_TYPES[6][4];
char alpha_num[36] = {'a','b','c','d','e','f','g','h','i','j','k','l',
                      'm','n','o','p','q','r','s','t','u','v','w','x',
                      'y','z','1','2','3','4','5','6','7','8','9','0'};


long i, j, k, K, N, M, W, NUM_CORE_TERM, END_CORE_TIME, final_1_ndat,
         final_2_ndat, final_3_ndat, NUM_BREAKS, loc_time[50],
         real_syn_data, num_rates, chk_local, cnt_bst;

long IBEST, JBEST, KBEST, MPT, IPT, KPT, MTOM, NDAT,
         NMPT, NNN, IITOM, KTOM[20], LOB, JACK, IJK,  IJKBST, fl_incr,
         START_PT, NCON1, NCON2, JTOM[4], int_K, int_NDAT, int_NMPT,
         NUM_PARA, QN_LOCO[4], set_min_flag, DEP_LOCO[8], NTIME,
         X_DAT[3][150], Y_DAT[150], PREV[150], X_PREV, LINES_PER_SEC,
         NUM_PTS, NUM_SPLINE, TERMS, BEST_FLAG;

int gdriver = DETECT, gmode, errorcode;

double SBEST, A1, A2, B1, PBMAX, SLOPE, WEIGHT, ZMAX, PBMIN,
         CHSQL, ABEST, BBEST, XBEST, YBEST, ZBEST, ZZZZ, SMCH1, PARA[3][3],
         BTOM[3][3], ATOM[20], SSSS, LLL, CHSQL1, BST_SLP, FLAMD,
         BEE, SSS0, SS00, DPTS, tmp[4], PBMAX_BST, WGT_BEST, DELTA_PARA[4],
         NUM_SPACE[2], SSS1[20], DSTP[20], DDST[20], DDYY[20], FF1[20],
         CHI_1, CHI_2, SUM_QN, DELTA_N, RESULT_TANH, DLT_PARA, DELTA_CHI,
         PREV_BTOM[3][4], BST_CHI, LINEAR_DEPTH, CONS_SRCM,
         DEP_TIM_RAW[2][50], DEP_TIM_CAL[2][50], loc_chi_time, X_maxx,
         Y_maxx, CURRENT_TIME, INITIAL_TIME, CURRENT_PB, INITIAL_PB, X0, Y0,
         CHI_PB_TIME, CHI, MINRATE, MAXRATE, loc_time_min[75], chi_mean[12][22],
         terms_chi[3][50], tmp_slp[25], LIN_CHI, Def_iso, MIN_YRS, MAX_YRS,
         DATA_ERR, TIME_ERR, time_slope, AVG_ERR_DATA, MAXCHI;


//*********************************************************************
//  Dynamically allocated variables
//*********************************************************************
```

```
double huge* *A, huge* *B, huge* *F, huge* *ANM, *CC, *E, *O, *COEF,
        *TBST, *PBST, *RBST, *DBST, *CBST, *DLTAZ, *DLTAT, *CKANM1,
        *CKANM2, *CKANM, *XMPT, *SRATE, *Pred_P,  *SZ, *SZV, *CHISQM,
        *TIME1, *SRCM, *SEDM, *DTDZ, *P, *PZ, *Z, *ZN, *BEST_COEF,
        *SUMPC, *SUMPE, *tmp_P, *tmp_DLTAZ, *tmp_Z, *PERCENT, *tmp_SRCM,
        huge* *PRN_COEF, *tmp_COEF, *QN, *ALPHA, *CONS_SEDM,
        *loc_min_time, *loc_min_COEF, *VAR_DATA, *VAR_TIME, *CORE_DATE,
        huge* *BLEND, *Xsm, *Ysm, *Ax, *Ay, *Pre_SPLINE_P, huge* *ONE_TERM,
        huge* *TWO_TERM, huge* *THREE_TERM, *ONE_CHI, *TWO_CHI, *THREE_CHI,
        *A_sub0, *A_sub1, *rate_adj, *rate_B, *TIME_VEC, *LOC_P, *ORG_DATA,
        *FLUX, *DEPTH_TIME, *LOC_SRC, *FLUX_COEF, *LOC_RATE, *LOC_FLUX;

long color, maxcolor, x1, y1;

char *CHI_tim_sta = "THE BEST CHI TIME VALUE IS ";
char *CHI_statement = "THE BEST CHI VALUE IS ";
char *lname[] = {
   "SOLID_LINE",
   "DOTTED_LINE",
   "CENTER_LINE",
   "DASHED_LINE",
   "USERBIT_LINE'
   };

char *dir[] = { 'HORIZ_DIR", "VERT_DIR" };

double first, second;
struct  time t;


//****************************************************************************
//  A procedure to calculate the chi squared value
//****************************************************************************
double Calc_CHI()
{

  double tmp_CHI;
  tmp_CHI = 0;
  for (i = 0; i < NDAT; i++)
          tmp_CHI += pow(Pred_P[i] - P[i] ,2);
  tmp_CHI /= NDAT;

  return(tmp_CHI);
}


//****************************************************************************
//  Set core date and reassign the source value.
//****************************************************************************
void SET_CORE_DATE()
{

  for (i = 1; i < NDAT; i++)
          CORE_DATE[i] = CORE_DATE[0] - TBST[i];
```

```
    for (i = 0; i < NDAT; i++)
          DBST[i] = exp(CBST[i]);
}


//***************************************************************************
//   The following procedures are used for graphics.
//***************************************************************************
//***************************************************************************
//***************************************************************************
void label_graph() // NOT Implemented yet!!!!!
{
    long y_lab[50], x_lab[50], y_min, y_max, x_min, x_max;
    unsigned long tmp_y, tmp_x;
    char *tmpstr;

    y_min = 0;                      // Find the min and max
    y_max = (long)ceil(ZMAX+25);
    x_min = (long)ceil(PARA[0][1]);
    x_max = (long)ceil(PARA[0][2]);

//75+(getmaxy()-150)*SZV[i]/ceil(SZV[NDAT-3]+25);

    for (i = 0; i <y_max/10; i++)
          y_lab[i] = i*10;

    for (i = 0; i <(x_max+25); i++)
          x_lab[i] = i*10;

    for (i = 0; i <=(100*y_max); i+=10)
          {
              itoa(y_lab[i],tmpstr,10);
              tmp_y = (long)(75+(getmaxy()-150)*i/(100*y_max));
              line (75, tmp_y,70, tmp_y );
          }

    for (i = 0; i <=(100*y_max); i+=10)
          {
              itoa(x_lab[i],tmpstr,10);
              tmp_y = (long)(75+(getmaxy()-150)*i/(100*y_max));
              line (75, tmp_y,70, tmp_y );
          }
}

void draw_data_graph()
{
    int dec, sign;
    double X_maxx, Y_maxx;
    char *loc_tmp;

    detectgraph(&gdriver, &gmode);
    initgraph(&gdriver, &gmode, "");

/* read result of initialization */
    errorcode = graphresult();
```

```
if (errorcode != grOk)   /* an error occurred */
        {
        printf("\nGraphics error: %s\n", grapherrormsg(errorcode));
        printf("Press any key to halt:");
        getch();
        closegraph();
        restorecrtmode();

        exit(1);                /* return with error code */
        }

setlinestyle(0, 1, 10);
setbkcolor(1);
setcolor(4);                    // set a rectangle around the entire screen
rectangle(0,0,getmaxx(), getmaxy());
setcolor(63);                   // set a rectangle interior of the screen
rectangle(40,40,getmaxx()-50, getmaxy()-50);
setlinestyle(0, 1, 3);
setcolor(62);
line(75,75,getmaxx()-75,75);   // the X and Y AXIS
line(75,75,75,getmaxy()-75);

X_maxx = P[0];
for (i = 0; i < NDAT; i++)
        if (X_maxx < P[i])
          X_maxx = (long)ceil(P[i]);

for (i= 0; i<NDAT; i++)
        {
          X_DAT[0][i] =(long)(75+(getmaxx()-150)*P[i]/(X_maxx)*0.5);
          X_DAT[1][i] =(long)(75+(getmaxx()-150)*PBST[i]/(X_maxx)*0.5);
          Y_DAT[i]    = (long)(75+(getmaxy()-150)*SZV[i]/ceil(SZV[NDAT-1]));
        }

//  label_graph();

setcolor(59);
for (i= 0; i<NDAT-1; i++)   // PLOT the SPLINED OR RAW DATA
        line(X_DAT[0][i],Y_DAT[i],X_DAT[0][i+1],Y_DAT[i+1]);


setcolor(60);
for (i= 0; i<NDAT-1; i++)   // PLOT the MODELED DATA
        line(X_DAT[1][i],Y_DAT[i],X_DAT[1][i+1],Y_DAT[i+1]);

loc_tmp = fcvt(BST_CHI,3,&dec, &sign);
strcat(CHI_statement,loc_tmp);
outtextxy(200,25,CHI_statement);

outtextxy(25,getmaxy()-25,
  "NOTE: If no decimal showing, then assume it precedes at least 3 numbers");

// DEFINATIONS OF THE PLOT
if (strcmp(SPLINE, "SPLINE") == 0)
```

```
        {
        for (i= 0; i<NDAT; i++)
        X_DAT[2][i]=(long)(75+(getmaxx()-150)*Pre_SPLINE_P[i]/(X_maxx)*0.5);
        setcolor(61);
        for (i= 0; i<NDAT-1; i++)
        line(X_DAT[2][i],Y_DAT[i],X_DAT[2][i+1],Y_DAT[i+1]);

        setcolor(61);
        line (25,15,50,15);
        outtextxy(60,12, "RAW DATA");

        setcolor(60);
        line (25,25,50,25);
        outtextxy(60,22, "MODEL DATA");

        setcolor(59);
        line (25,35,50,35);
        outtextxy(60,32, "SPLINED DATA");

        setcolor(YELLOW);
        settextstyle(0,0,0);
        outtextxy((getmaxx()-75)/2,60, "ACTIVITY");

        setcolor(YELLOW);
        settextstyle(0, 1,0);
        outtextxy(65,(getmaxy()-75)/2, "DEPTH");
        }
   else
        {
        setcolor(59);
        line (25,15,50,15);
        outtextxy(60,12, "RAW DATA");

        setcolor(60);
        line (25,25,50,25);
        outtextxy(60,22, "MODEL DATA");

        setcolor(YELLOW);
        outtextxy((getmaxx()-70)/2-25,60, "ACTIVITY");

        setcolor(YELLOW);
        settextstyle(0,1,0);
        outtextxy(65,(getmaxy()-75)/2, "DEPTH");
        }
}

void SET_GRAPH()
{
// The graphics initialization for the data plot

  X_maxx = P[0];
  for (i = 0; i < NDAT; i++)
        if (X_maxx < P[i])
           X_maxx = (long)ceil(P[i]);
```

```
clearviewport();
setlinestyle(0, 1, 10);
setbkcolor(BLUE);
setcolor(RED);
rectangle(0,0,getmaxx(), getmaxy());
setcolor(WHITE);
rectangle(40,40,getmaxx()-50, getmaxy()-50);
setlinestyle(0, 1, 3);
setcolor(YELLOW);
line(75,75,getmaxx()-75,75);
line(75,75,75,getmaxy()-75);

for (i= 0; i<NDAT; i++)
       {
          X_DAT[0][i]  =(long)(75+(getmaxx()-150)*P[i]/(X_maxx)*0.5);
          Y_DAT[i]   = (long)(75+(getmaxy()-150)*SZV[i]/ceil(SZV[NDAT-1]));
       }
if (strcmp(SPLINE, "SPLINE") == 0)
       {
       for (i= 0; i<NDAT; i++)
       X_DAT[2][i]  =(long)(75+(getmaxx()-150)*Pre_SPLINE_P[i]/(X_maxx)*0.5);
       setcolor(61);
       for (i= 0; i<NDAT-1; i++)
       line(X_DAT[2][i],Y_DAT[i],X_DAT[2][i+1],Y_DAT[i+1]);

       setcolor(61);
       line (25,15,50,15);
       outtextxy(60,12, "RAW DATA");

       setcolor(60);
       line (25,25,50,25);
       outtextxy(60,22, "MODEL DATA");

       setcolor(59);
       line (25,35,50,35);
       outtextxy(60,32, "SPLINED DATA");

       setcolor(YELLOW);
       outtextxy((getmaxx()-70)/2-25,60, "ACTIVITY");

       setcolor(YELLOW);
       settextstyle(0,1,0);
       outtextxy(65,(getmaxy()-75)/2, "DEPTH");
       }
else
     {
       setcolor(59);
       line (25,15,50,15);
       outtextxy(60,12, "RAW DATA");

       setcolor(60);
       line (25,25,50,25);
       outtextxy(60,22, "MODEL DATA");
       setcolor(YELLOW);
```

```
       outtextxy((getmaxx()-70)/2-25,60, "ACTIVITY");

       setcolor(YELLOW);
       settextstyle(0,1,0);
       outtextxy(65,(getmaxy()-75)/2, "DEPTH");
      }

  setcolor(59);
  for (i= 0; i<NDAT-1; i++)
         line(X_DAT[0][i],Y_DAT[i],X_DAT[0][i+1],Y_DAT[i+1]);

}

void draw_updated_data_graph(long k)
{
  int dec, sign;
  char *tmpstr;
// SIMIPLILY UPDATES THE PLOT (redraw)

  for (i= 0; i<NDAT; i++)
         X_DAT[1][i]   =(long)(75+(getmaxx()-150)*PBST[i]/(X_maxx)*0.5);

  if (k != 1)
         {
          setcolor(BLUE);
          for (i= 0; i<NDAT-1; i++)
          line(PREV[i],Y_DAT[i],PREV[i+1],Y_DAT[i+1]);
          line(X_PREV,Y_DAT[0],X_DAT[0][1],Y_DAT[1]);
         }
  if(strcmp(SPLINE, "SPLINE") == 0)
         {
          setcolor(61);
          for (i= 0; i<NDAT-1; i++)
          line(X_DAT[2][i],Y_DAT[i],X_DAT[2][i+1],Y_DAT[i+1]);
         }

  setlinestyle(0, 1, 10);
  setbkcolor(BLUE);
  setcolor(RED);
  rectangle(0,0,getmaxx(), getmaxy());
  setcolor(WHITE);
  rectangle(40,40,getmaxx()-50, getmaxy()-50);
  setcolor(YELLOW);
  line(75,75,getmaxx()-75,75);
  line(75,75,75,getmaxy()-75);

  setcolor(59);
  for (i= 0; i<NDAT-1; i++)
         line(X_DAT[0][i],Y_DAT[i],X_DAT[0][i+1],Y_DAT[i+1]);

  setcolor(60);
  for (i= 0; i<NDAT-1; i++)
         line(X_DAT[1][i],Y_DAT[i],X_DAT[1][i+1],Y_DAT[i+1]);
  for (i= 0; i<NDAT; i++)
```

```
                {
                PREV[i]   =(long)(75+(getmaxx()-150)*PBST[i]/(X_maxx)*0.5);
                X_PREV = X_DAT[0][0];
                }
     X_DAT[0][0] = X_DAT[1][0];
}
//***********************************************************************
//***********************************************************************
//***********************************************************************

//***********************************************************************
//***********************************************************************
//***********ALLCCATE DATA SETS BASED ON THE PPP AND DATA FILE **********
//***********************************************************************
//***********************************************************************
double huge ***ALLOC_3D_MATRIX(long size, long size2, long size3)
{
  long i;
  double huge ***m;
  m=(double huge ***) calloc((size),sizeof(double huge **)); //ALLOCATE X
  if (m== NULL)
          {
            printf("allocation failure 1 in ALLOC_3D_MATRIX()\n");
            printf("Press any key to halt:");
            getch();
            closegraph();
            restorecrtmode();
            exit(0);
            }
  for (i=0;i<size;i++)
         {                                      //ALLOCATE Y
         m[i]=(double huge **) calloc((unsigned) (size2),sizeof(double *));
         if (m[i]== NULL)
                 {
                 printf("allocation failure 2 in ALLOC_3D_MATRIX()\n");
                 printf("Press any key to halt:");
                 getch();
                 closegraph();
                 restorecrtmode();
                 exit(0);
                 }
         for (j = 0; j < size2; j++)
             {                                      //ALLOCATE Z
             m[i][j]=(double huge *) calloc((unsigned) (size3),sizeof(double));
             if (m[i][j]== NULL)
                   {
                   printf("allocation failure 3 in ALLOC_3D_MATRIX()\n");
                   printf("Press any key to halt:");
                   getch();
                   closegraph();
                   restorecrtmode();
                   exit(0);
                    }
             }
```

```
      }
   return m;
}


double huge **ALLOC_HUGE_MATRIX(long size)  //ALLOCATE A SQUARE MATRIX
{
   long i;
   double huge **m;
   m=(double huge **) malloc((size)*sizeof(double huge *)); //ALLOCATE X
   if (m== NULL)
           {
           printf("allocation failure 1 in ALLOC_HUGE_MATRIX()\n");
           printf("Press any key to halt:");
           getch();
           closegraph();
           restorecrtmode();
           exit(0);
           }
   for(i=0;i<size;i++)
   {                                    //ALLOCATE Y
           m[i]=(double huge *) malloc((unsigned) (size)*sizeof(double ));
           if (m[i]== NULL)
           {
           printf("allocation failure 2 in ALLOC_HUGE_MATRIX()\n");
           printf("Press any key to halt:");
           getch();
           closegraph();
           restorecrtmode();
           exit(0);
           }
           else
           memset(m[i], '\0' ,(size)*sizeof(double));
   }
   return m;
}

double huge **ALLOC_MATRIX(long size,long size2)
                                    //ALLOCATE A NON-SQUARE MATRIX
{
   long i;
   double huge **m;                         //ALLOCATE X
   m=(double huge **) calloc((size),sizeof(double huge *));
   if (m== NULL)
           {
           printf("allocation failure 1 in ALLOC_MATRIX()\n");
           printf("Press any key to halt:");
           getch();
           closegraph();
           restorecrtmode();
           exit(0);
           }
   for(i=0;i<size;i++)
   {                              //ALLOCATE Y
```

```
      m[i]=(double huge *) calloc((unsigned) (size2),sizeof(double ));
      if (m[i]== NULL)
        {
        printf("allocation failure 2 in ALLOC_MATRIX()\n");
        printf("Press any key to halt:");
        getch();
        closegraph();
        restorecrtmode();
        exit(0);
        }
  }
  return m;
}
long **ALLOC_int_MAT(long size, long size2)   //ALLOCATE int NON-SQUARE MATRIX
{
  long i;
  long **m;
  m=(long **) calloc (size),sizeof(long *)); //ALLOCATE X
  if (m== NULL)
        {
        printf("allocation failure 1 in ALLOC_int_MATRIX()\n");
        printf("Press any key to halt:");
        getch();
        closegraph();
        restorecrtmode();
        exit(0);
        }
  for(i=0;i<size;i++)
  {                                         //ALLOCATE Y
        m[i]=(long *) calloc((unsigned) (size2),sizeof(long));
        if (m[i]== NULL)
          {
        printf("allocation failure 2 in ALLOC_int_MATRIX()\n");
         printf("Press any key to halt:");
         getch();
         closegraph();
         restorecrtmode();
        exit(0);
          }
  }
  return m;
}



double *ALLOC_VECTOR(long size) //ALLOCATE A VECTOR
{
  long i;
  double *m;                      //ALLOCATE X
  m=(double *) malloc((size)*sizeof(double));
  if (!m)
        {
        printf("allocation failure 1 in ALLOC_VECTOR()\n");
        printf("Press any key to halt:");
```

```
            getch();
            closegraph();
            restorecrtmode();
            exit(0);
            }
  else
          memset(m,'\0',(size)*sizeof(double));

  return(m);
}
//****************************************************************************
//****************************************************************************
//****************FREE ALLOCATED DATA SETS ***********************************
//****************************************************************************
//****************************************************************************
//****************************************************************************

void FREE_HUGE_MATRIX(double huge **a1,long size)
{
 long i;
 for(i=0;i<size;i++)
          free(a1[i]);
 free(a1);
}

void FREE_VECTOR(double *a1)
{
 free(a1);
}

void FREE_int_K_DATA()
{
 FREE_HUGE_MATRIX(A,(2*int_K+1));
 FREE_HUGE_MATRIX(B,(2*int_K+1));
 FREE_HUGE_MATRIX(F,(2*int_K+1));
}

void FREE_DATA_VARS()
{
 FREE_VECTOR(TBST);
 FREE_VECTOR(PBST);
 FREE_VECTOR(RBST);
 FREE_VECTOR(DBST);
 FREE_VECTOR(CBST);
 FREE_VECTOR(DLTAZ);
 FREE_VECTOR(DLTAT);
 FREE_VECTOR(PZ);
 FREE_VECTOR(Z);
 FREE_VECTOR(ZN);
 FREE_VECTOR(CKANM2);
 FREE_VECTOR(CKANM1);
 FREE_VECTOR(CKANM);
       FREE_VECTOR(SRATE);
 FREE_VECTOR(Pred_P);
```

```
FREE_VECTOR(SZ);
FREE_VECTOR(SZV);
FREE_VECTOR(TIME1);
FREE_VECTOR(SRCM);
FREE_VECTOR(SEDM);
FREE_VECTOR(DTDZ);
FREE_VECTOR(P);
FREE_VECTOR(CC);
FREE_VECTOR(E);
FREE_VECTOR(O);
FREE_VECTOR(COEF);
FREE_VECTOR(BEST_COEF);
FREE_VECTOR(XMPT);
FREE_VECTOR(tmp_Z);
FREE_VECTOR(tmp_P);
FREE_VECTOR(tmp_DLTAZ);
FREE_VECTOR(SUMPE);
FREE_VECTOR(SUMPC);
}
//********************************************************************
//********************************************************************
//********************************************************************
//***Initialize array variables to zero******************************
//********************************************************************
void init()
{
 for (i = 0; i < 20; i++)
        SSS1[i] = DSTP[i] = DDST[i] = DDYY[i] = FF1[i] = ATOM[i] = 0;
}



//********************************************************************
//*****SPLINING VARIABLES ALLOCATED **********************************
//********************************************************************
//********************************************************************
//********************************************************************

void SPLINE_VAR_INIT()
{
 BLEND = ALLOC_MATRIX(5,MAX_NUM_LINES);
 Xsm = ALLOC_VECTOR(5);
 Ysm = ALLOC_VECTOR(5);
 Ax  = ALLOC_VECTOR(NDAT+2);
        Ay  = ALLOC_VECTOR(NDAT+2);
 Pre_SPLINE_P = ALLOC_VECTOR(NDAT+2);
}

//********************************************************************
//******VARIABLES ALLOCATED BASED ON THE NUMBER OF TERMS*************
//********************************************************************
//********************************************************************
//********************************************************************

void ALLOC_int_K_DATA()
```

```
{
  // ALLOC THE MATRICES FOR THE PROGRAM
  A = ALLOC_HUGE_MATRIX(2*int_K+1);
  B = ALLOC_HUGE_MATRIX(2*int_K+1);
  F = ALLOC_HUGE_MATRIX(2*int_K+1);
  CKANM2 = ALLOC_VECTOR(2*int_K+2);
  CKANM1 = ALLOC_VECTOR(2*int_K+2);
  CKANM = ALLOC_VECTOR(2*int_K+2);
  SUMPC = ALLOC_VECTOR(2*int_K+2);
  SUMPE = ALLOC_VECTOR(2*int_K+2);
        CC = ALLOC_VECTOR(2*int_K+2);
  E = ALLOC_VECTOR(2*int_K+2);
  O = ALLOC_VECTOR(2*int_K+2);
  COEF = ALLOC_VECTOR(2*int_K+2);
  tmp_COEF = ALLOC_VECTOR(2*int_K+2);
  BEST_COEF = ALLOC_VECTOR(2*int_K+2);
  QN = ALLOC_VECTOR(5);
  ALPHA = ALLOC_VECTOR(5);
}


//**********************************************************************
//*****VARIABLES ALLOCATED BASED ON THE NUMBER OF DATA POINTS*************
//**********************************************************************
//**********************************************************************
//**********************************************************************
void ALLOC_VECTOR_DATA()
{
  // ALLOC THE VECTOR FOR THE PROGRAM
  CORE_DATE = ALLOC_VECTOR(int_NDAT+2);
  TBST = ALLOC_VECTOR(int_NDAT+2);
  PBST = ALLOC_VECTOR(int_NDAT+2);
  RBST = ALLOC_VECTOR(int_NDAT+2);
  DBST = ALLOC_VECTOR(int_NDAT+2);
  CBST = ALLOC_VECTOR(int_NDAT+2);
  DLTAZ = ALLOC_VECTOR(int_NDAT+2);
  DLTAT = ALLOC_VECTOR(int_NDAT+2);
  PZ = ALLOC_VECTOR(int_NDAT+2);
  Z = ALLOC_VECTOR(int_NDAT+2);
  ZN = ALLOC_VECTOR(int_NDAT+2);
  SRATE = ALLOC_VECTOR(2*int_NDAT+2);
  VAR_DATA = ALLOC_VECTOR(int_NDAT+2);
  Pred_P = ALLOC_VECTOR(int_NDAT+2);
  LOC_P = ALLOC_VECTOR(int_NDAT+2);
  LOC_RATE = ALLOC_VECTOR(int_NDAT+2);
  LOC_FLUX = ALLOC_VECTOR(int_NDAT+2);
  SZ = ALLOC_VECTOR(int_NDAT+2);
  SZV = ALLOC_VECTOR(int_NDAT+2);
  TIME1 = ALLOC_VECTOR(int_NDAT+2);
  SRCM = ALLOC_VECTOR(int_NDAT+2);
  SEDM = ALLOC_VECTOR(int_NDAT+2);
  DTDZ = ALLOC_VECTOR(int_NDAT+2);
  P = ALLOC_VECTOR(int_NDAT+2);
  tmp_SRCM = ALLOC_VECTOR(int_NDAT+2);
```

```
PERCENT = ALLOC_VECTOR(int_NDAT+2);
CONS_SEDM = ALLOC_VECTOR(int_NDAT+2);
FLUX = ALLOC_VECTOR(int_NDAT+2);
FLUX_COEF = ALLOC_VECTOR(int_NDAT+2);

XMPT = ALLOC_VECTOR(20*int_NDAT+2);
tmp_Z = ALLOC_VECTOR(20*int_NDAT+2);
tmp_P = ALLOC_VECTOR(20*int_NDAT+2);
tmp_DLTAZ = ALLOC_VECTOR(20*int_NDAT+2);

 init();
}

//*************************************************************************
//*************************************************************************
//*****WRITE A XYZ FILE FOR PLOTTING PROGRAMS******************************
//********WITH AND WITHOUT THE AGE PRINTED TO THE FILE*********************
//*************************************************************************
void write_xyz_file_w()
{
 FILE *out_put_file;

  out_put_file = fopen(XYZFL, "w");

  if (strcmp(SPLINE, "SPLINE") != 0)
      {
       fprintf(out_put_file," DEPTH AGE TIME RATE SRCS FLUX DATA MODEL\n");
       for (i = 0; i < NDAT; i++)
         fprintf (out_put_file, "%6.2lf %6.2lf %6.2lf %6.4lf %6.2lf %6.2lf
                                                %7.3lf  %8.3lf\n",
         SZV[i],CORE_DATE[i],TBST[i],RBST[i], DBST[i], FLUX[i],P[i],PBST[i]);
      }
  else
      {
       Pre_SPLINE_P[0] = P[0];
       fprintf(out_put_file,"\n                        ORG. SPLINE");
       fprintf(out_put_file,"\n DEPTH AGE TIME RATE SRCS FLUX DATA
                                                DATA MODEL\n");
       for (i = 0; i < NDAT; i++)
         fprintf (out_put_file, "%6.2lf %6.2lf %6.2lf %6.4lf %6.2lf %6.2lf
                                             %7.3lf %7.3lf %8.3lf\n",
         SZV[i],CORE_DATE[i],TBST[i],RBST[i],DBST[i],FLUX[i],Pre_SPLINE_P[i],
                                                P[i],PBST[i]);
      }
  fclose(out_put_file);
}

void write_xyz_file_wo()
{
 FILE *out_put_file;

  out_put_file = fopen(XYZFL, "w");

  if (strcmp(SPLINE, "SPLINE") != 0)
```

```
      {
      fprintf(out_put_file," DEPTH TIME RATE SRCS FLUX DATA MODEL\n");
      for (i = 0; i < NDAT; i++)
       fprintf (out_put_file, "%6.2lf %6.2lf %6.4lf %6.2lf %6.2lf %7.3lf
            %8.3lf\n", SZV[i],TBST[i],RBST[i],DBST[i],FLUX[i],P[i],PBST[i]);
      }
   else
      {
      Pre_SPLINE_P[0] = P[0];
      fprintf(out_put_file,"\n              ORG. SPLINE");
      fprintf(out_put_file,"\n DEPTH TIME RATE SRCS FLUX DATA DATA MODEL\n");
      for (i = 0; i < NDAT; i++)
       fprintf (out_put_file, "%6.2lf %6.2lf %6.4lf %6.2lf %6.2lf %7.3lf
                                        %7.3lf %8.3lf\n",
          SZV[i],TBST[i],RBST[i],DBST[i],FLUX[i],Pre_SPLINE_P[i],P[i],PBST[i]);
      }


   fclose(out_put_file);
}

void FILE_PROB_TYPES(long l, long m)
{
   int ii, jj;
   for (jj = 0; jj < 5; jj++)
    {
        if (strcmp(strupr(PROB_TYPES[jj]), "P") == 0)
          {
               for (ii = 0; ii < NDAT+1; ii++)
                 {
                        if (ii == 0)
                         fprintf (out_time,"%ld %ld", (l+1), (m+1));
                        else if (jj == 0 && ii > 0)
                         fprintf (out_time,"%1.6E",TIME_VEC[ii-1]);
                        else if (jj == 1 && ii > 0)
                         fprintf (out_time,"%1.6E", LOC_RATE[ii-1]);
                        else if (jj == 2 && ii > 0)
                         fprintf (out_time,"%1.6E", LOC_SRC[ii-1]);
                        else if (jj == 3 && ii > 0)
                         fprintf (out_time,"%1.6E", LOC_FLUX[ii-1]);
                        else if (jj == 4 && ii > 0)
                         fprintf (out_time,"%1.6E ", LOC_P[ii-1]);
                 }
               fprintf (out_time,"\n");
          }
    }
}

void SET_PROB_TYPES()
{
   fprintf(out_time, "%ld %ld", cnt_bst, NDAT);
   for (i = 0; i < 5; i++)
        if (strcmp(strupr(PROB_TYPES[i]),"P") == 0)
          {
```

```
                    switch(i){
                            case 0: fprintf(out_time,"TIME "); break;
                            case 1: fprintf(out_time,"RATE "); break;
                            case 2: fprintf(out_time,"SOURCE "); break;
                            case 3: fprintf(out_time,"FLUX "); break;
                            case 4: fprintf(out_time,"MODEL "); break;
                            }
            }
}


//*********************************************************************************
//*********************************************************************************
//*********************************************************************************
//*********************************************************************************
//*****SET THE VARIABLE FLAMD, TO THE PROPER VALUE OF DPM AND SET     ********
//*****THE END OF THE CORE TO BE SIX TIMES THE HALF LIFE OR SPECIFIED********
//*****IN THE PARAMETER FILE (SIT.PPP) TO ANOTHER VALUE.             ********
//*********************************************************************************
void SET_ISOTOPE()
{
  strupr(ISOTOPE);
  if (strcmp(ISOTOPE, "PB210") == 0)
            {
                    if (strcmp(strupr(max_time),"YES") == 0)
                      END_CORE_TIME = MAX_YRS;
                    else
                      END_CORE_TIME = (long)6*22.26 ;
                    FLAMD = M_LN2/22.26;
            }
  else if (strcmp(ISOTOPE, "TH230") == 0)
            {
                    if (strcmp(strupr(max_time),"YES") == 0)
                      END_CORE_TIME = MAX_YRS;
                    else
                    END_CORE_TIME = (long)6*7.54e4 ;
                    FLAMD = M_LN2/(7.54e4);
            }
  else if (strcmp(ISOTOPE, "U234") == 0)
            {
                    if (strcmp(strupr(max_time),"YES") == 0)
                      END_CORE_TIME = MAX_YRS;
                    else
                      END_CORE_TIME = (long)6*2.4e5 ;
                    FLAMD = M_LN2/2.45e5;
            }
  else if (strcmp(ISOTOPE, "RA226") == 0)
            {
                    if (strcmp(strupr(max_time),"YES") == 0)
                      END_CORE_TIME = MAX_YRS;
                    else
                      END_CORE_TIME = (long)6*1.599e3 ;
                    FLAMD = M_LN2/1.599E3;
            }
```

```
else if (strcmp(ISOTOPE, "PA231") == 0)
        {
                if (strcmp(strupr(max_time),"YES") == 0)
                  END_CORE_TIME = MAX_YRS;
                else
                END_CORE_TIME = (long)6*3.25e4 ;
                FLAMD = M_LN2/3.25e4;
        }
else if (strcmp(ISOTOPE, "CS137") == 0)
        {
                if (strcmp(strupr(max_time),"YES") == 0)
                  END_CORE_TIME = MAX_YRS;
                else
                END_CORE_TIME = (long)6*30.17 ;
                FLAMD = M_LN2/30.17;
        }
else if (strcmp(ISOTOPE, "PU239") == 0)
        {
                if (strcmp(strupr(max_time),"YES") == 0)
                  END_CORE_TIME = MAX_YRS;
                else
                  END_CORE_TIME = (long)6*2.411e4 ;
                FLAMD = M_LN2/2.411E4;
        }
else if (strcmp(ISOTOPE, "PU240") == 0)
        {
                if (strcmp(strupr(max_time),"YES") == 0)
                  END_CORE_TIME = MAX_YRS;
                else
                  END_CORE_TIME = (long)6*6.537e3;
                FLAMD = M_LN2/6.537E3;
        }
else if (strcmp(ISOTOPE, "AM241") == 0)
        {
                if (strcmp(strupr(max_time),"YES") == 0)
                  END_CORE_TIME = MAX_YRS;
                else
                  END_CORE_TIME = (long)6*432.2;
                FLAMD = M_LN2/432.2;
        }
else if (strcmp(ISOTOPE, "BE7") == 0)
        {
                if (strcmp(strupr(max_time),"YES") == 0)
                  END_CORE_TIME = MAX_YRS;
                else
                  END_CORE_TIME = (long)6*0.1459726027;
                FLAMD = M_LN2/0.1459726027;
        }
else if (strcmp(ISOTOPE, "BE10") == 0)
        {
                if (strcmp(strupr(max_time),"YES") == 0)
                  END_CORE_TIME = MAX_YRS;
                else
                  END_CORE_TIME = (long)6*1.52e6;
```

```
                        FLAMD = M_LN2/1.52E6;
              }
  else if (strcmp(ISOTOPE, "PO210") == 0)
              {
                        if (strcmp(strupr(max_time),"YES") == 0)
                          END_CORE_TIME = MAX_YRS;
                        else
                          END_CORE_TIME = (long)6*0.3791780822;
                        FLAMD = M_LN2/0.3791780822;
              }
  else
              {
                        if (strcmp(strupr(max_time),"YES") == 0)
                          END_CORE_TIME = MAX_YRS;
                        else
                          END_CORE_TIME = (long)6*(M_LN2/Def_iso);
                        FLAMD = M_LN2/Def_iso;
              }
}


//*********************************************************************
//*********************************************************************
//*****REASSIGN THE USER SPECIFED SEDIMENTATION RATE TO A SLOPE VALUES*******
//*********************************************************************
//*********************************************************************
double reset_Slope(double *reset, long k)
{
  double temp;
  temp = FLAMD/reset[k]*100*ZMAX;
  return(temp);
}


//*********************************************************************
//*************NORMALIZE THE MODIFIED RATES.********************************
//*********************************************************************
double chg_R_to_B(long t)
{
  double X;

  X =    (rate_B[num_rates-t-1] - BTOM[1][1])/(BTOM[2][1] - BTOM[1][1]);
  return (X);
}


//*********************************************************************
//*****SCAN FOR COMMA'S IN THE DATA FILE AND REPLACE THEM WITH***************
//*****A SPACE. *******************************************************
//*********************************************************************
//*********************************************************************
void scan_comma()
{
        char *ptr, comma = ',', space = ' ';
        char tmp_str[80];
        long com_loc;
```

```
          ptr = strchr(tempstr, comma);
          do
           {
           if (ptr)  // FIND THE COMMA
           {
            gotoxy(15,8);
            printf("Correcting for comma's in the data file %s.\n", DATFL);
            strcpy(tmp_str,tempstr);
            com_loc = (long)(ptr-tempstr);
            tmp_str[com_loc] = space;
            strcpy(tempstr,tmp_str); // REPLACE THE COMMA WITH A SPACE
           }
           ptr = strchr(tempstr, comma);
           }
          while(ptr);
}


//**********************************************************************************
//***Read the Input File and Parameter File ***************************************
//**********************************************************************************
void input()
{
 double DATMIN, int_date, tmpdouble;
 long strlengh, tmpint;
 struct time t;
 struct date d;
 char buffer[MAXPATH];

 getcwd(buffer, MAXPATH);

 getdate(&d); // FIND THE DATE AND TIME THE CORE WAS PROCESSED
 gettime(&t);

 fgets (tempstr, 80, in2);
 sscanf(tempstr,"%ld %s %lf", &NMPT, &ISOTOPE, &Def_iso);
 int_NMPT = NMPT;

 ANM = ALLOC_HUGE_MATRIX(int_NMPT+1);//ALLOCATE DATA BASED ON # OF PARAMETERS
 tmp_COEF = ALLOC_VECTOR(int_NMPT+1);
 PRN_COEF = ALLOC_HUGE_MATRIX(2*int_NMPT+1);
 CHISQM = ALLOC_VECTOR(int_NMPT+1);
 loc_min_time = ALLOC_VECTOR(int_NMPT+1);

//**********************************************************************************
//**SCAN THE PARAMETER FILE      **************************************************
//**********************************************************************************
    fgets (tempstr, 80, in2);
    sscanf(tempstr,"%ld %s %s %ld", &MTOM, &GRAPHICS, &SPLINE, &NUM_SPLINE);


    fgets (tempstr, 80, in2);
    sscanf(tempstr,"%ld %s %ld %ld %lf %lf", &TERMS,&CONTOUR,&NDAT,&START_PT,
                                                    &DPTS, &int_date);
```

```
    strupr(TERM);
    strupr(CONTOUR);

    for (i = 0, i < NNTOM; i++)
          {
             fgets (tempstr, 80, in2);
             sscanf(tempstr,"%s  %lf %lf %lf %lf", &LTOM[i],
                      &PARA[0][i], &PARA[1][i], &PARA[2][i], &DELTA_PARA[i]);
          }

    strupr(LTOM);
    tmpdouble = (PARA[2][1]-PARA[1][1])/DELTA_PARA[1]+1;
    num_rates = tmpint = (long)(tmpdouble+0.001);

    if (MTOM == 3)
          rate_B = ALLOC_VECTOR(tmpint+1);
    else
          {
             rate_B = ALLOC_VECTOR(tmpint+1);
             for (i = 0, i < num_rates+1; i++)
                    {
                     rate_B[i] = PARA[1][1] + i*DELTA_PARA[1];
                     if (i == 0) PARA[1][1] = rate_B[i];
                     if (i == num_rates-1) PARA[2][1] = rate_B[i];
                    }

          }

    fgets (tempstr, 80, in2);
    sscanf (tempstr,"%ld %s %s %s %s %s",&MAXCHI,&PROB_TYPES[0],&PROB_TYPES[1],
                           &PROB_TYPES[2], &PROB_TYPES[3], &PROB_TYPES[4]);

    fgets (tempstr, 80, in2);
    sscanf (tempstr,"%ld %lf %lf %ld %s", &real_syn_data, &DATA_ERR, &TIME_ERR,
                                        &NUM_BREAKS, &timefile);

    fgets (tempstr, 80, in2);
    sscanf(tempstr, "%s %lf %s %lf", &min_time, &MIN_YRS, &max_time, &MAX_YRS);

    fgets (tempstr, 80, in2);
    sscanf (tempstr,"%ld", &NTIME);

    VAR_TIME = ALLOC_VECTOR(NTIME+1);
    DEPTH_TIME = ALLOC_VECTOR(NTIME+1);
    for (i = 0; i < NTIME; i++)
     {
       fgets (tempstr, 80, in2);
       sscanf (tempstr, '%lf %lf %lf', &DEP_TIM_RAW[0][i], &DEP_TIM_RAW[1][i],
                                        &VAR_TIME[i]);
     }
//******************************************************************************
//******************************************************************************

    DATMIN = 1.E5;
```

```
//*********************************************************************************
//**SCAN THE DATA FILE ***********************************************************
//*********************************************************************************
    fgets(tempstr,80,in1);
    sscanf (tempstr, "%[^\n]s", FNAME);

    fgets(tempstr,80,in1);

    if (NDAT > 150)
            {
              gotoxy(15,7);
              printf ("NOTE:  TO MANY DATA POINTS TO PROCESS GRAPHICS");
              delay(1500);
              strcpy(GRAPHICS, "OFF");
            }

    strupr(SPLINE);

    if (strcmp(SPLINE, "SPLINE") == 0)
            SPLINE_VAR_INIT();
    int_NDAT = NDAT;
    if (MTOM == 3)
            {
              out_time = fopen (timefile,"w+");
              LTOM[0] = LTOM[1] = 'V';
//            DELTA_PARA[0] = 10;
//            DELTA_PARA[1] = 20;
            }

   if (MTOM == 4)
   {
            TIME_VEC = ALLOC_VECTOR(int_NDAT);
            LOC_SRC = ALLOC_VECTOR(int_NDAT);
            out_time = fopen (timefile,"w+");
   }

   if (real_syn_data == 1)
            ORG_DATA = ALLOC_VECTOR(int_NDAT);


    ALLOC_VECTOR_DATA();
    CORE_DATE[0] = int_date;
// This Condition is for reading from the depth of DPTS to the bottom.
    if (START_PT == 1)
            {
             for (i = 0; i < NDAT; i++)
             {
               fgets (tempstr, 80, in1);
               scan_comma();
               sscanf (tempstr, "%lf %lf %lf",&tmp[0], &tmp[1], &tmp[3]);
                      if ((long)(10*DPTS) == (long)(10*tmp[0]))
                      {
                        NDAT -= (i);
                        ZN[1] = (tmp[0] - tmp[2])/100;
```

```
                        PZ[1] = tmp[1];
                        P[1] = PZ[1];
                        VAR_DATA[0] = 1;
                        VAR_DATA[1] = tmp[3];
//   tmp[2] = tmp[0];
                        break;
                        }
                P[0] = tmp[1];
                tmp[2] = tmp[0];
            }
        j = 0;
        for (i = 2; i <= NDAT; i++)
        {
            fgets(tempstr, 80,in1);
            scan_comma();
            sscanf(tempstr,"%lf %lf %lf %lf",&ZN[i],&PZ[i],&FLUX_COEF[i],
                                                        &VAR_DATA[i]);

            P[i] = PZ[i];
            ZN[i] = (ZN[i] - tmp[2])/100;

            if (PZ[i] <= DATMIN && PZ[i] > 0)
                    DATMIN = PZ[i];
            if (P[i] <= 0)    P[i] = DATMIN;

            if (fabs(ZN[i]*100 - DEP_TIM_RAW[1][j]) < 0.001)
                    DEP_LOCO[j++] = i;
        }
    }
    else
        {
        j =0;
        for (i = 0; i < NDAT; i++)
         {
                if (real_syn_data == 1)
                   {
                    fgets(tempstr, 80, in1);
                    scan_comma();
                    sscanf(tempstr,"%lf %lf %lf %lf %lf",&ZN[i+1],&PZ[i],
                            &FLUX_COEF[i+1],&VAR_DATA[i+1],&ORG_DATA[i+1]);
                   }
                else
                   {
                    fgets(tempstr, 80, in1);
                    scan_comma();
                    sscanf(tempstr,"%lf %lf %lf %lf", &ZN[i+1],&PZ[i],
                                        &FLUX_COEF[i+1],&VAR_DATA[i+1]);
                   }

                if ((long)(100*DPTS) == (long)(100*ZN[i]))
                   {
                    P[i+1] = PZ[i];
                    ZN[i+1] = ZN[i+1]/100;
                    if (PZ[i] < DATMIN && PZ[i] >= 0) DATMIN = PZ[i];
                    if (P[i+1] < 0)    P[i+1] = DATMIN;
```

```
                        NDAT = i+1;
                        break;
                        }

                if (fabs(ZN[i]*100-DEP_TIM_RAW[0][j]) < 0.001)
                    DEP_LOCO[j++] = i;

                P[i+1] = PZ[i];
                ZN[i+1] = ZN[i+1]/100;

                if (PZ[i] < DATMIN && PZ[i] >= 0) DATMIN = PZ[i];
                if (P[i+1] < 0)    P[i+1] = DATMIN;

            }
            VAR_DATA[0] = VAR_DATA[1];
            }
//*****************************************************************************
//*****************************************************************************
    if (MTOM == 3)
            {
            #undef MIN_NUM_2_TERMS
            #undef MIN_NUM_3_TERMS
            #define MIN_NUM_2_TERMS 12
            #define MIN_NUM_3_TERMS 12
            }


        SET_ISOTOPE(); // set FLAMD to a value

//*****************************************************************************
//**OUTPUT ALL OF THE IMPORTANT IMFORMATION TO THE *.OUT FILE*****************
//*****************************************************************************
    fprintf (out1, "DATA TITLE = %s.\n\n", FNAME);
    fprintf (out1, "PROCESSED ON %d/%d/%d",d.da_mon,d.da_day,d.da_year);
    fprintf (out1, "AT %2d:%02d. \n",t.ti_hour,t.ti_min);
    fprintf (out1, "THE DATA FILE NAME AND LOCATION: %s\\ %s \n\n", buffer,
                                                        strupr(DATFL));

// THIS WRITES VERBATUM THE PARAMETER FILE
//*****************************************************************************
    fprintf(out1, "*************************************************\n");
    rewind(in2);
    fgets(tempstr, 80, in2);
    fputs(tempstr,out1);
    fgets(tempstr, 80, in2);
    fputs(tempstr,out1);
    fgets(tempstr, 80, in2);
    fputs(tempstr,out1);
    fgets(tempstr, 80, in2);
    fputs(tempstr,out1);
    fgets(tempstr, 80, in2);
    fputs(tempstr,out1);
    fgets(tempstr, 80, in2);
    fputs(tempstr,out1);
```

```
    fgets(tempstr, 80, in2);
    fputs(tempstr,out1);
    fgets(tempstr, 80, in2);
    fputs(tempstr,out1);
    fgets(tempstr, 80, in2);
    fputs(tempstr,out1);
    fprintf(out1, "*****************************************************\n\n");


    fprintf(out1,"              PARAMETER FILE INPUT: \n\n");


    fprintf (out1, "MTOM SET TO %ld.\n", MTOM);
    fprintf (out1, "MAXIUM CHI SQUARE VALUE IS %5.3lf.\n", MAXCHI);
//  fprintf (out1, "FLUX COEFFIENT WAS SET TO %5.3lf.\n", FLUX_COEF);
    fprintf (out1, "THERE ARE %ld DATA POINTS IN THIS CORE.\n", NDAT);
    if (START_PT == 0)
      fprintf (out1, "THIS CORE WAS PROCESSED FROM THE SURFACE.\n");
    else
      fprintf (out1, "THIS CORE WAS PROCESSED FROM %5.3lf.\n",DPTS);

    if (strcmp(CONTOUR,"YES") == 0)
      fprintf (out1, "THIS CORE WAS PROCESSED FOR A CONTOUR (ACCEPT ALL
                                                   ANSWERS).\n");
    else
      fprintf (out1, "CONTOURING FILE (IS_R_CHI.DAT) IS NOT COMPLETE.\n");


    fprintf (out1, "NUMBER OF ITERATIONS FOR ADJUSTING COEF IS %ld.\n", NMPT);
    fprintf (out1, "DECAY COEFFICIENT FOR THE ISOTOPE %s IS:  %10.6E.\n",
                                              strupr(ISOTOPE), FLAMD);


    fprintf (out1, "THE YEAR THE CORE WAS TAKEN: %5.0lf.\n", CORE_DATE[0]);

    if (strcmp(SPLINE, "SPLINE") == 0)
    fprintf (out1, "NOTE: THIS DATA HAS BEEN SPLINED WITH A TENSION FACTOR
                                              OF %ld.\n",NUM_SPLINE);
    else
    fprintf (out1, "NO DATA SPLINING.\n");

    if (MTOM == 3)
      if (TERMS == 10)
          fprintf (out1,"THIS DATA HAS BEEN BROKEN INTO PARTS WITH 1, 2, AND
                                                   3 TERMS.\n");
      else
          fprintf (out1,"THIS DATA HAS BEEN BROKEN INTO PARTS WITH %ld TERMS.
                                                   \n", TERMS);

    if (MTOM == 0 || MTOM == 3 || MTOM == 4)
    fprintf(out1,"\nTHIS IS A LINEAR SEARCH TASK WITH THE FOLLOWING PARAMETERS:
                                                            \n\n");
    else
    fprintf(out1,"\nTHIS IS A TOMOGRAPHY SEARCH TASK WITH THE FOLLOWING
                                              PARAMETERS: \n\n");
```

```
    for (i = 0; i < NNTOM; i++)
      fprintf (out1,"%c %10.4lf %10.4lf %10.4lf %10.4lf\n", LTOM[i],
                          PARA[0][i], PARA[1][i], PARA[2][i], DELTA_PARA[i]);
    fprintf (out1, "\n\n");

    if (NTIME == 0)
      fprintf (out1, "THERE ARE NO KNOWN TIME TO DEPTH OR VARIANCE VALUES
                                                    INDICATED.\n\n");
    else
    {
      fprintf (out1, "THESE ARE THE KNOWN TIME TO DEPTH VALUES
                                          WITH A VARIANCE(S).\n");
      for (i = 0; i < NTIME; i++)
        fprintf (out1, "  %4.2lf cm   %4.2lf years   * = %4.3lf\n",
                          DEP_TIM_RAW[0][i], DEP_TIM_RAW[1][i], VAR_TIME[i]);
      fprintf (out1, "\n\n");
    }

    if (START_PT != 1)
    P[0] = PARA[0][0];//must set the initial SURFACE VALUE.
    else
    {
// PARA[0][0] = P[0];
// PARA[1][0] = 0.80*P[0];
// PARA[2][0] = 1.20*P[0];

    fprintf (out1, "NOTE: This output file is Normalized to the surface.\n");
    fprintf (out1, "  These are the new parameters for the decay values\n\n");

    fprintf (out1,"%10.4lf %10.4lf %10.4lf\n\n", PARA[0][0], PARA[1][0],
                                                          PARA[2][0]);

    fprintf (out1, "Original depth was %5.2lf is now the surface depth
                                          (0 units)\n", tmp[2]);
    fprintf (out1, "and %5.2lf is the first step below the surface.\n", DPTS);
    fprintf (out1, "\n\n");
    }
//*********************************************************************
//*********************************************************************
//*********************************************************************
}

//*********************************************************************
//**THIS PROCEDURE FINDS EITHER THE *.DAT FILES OR THE *.* FILE WHEN ********
//**INPUTING THE DATA FILE NAME   *********************************************
//*********************************************************************
void FILE_FIND()
{
  struct ffblk ffblk;
  char *exten, tmp_chr[3], *tmpstr;
  long done, i, j, k;
  char buffer[MAXPATH];

  getcwd(buffer, MAXPATH);
```

```
ptr = strchr(DATFL,point);
i = 0;
strcpy (tmp_chr, "*.*");

if (ptr == NULL)
        exten = "*.*";
else
        {
         for (j = ptr-DATFL+1; j < strlen(DATFL) ; j++)
         tmp_chr[i++] = DATFL[j];

         strcpy(exten,"*.");
         strcat(exten,tmp_chr);
         strupr(exten);
        }

i = 1;
j = 15;
k = 0;
done = findfirst(exten,&ffblk,0);
if (done)
        {
         strcpy(exten,"*.*");
         done = findfirst(exten,&ffblk,0);
        }

gotoxy(5,11);
printf("The current directory is: %s\n", buffer);
gotoxy(5,13);
printf("Please select a file listing of %s below:\n",exten);

while (!done)
{
        gotoxy(5+k*15,j);
        printf("%s", ffblk.ff_name);
        done = findnext(&ffblk);
        k++;
        if (i++ % 5 == 0)
          {
          printf("\n");
          gotoxy(5,++j);
          k = 0;
          }
}
}

//*******************************************************************
//***Determine the Primary Data Name*********************************
//*******************************************************************
void UPDATE_DATA_FILE()
{
  char tmpc;

  gotoxy(15,1);
```

```c
  printf("************************************************");
  gotoxy(15,2);
  printf("*** SEDIMENTARY ISOTOPE TOMOGRAPHY MODEL (SIT) ***");
  gotoxy(15,3);
  printf("************************************************");
  gotoxy(25,5);
  printf("INPUT FILE IS: %s", DATFL);
  gotoxy(22,7);
  printf("Do you want to change the input file? (Y/N)");
  tmpc = getch();

  if (tmpc == 'Y' || tmpc == 'y')
    {
    clrscr();

    do
     {
    gotoxy(15,1);
    printf("************************************************");
    gotoxy(15,2);
    printf("*** SEDIMENTARY ISOTOPE TOMOGRAPHY MODEL (SIT) ***");
    gotoxy(15,3);
    printf("************************************************");
    gotoxy(25,5);
    printf("INPUT NEW DATA FILE IS: ");
    gets(tempstr);
    sscanf(tempstr,"%s", DATFL);

    in1 = fopen(DATFL, "r");
    clrscr();
    if (in1 == NULL)
      {
       strupr(DATFL);
       gotoxy(5,9);
       printf ("\a\a\a\aThe data file %s was not found in current directory.
                                                    \n", DATFL);
       FILE_FIND();
      }
    fl_incr = 1;
     }while (in1 == NULL);
    }
  else
    in1 = fopen(DATFL, "r");
}


//***********************************************************************
//****THIS IS A SAFeTY VALUE IF THE SIT.CFG DOES NOT EXIST ******************
//****IT WILL CREATE IT BY FINDING THE FIRST *.DAT FILE ON THE DISK**********
//****AND WRITING IT TO THE SIT.CFG FILE**********************************
//***********************************************************************
void FILE_FIND_FIRST()
{
  struct ffblk ffblk;
```

```
  findfirst("*.dat",&ffblk,0);
  fprintf(CFG, " %s %ld\n", ffblk.ff_name, fl_incr);
  strcpy (DATFL,ffblk.ff_name);
}


//******************************************************************
//******************************************************************
//***********INVERSION PROCESS AND AIDS FOR THE PROCESS*************
//******************************************************************
//******************************************************************
void print_it(double huge **a1, long size)   // PRINT TO THE SCREEN
{
  long i,j;
  printf("\n");
  for (i=0;i< size;i-+)
  {
        for (j=0;j< size;j++)
         printf("%6 31f  ",a1[i][j]);
        printf("\n");
  }
}

void copymatrix(double huge **a1, double huge **b1,long size)
{
  long i,j;                      // SIMPLY COPY A MATRIX
  for (i=0;i< size;i-+)
        for (j=0;j< size;j++)
         b1[i][j] = a1[i][j];
}


void multiply(double huge **a1, double huge **b1, double huge **C, long size)
{
  long i,j,k;                         // SIMPLY MULTIPLY A MATRIX
  for (i=0;i< size;i++)
        for (j=0;j< size;j++)
        {
         C[i][j] = 0.0;
         for (k=0;k< size;k++)
        C[i][j] = a1[i][k] * b1[k][j] + C[i][j];
        }
}


void trace(long l, double huge **a1, long size, double *det)
{
  long m;                            // CREATE A TRACE OF MATRIX (DET OF A)
  *det = 0.0;
  for (m=0;m < size;m++)
        *det = a1[m][m]/l + *det;
}

void russian(double huge **a1,double huge **b1, long size)
```

```
{
    long i,j;                              // INVERSION OF A MATRIX
    double det;                            // RETURNED AS B1 ==> A
    copymatrix(a1,b1,size);
    for (i=0;i<size-1;i++)
    {
        copymatrix(b1,F,size);
        trace(i+1,F,size,&det);
        for (j=0;j<size;j++)
          F[j][j] -=  det;
        multiply(a1,F,b1,size);
    }
    trace(size,b1,size,&det);
    multiply(a1,F,b1,size);

    for (i=0;i<size;i++)
        for (j=0;j<size;j++)
        b1[i][j] = F[i][j]/det;
}


//*************************************************************************
//***THIS PROCEDURE UN-NORMALIZES THE DATA FOR THE LEAD FUNCTION*************
//*************************************************************************
void DENOMO(long j, double *X)
{
        double X1, X0, Xm;

        if (JTOM[j] == 0)
        *X  = BTOM[0][j];
        else
        {
          X1 = ATOM[JTOM[j]-1];
          X0 = BTOM[2][j];
          Xm = BTOM[1][j];
          *X = X1*(X0-Xm)+Xm;
          if (*X < Xm)
          {
                  *X = 1.1*Xm;
                  set_min_flag = 1;
          }
          if (*X > X0)
          {
                  *X = 0.9*X0;
                  set_min_flag = 1;
          }
          BTOM[0][j] = *X;
        }
}


//*************************************************************************
//***THIS PROCEDURE NORMALIZES THE DATA OUTSIDE THE LEAD FUNCTION************
//*************************************************************************
```

```
void NORMALIZE()
{
  double tmp[1];

  i = 0;
  for (j = 0; j < NNTOM; j++)
          {
              JTOM[j] = 0;
              if (LTOM[j] == 'F')
               continue;

              if (LTOM[j] == 'V')
            {
              DDST[i] = BETA_0;
              JTOM[j] = i+1;
              KTOM[i] = j;
              switch (KTOM[i])
              {
               case 0: {strcpy(PARA_NAME[i],"Surface");QN_LOCO[i] = 0;break;}
               case 1: {strcpy(PARA_NAME[i],"Rate");QN_LOCO[i] = 1;break;}
               case 2: {strcpy(PARA_NAME[i],"Weight");QN_LOCO[i] = 2;break;}
               default:  printf("Bug!\n");
              }
// if (MTOM == 3)
//                    NUM_SPACE[i] = DELTA_PARA[j];
//            else
//                    {
                      if (strcmp(strupr(PARA_NAME[j]),"RATE")== 0)
                      NUM_SPACE[i] = num_rates;
                      else
                      NUM_SPACE[i] = (BTOM[2][j]-BTOM[1][j])/DELTA_PARA[j];
//                    }
              ATOM[i] = (BTOM[0][j] - BTOM[1][j])/(BTOM[2][j] - BTOM[1][j]);
              i++;
            }
          }
}



//**************************************************************************
//***Set up a portion of Rk and Uk  ***************************************
//***THIS IS TO ASSIGN THE EVEN AND ODD FUNCTIONS FOR THE *****************
//***THE A-MATRIX**********************************************************
//**************************************************************************
void calc_E_O()
{
  double OR, ER, E1, E2, E3, O1, O2, O3, B, slp;
  double E11, E21, E31, O11, O21, O31;
  double E12, E22, E32, O12, O22, O32;

  for (j = 0; j <= N; j++)
   {
     OR = ER = E1 = E2 = E3 = O1 = O2 = O3 = 0;
     E11 = E21 = E31 = O11 = O21 = O31 = 0;
```

```
E12 = E22 = E32 = O12 = O22 = O32 = 0;
for (M = 0; M < NDAT-1; M++)    //This is to calculate the integral
{                               //portion R(k) and U(k)  A.16c and A.18c equations
  B = slp = 0;
  slp = (P[M+1]-P[M])/(Z[M+1]-Z[M]);
  B = P[M]-slp*Z[M];

  if (j == 0)
    {
    E11 = pow(slp,2)*pow(Z[M],3)/3 + slp*B*pow(Z[M],2) + pow(B,2)*Z[M];
    E12 = pow(slp,2)*pow(Z[M+1],3)/3+slp*B*pow(Z[M+1],2)+pow(B,2)*Z[M+1];
    E1 = E12-E11;
    }
  else
    {
    E11 = pow(B,2)*sin(j*PI*Z[M])/(j*PI);
    E21 = 2*slp*B*(PI*j*Z[M]*sin(j*PI*Z[M])+cos(j*PI*Z[M]))/pow(j*PI,2);
    E31 = pow(slp,2)*(pow(j*PI*Z[M],2)*sin(j*PI*Z[M]) +
              2*j*PI*Z[M]*cos(j*PI*Z[M]) - 2*sin(j*PI*Z[M]))/pow(j*PI,3);

    O11 = -pow(B,2)*cos(j*PI*Z[M])/(j*PI);
    O21 = 2*slp*B*(-PI*j*Z[M]*cos(j*PI*Z[M])+sin(j*PI*Z[M]))/pow(j*PI,2);
    O31 = pow(slp,2)*(-pow(j*PI*Z[M],2)*cos(j*PI*Z[M]) +
              2*j*PI*Z[M]*sin(j*PI*Z[M]) + 2*cos(j*PI*Z[M]))/pow(j*PI,3);

    E12 = pow(B,2)*sin(j*PI*Z[M+1])/(j*PI);
    E22 = 2*slp*B*(PI*j*Z[M+1]*sin(j*PI*Z[M+1])+
                                       cos(j*PI*Z[M+1]))/pow(j*PI,2);
    E32 = pow(slp,2)*(pow(j*PI*Z[M+1],2)*sin(j*PI*Z[M+1]) +
        2*j*PI*Z[M+1]*cos(j*PI*Z[M+1]) - 2*sin(j*PI*Z[M+1]))/pow(j*PI,3);

    O12 = -pow(B,2)*cos(j*PI*Z[M+1])/(j*PI);
    O22 = 2*slp*B*(-PI*j*Z[M+1]*cos(j*PI*Z[M+1])+
                                       sin(j*PI*Z[M+1]))/pow(j*PI,2);
    O32 = pow(slp,2)*(-pow(j*PI*Z[M+1],2)*cos(j*PI*Z[M+1]) +
        2*j*PI*Z[M+1]*sin(j*PI*Z[M+1]) + 2*cos(j*PI*Z[M+1]))/pow(j*PI,3);

    E1 = E12-E11;
    E2 = E22-E21;
    E3 = E32-E31;

    O1 = O12-O11;
    O2 = O22-O21;
    O3 = O32-O31;
    }

  ER += (E1+E2+E3);      // PP2*cos((j)*PI*Z[M])*DLTAZ[M];
  OR += (O1+O2+O3);      //PP2*sin((j)*PI*Z[M])*DLTAZ[M];
  }
  E[j] = ER;
  O[j] = OR;
  }

}
```

```
//****************************************************************************
//***Calculate Delta Rk and Uk--Complete Formula ****************************
//***Rk is the first half of CC[] and Uk is the second half of CC[]**********
//****************************************************************************

void Calc_COEF_GT_0(
{
    long t;
    double  ANX, SUME1, P2P, SMO, SME, CM21, CM24, B1, slp;
    double num_dlt_z;

        t=j=num_dlt_z=0;
        k=1;
        for (i = 0; i < NDAT-1; i++)
          {
          num_dlt_z = (Z[i+1] - Z[i])/5;
          k--;
          do {
             tmp_Z[k] = Z[i] + num_dlt_z*j++;
          } while( tmp_Z[k++] < Z[i+1]);
          j = 0;
          }

        k--;
        j = 0;
        t = 1;
        for (i = 0 : i < NDAT-1; i++)
          {
          B1 = slp = 0;
          t--;
          slp = (P[(i+1)]-P[i])/(Z[(i+1)]-Z[i]);
          B1 = P[i]-slp*Z[i];

          do{tmp_P[t] = slp*tmp_Z[t] + B1; } while(tmp_Z[t++] < Z[i+1]);
          }

        for (i = 0; i < k; i++)
        tmp_DLTAZ[i+1] = tmp_Z[i+1] - tmp_Z[i];



        for (IPT = 0; IPT < K; IPT ++)
          {
          ANX = 0;
          for (KPT = 0; KPT < k; KPT++)
                  ANX += COEF[IPT]*cos((IPT+1)*PI*tmp_Z[KPT]);

          XMPT[IPT] = ANX/SLOPE;   // this is an approximation
          }                        // for equ A.19b

        for (KPT = 0; KPT < K; KPT++)
          {
          SME = SMO = SUME1 = 0;
          for (IPT = 0; IPT < k; IPT++)
```

```
                {
                SUME1 = sqrt(1 + 2*fabs(XMPT[KPT]))-1-XMPT[KPT];//delta R(k)
                P2P = tmp_P[IPT]*tmp_P[IPT];
                SUME1 *= P2P;
                SME += SUME1*cos((KPT+1)*PI*tmp_Z[IPT])*tmp_DLTAZ[IPT];
                SMO += SUME1*sin((KPT+1)*PI*tmp_Z[IPT])*tmp_DLTAZ[IPT];
                }
            SUMPE[KPT] = SME;     // portion of equation A.22
            SUMPC[KPT] = SMO;     // portion of equation A.23
            }

        for (M = 0; M < K; M++)
            {
            CM21 = PBMIN*PBMIN*pow(W,(M+1))-PBMAX*PBMAX;
            CM24 = PI*(M+1)*O[M+1]+2*SLOPE*E[M+1];
            CC[M] = CM21 + CM24 - 2*SLOPE*SUMPE[M]; // Rk + DelRk

            CM21 = (M+1)*PI*E[M+1];
            CM24 = 2*SLOPE*O[M+1];
            CC[M+K] = -CM21 + CM24 - 2*SLOPE*SUMPC[M]; // Uk + DelUk
            }
}


//****************************************************************************
//***Calculate the initial Rk and Uk*****************************************
//****************************************************************************
void Calc_COEF_0()
{
    double CM21, CM24;
    for (M = 0; M < K; M++)            // calculating rest of the
        {                             // formula A.16c
        CM21 = PBMIN*PBMIN*pow(W,(M+1))-PBMAX*PBMAX;
        CM24 = PI*(M+1)*O[M+1] + 2*SLOPE*E[M+1];
        CC[M] = CM21 + CM24;        //CC[i] = R(k) real
        CC[M+K] =-1*(M+1)*PI*E[M+1]+2*SLOPE*O[M+1];//CC[i+k] = U(k) complex
        }
}


//****************************************************************************
//***Create a matrix from the parts of Rk/Uk and the complete Rk and Uk******
//****************************************************************************
//************************_                    _***********************************
//***********************|E/R(k) O/R(k) |*********************************
//******************  A = |                    |********************************
//***********************|O/U(k) E/U(k) |********************************
//************************_                    _***********************************
//****************************************************************************
void Calc_A_Matrix()
{

    for (M = 1; M <= K; M++)    // k
        for (j = 1; j <= K; j++) // n
```

```
                {
            if (M == j)
              {
                    A[M-1][j-1]     = -(E[abs(M+j)] + E[abs(j-M)])/CC[M-1];
                    A[M-1][K+j-1]   = (O[abs(M+j)])/CC[M-1];
                    A[K+M-1][j-1]   = -O[abs(M+j)]/CC[M+K-1];
                    A[K+M-1][j+K-1] = (E[abs(j-M)] - E[abs(M+j)])/CC[M+K-1];
              }
            else if (M < j)
              {
                    A[M-1][j-1]     = -(E[abs(M+j)] + E[abs(j-M)])/CC[M-1];
                    A[M-1][K+j-1]   = (O[abs(M+j)] + O[abs(j-M)])/CC[M-1];
                    A[K+M-1][j-1]   = -(O[abs(M+j)] - O[abs(j-M)])/CC[M+K-1];
                    A[K+M-1][j+K-1] = (E[abs(j-M)] - E[abs(M+j)])/CC[M+K-1];
              }
            else
              {
                    A[M-1][j-1]     = -(E[abs(M+j)] + E[abs(M-j)])/CC[M-1];
                    A[M-1][K+j-1]   = (O[abs(M+j)] - O[abs(M-j)])/CC[M-1];
                    A[K+M-1][j-1]   = -(O[abs(M+j)] + O[abs(M-j)])/CC[M+K-1];
                    A[K+M-1][j+K-1] = (E[abs(M-j)] - E[abs(M+j)])/CC[M+K-1];
              }
            }

}

//*********************************************************************************
//***Initialize Btom for LEAD Function
//*********************************************************************************
void SET_BTOM()
{
  for (j = 0; j < NNTOM; j++)
          {
            if ((MTOM == 0 || MTOM == 3 || MTOM == 4) && j == 0)
                  PARA[0][j] = PARA[1][0];
            if ((MTOM == 0 || MTOM == 3 || MTOM == 4) && j == 1)
                  PARA[0][j] = PARA[1][1];


            if (j == 1)
                  {
                    BTOM[0][j] = FLAMD/PARA[2][j]*ZMAX*100;
                    BTOM[1][j] = FLAMD/PARA[2][j]*ZMAX*100;

                    BTOM[2][j] = FLAMD/PARA[1][j]*ZMAX*100;
                  }
            else
                  {
                    BTOM[0][j] = PARA[0][j];
                    BTOM[1][j] = PARA[1][j];
                    BTOM[2][j] = PARA[2][j];
                  }
          }
}
```

```
//*************************************************************************
//***Lead procedure to calculate the Predicted Pb values for Tomography *****
//***for either the Linear Search or Core Cutting      *********************
//*************************************************************************
void LEAD(long CHK_TIME)
{
  long JI, ii, cnt_chi, rate_CHK, chk_time_values, chk_data_err,
          chk_time_loc, chk_date_mark, chk_con_dat_tim;
  double CHKA1, CHKA2, ABSA1, ABSA2, FM, SRS, CS14, PIM, SR1,
          SR2, CS12, CS13, TIMECON, TIME1ON, SQ11, SQ12, CHECK,
          CHI, PPDIF, EPD, EXPSED, EXPSRC, tmp_TIMECON,
          tmp_CS14, tmp_SQ11, tmp_CHI, tmp_SUM_QN,
          tmp_FF1, loc_chi, loc_data_chi, CHK_MAXCHI;

  double chk_ratio;

  rate_CHK = 0;
  loc_chi = 100000;

  DENOMO (0,&PBMAX);    // Un-Normalized data for the LEAD Routine
  DENOMO (1,&SLOPE);
  DENOMO (2,&WEIGHT);

  W = -1;

  for (MPT = 0; MPT < NMPT; MPT++)
                          //The Coefficient Loop set in parameter file
        {
          if (MPT > 0)
              Calc_COEF_GT_0();
          else
              Calc_COEF_0();//Initial Coefficients for Calculating A-Matrix

          Calc_A_Matrix();   //Calculate the A-Matrix

          russian(A, B, N); //Calculate the invere of The A-Matrix

          for (j = 0; j < N; j++)
                COEF[j] = 0;

          for (j = 0; j < N; j++)
              for (M = 0; M < N; M ++)
                  COEF[j] += B[j][M];

          for (j = 0; j < N; j++)     // Maintain all coef
                ANM[MPT][j] = COEF[j];

          if (MPT != 0)
                {

//*************************************************************************
//This section determines the direction and magnitude for the Coefficients  *
//*************************************************************************
                for (i = 0; i < N; i++)
```

```
                             {
                        CHKA1 = ANM[MPT][i];
                        CHKA2 = ANM[MPT-1][i];
                        ABSA1 = fabs(CHKA1);
                        ABSA2 = fabs(CHKA2);
                        CKANM1[i] = fabs(CHKA1 - CHKA2);
                        CKANM2[i] = fabs(CHKA1 + CHKA2);
                        CKANM[i] = CKANM2[i]*FINC;
                        if (CKANM1[i] > CKANM[i])
                             {
                           FM = fabs(FINC*ANM[MPT-1][i]);
                           if (ABSA1 < ABSA2)
                                  if (CHKA2 < 0)
                                COEF[i] = ANM[MPT-1][i]-FM;
                                 else
                                COEF[i] = ANM[MPT-1][i]+FM;
                            else
                                  if (CHKA2 < 0)
                                COEF[i] = ANM[MPT-1][i]+FM;
                                 else
                                COEF[i] = ANM[MPT-1][i]-FM;
                                 }
                        ANM[MPT][i] = COEF[i];
                        }
//*******************************************************************************
                   }

      for (j = 0; j < NDAT; j++)
       {
       SRS = CS14 = 0;
       for (M = 0; M < K; M++)
       {
       SR1 = COEF[M+K]/((M+1)*PI);//COEF[M] was COE[M+K]
       SR2 = 1 - cos (M+1)*PI*Z[j]);
       SR1 *= SR2;
       SRS += SR1;                // SRS = F(z) Equ A.9

       if (MPT == 0)
          {
          CS12 = sin( M+1)*Z[j]*PI);//Use a SIN func. on the first run in LEAD
          CS13 = COEF M]*CS12;
          CS14 += CS13/((M+1)*PI);//CS14 is the derivitive of t (equ A.19a)
          }
       else
          {
          CS12 = cos( M+1)*Z[j]*PI);//Use a COS func the any other time
          CS13 = COEF M]*CS12;
          CS14 += CS13;//CS14 is the derivitive of t (equ A.8)
          }
       }

       if (MPT == 0)
          TIMECON = SLOPE*Z[j] + CS14;
       else
```

```
          {
           DTDZ[j] = CS14/SLOPE;
           TIMECON =  0;

           for (JI = 0; JI < j+1 && NDAT; JI++) // TIMECON = tau(z)
           {
                   SQ11 = 1 + 2*fabs(DTDZ[JI]);
                   TIMECON += SLOPE*sqrt(SQ11)*DLTAZ[JI]; //tau(z)
           }
          }


        SEDM[j] = TIMECON*(1-WEIGHT); // -FLAMD*SZV[j+1]-tmp_sedm;
        SRCM[j] = SRS*(1+WEIGHT);
        TIME1[j] = SEDM[j]/FLAMD;   //t=tau(z)/lambda
        CHECK = SRCM[j]-SEDM[j] ;


        if (fabs(CHECK) > 20)
          Pred_P[j] = 3*PBMAX;
        else
           Pred_P[j] = PBMAX*exp(CHECK); //Equation A.10
        }

  CHI_PB_TIME = CURRENT_PB = CURRENT_TIME = i = 0;

  TIME1[0] = 0;
  for (i = 0; i < NDAT; i++)
      {
        SZ[i] = SZV[i+1] - SZV[i];
        DLTAT[i] = TIME1[i+1] - TIME1[i];
        SRATE[i+1] = SZ[i]/DLTAT[i];
      }

//*****************************************************************************
//**Define restrictions within the parameter file, such as Time to the   ****
//**bottom of the core (Min and Max), Relative fit, Chi fit and any       ****
//**date markers that might be available, and negative rates (due to SIN ****
//**Function).*****************************************************************
//*****************************************************************************
    for (i = 0; i < NDAT; i++)
          if (SRATE[i] < 0)
              {
              rate_CHK = 1;
              break;
              }
           else
              rate_CHK = 0;

    chk_time_loc = 1;
    if (strcmp(strupr(min_time),"YES") == 0 && TIME1[NDAT-1]< MIN_YRS)
          chk_time_loc = 0;

    chk_con_dat_tim = chk_date_mark = chk_time_values = chk_data_err = 1;
```

```
   i = 0;
   for (ii = 0; ii < NDAT; ii++)
     {
     if (i >= NTIME || rate_CHK)
             break;

     if (TIME1[ii]>:DEP_TIM_RAW[1][i]+VAR_TIME[i]+TIME_ERR*VAR_TIME[:]/100))
             chk_time_values = 0;

     if (fabs(100*ZN[ii]- DEP_TIM_RAW[0][i]) < 0.001)
             i++;
     }

   i = 0;
   for (ii = 1; ii < NDAT; ii++)
     {
      if (real_syn_data == 0  &&
             fabs(Pred_P[ii]-P[ii])>VAR_DATA[ii]+DATA_ERR*VAR_DATA[:i]/100)
           {
            chk_data_err = 0;
            break;
           }
      else if (real_syn_data == 1  &&
              fabs Pred_P[ii]-ORG_DATA[ii])>DATA_ERR*ORG_DATA[i:]/100)
           {
            chk_data_err = 0;
            break;
           }
      else if (real_syn_data == 2  &&
              fabs(Pred_P[ii]-P[ii])>VAR_DATA[ii]+DATA_ERR*VAR_DATA[:i]/100)
           {
            chk_data_err = 0;
            break;
           }
      else if (real_syn_data == 3)
           {
            chk_data_err = 1;
            break;
           }
     }

   for ( ii = 0; ii < NTIME; ii++)
     if (fabs(TIME1[DEP_LOCO[ii]] - DEP_TIM_RAW[1][ii]) >
         VAR_TIME[i:]+TIME_ERR*VAR_TIME[ii]/100&&NDAT-1>= DEP_LOCO[i:])
         chk_date_mark = 0;

//**************************************************************=***=*******
//**************************************************************=***=*******
//**Set the Global BEST values of the model*********************=***=*******
//**************************************************************=***=*******
         loc_data_chi = Calc_CHI();
         if (loc_data_chi  <= CHSQL && !rate_CHK && chk_time_values  &&
                 ·TIME1[NDAT-1] < END_CORE_TIME) && chk_time_loc  &&
                chk_data_err && chk_date_mark && loc_data_chi < MAXCHI)
```

```
                    {
                              CHSQL = loc_data_chi ;
                              CHSQL1 = CHSQL;
                              BEST_FLAG=1;

                              for (j = 0; j < NDAT; j++)
                                  {
                          P[0] = Pred_P[0];
                          TBST[j] = TIME1[j];
                          RBST[j] = SRATE[j];
                          PBST[j] = Pred_P[j];
                          DBST[j] = SEDM[j];
                          CBST[j] = SRCM[j];
                          tmp_SRCM[j] = PBMAX*SRCM[j];
                          PERCENT[j] = tmp_SRCM[j]/Pred_P[j];
                          if (j == 0)
                                  {
                                  for (i = 0; i < N; i++)
                                          BEST_COEF[i] = COEF[i];
                                  BST_CHI = CHSQL1;
                                  PBMAX_BST = Pred_P[0];
                                  WGT_BEST = WEIGHT;
                                  BST_SLP   = SLOPE;
                                  }
                                  }
                    }
//****************************************************************************
//****************************************************************************
          if ((strcmp(CONTOUR,"YES") == 0) || (chk_time_values  &&
                          (TIME1[NDAT-1] < END_CORE_TIME) && chk_time_loc  &&
                                          chk_data_err && chk_date_mark))
                    chk_con_dat_tim = 1;
          else
                    chk_con_dat_tim = 0;

          if (strcmp(CONTOUR,"YES") == 0)
                    CHK_MAXCHI = 1;
          else if (loc_data_chi < MAXCHI)
                    CHK_MAXCHI = 1;
          else
                    CHK_MAXCHI = 0;

//****************************************************************************
//**Set the Local BEST values of the model***********************************
//****************************************************************************
   if ((loc_data_chi <= loc_chi && !rate_CHK) && chk_con_dat_tim &&
                                                            CHK_MAXCHI)
     {
       chk_local = 1;
       for (JI = 0; JI < NDAT; JI++)
         {
                  if (MTOM == 4)
                      {
                          TIME_VEC[JI] = loc_time_min[JI] = TIME1[JI];
```

```
                    LOC_SRC[JI] = exp(SRCM[JI]);
                    LOC_SRC[0] = 0;
              }
          else
            loc_time_min[JI] = TIME1[JI];
          LOC_P[JI] = Pred_P[JI];
    }
  if (MTOM == 4)
    for (JI = 0; JI < NDAT; JI++)
      {
        LOC_RATE[JI+1] = (SZV[JI+1]-SZV[JI])/(TIME_VEC[JI+1]-TIME_VEC[JI]);
        LOC_FLUX[JI] = (LOC_P[JI]*exp(FLAMD*TIME_VEC[JI]))*FLUX_COEF[JI]
                      *LOC_RATE[JI];
      }
  loc_chi = loc_data_chi;
  LIN_CHI = loc_data_chi;

  if (TIME1[NDAT-1] > END_CORE_TIME && strcmp(CONTOUR,"YES") == 0
        LIN_CHI = MAXCHI;

// if (LIN_CHI > MAXCHI)
// LIN_CHI = MAXCHI;
    for (JI = 0; JI < NTIME; JI++)
          DEPTH_TIME[JI] = TIME1[DEP_LOCO[JI]];
  }
//*****************************************************************************
//*****************************************************************************

                if (rate_CHK)
                  CHISQM[MPT] = 100000;
                else
                  CHISQM[MPT] = CHI_PB_TIME;//CHI/NDAT;//final CHI Squared

          for (i = 0; i < N; i++)
                PRN_COEF[MPT][i] = COEF[i];
          }

//*****************************************************************************
//**FIND THE LOCAL COEFFICIENTS FOR THE MODEL   ******************************
//*****************************************************************************

          SMCH1 = 1.E20;
          for (IPT = 0; IPT < NMPT; IPT++)
                if (CHISQM[IPT] < SMCH1 )
            {
                SMCH1 = CHISQM[IPT];

                if (SMCH1 == CHSQL1)
                  {
                        for (i = 0; i < N; i++)
                        tmp_COEF[i] = BEST_COEF[i];
                        break;
                  }
                else
```

```
                               {
                       for (i = 0; i < N; i++)
                              tmp_COEF[i] = PRN_COEF[IPT][i];
                       }
                }
              SSSS = SMCH1;
}

//**********************************************************************
//**GRAPH THE FINAL BEST ANSWER OF THE MODELING PROGRAM******************
//**********************************************************************
void grph_fin_vals(long NNN)
{
  int dec, sign;
  char *loc_tmp, str_grph[80];
  settextstyle(0,0,0);

  if (NNN == 1)
          {
                  setcolor(YELLOW);
                  loc_tmp = fcvt(SZV[NDAT-1],3,&dec, &sign);
                  strcpy(str_grph,"DEPTH TO BOTTOM = ");
                  strcat(str_grph, loc_tmp);
                  outtextxy(350,100, str_grph);

                  loc_tmp = fcvt(TBST[NDAT-1],3,&dec, &sign);
                  strcpy(str_grph,"TIME (yrs) TO BOTTOM = ");
                  strcat(str_grph, loc_tmp);
                  outtextxy(350,110, str_grph);

                  loc_tmp = fcvt(PBST[0],3,&dec, &sign);
                  strcpy(str_grph,PARA_NAME[0]);
                  strcat(str_grph, " = ");
                  strcat(str_grph, loc_tmp);
                  outtextxy(350,120, str_grph);

          }
  else
          {

                  setcolor(YELLOW);
                  loc_tmp = fcvt(SZV[NDAT-1],3,&dec, &sign);
                  strcpy(str_grph,"DEPTH TO BOTTOM = ");
                  strcat(str_grph, loc_tmp);
                  outtextxy(350,100, str_grph);

                  loc_tmp = fcvt(TBST[NDAT-1],3,&dec, &sign);
                  strcpy(str_grph,"TIME (yrs) TO BOTTOM = ");
                  strcat(str_grph, loc_tmp);
                  outtextxy(350,110, str_grph);

                  loc_tmp = fcvt(PBST[0],3,&dec, &sign);
                  strcpy(str_grph,PARA_NAME[0]);
                  strcat(str_grph, " = ");
```

```
                    strcat(str_grph, loc_tmp);
                    outtextxy(350,120, str_grph);

                    loc_tmp = fcvt(YBEST,3,&dec, &sign);
                    strcpy(str_grph,PARA_NAME[1]);
                    strcat(str_grph, " = ");
                    strcat(str_grph, loc_tmp);
                    outtextxy(350,130, str_grph);
            }
}

//******************************************************************************
//***Procedure to do the linear search tomography***************************
//******************************************************************************
void LINEAR_SEARCH()
{
      int dec, sign;
      long i, j, tot, chk_i_j, ii, jj, chk_time_values,
              chk_date_mark, chk_data_err;
      double *XTOM, *YTOM, tmp_dbl, CHI_PB_TIME, CURRENT_PB,
              CURRENT_TIME, BEST_REL, *BEST_RATE;
      double huge **sort_best,huge **DATA_MAT,huge **COEF_MAT,huge **SRC_MAT;
      char *tmpstr, str_grph[80], tmp_2_str[5];
      long  **I_J_LOC;

      FILE *OUT_B;

   OUT_B = fopen("IS_R_CHI.DAT","w");


   LOB = chk_local = 0;
   LLL = 0;
   SBEST = 1.0E7;
   cnt_bst = BEST_FLAG = 0;

 if (strcmp(strupr(GRAPHICS),"ON") == 0)
            {
            detectgraph(&gdriver, &gmode);
            initgraph(&gdriver, &gmode, "");

            // read result of initialization
            errorcode = graphresult();

            if (errorcode != grOk)  // an error occurred
              {
             printf("Graphics error: %s\n", grapherrormsg(errorcode));
             printf("Press any key to halt:");
             getch();
             closegraph();
             restorecrtmode();
             exit(1);               // return with error code
              }
            SET_GRAPH();
            }
```

```
fprintf (out1,"****************************************************\n");
fprintf (out1,"**                                                **\n");
fprintf (out1,"**            LINEAR SEARCH TOMOGRAPHY OUTPUT      **\n");
fprintf (out1,"**                                                **\n");
fprintf (out1,"****************************************************\n\n");

NCON1 = (long)(NUM_SPACE[0]+0.00001);
if (NNN == 2)
        NCON2 = (long)(NUM_SPACE[1]+0.00001);

if (NNN == 1) NCON2 = 0;


//*************************************************************************
//***Allocate the necessary variables for linear search *****************
//*************************************************************************

if (NNN == 0)
        {
          NCON1 = NCON2 = 1;
          strcpy(PARA_NAME[0],"SURFACE");
          strcpy(PARA_NAME[1],"RATE");
        }


if (NCON1 > NCON2)
        {
          XTOM = ALLOC_VECTOR(NCON1+2);
          YTOM = ALLOC_VECTOR(NCON1+2);
          rate_adj = ALLOC_VECTOR(NCON1+2);
        }
else
        {
          XTOM = ALLOC_VECTOR(NCON2+2);
          YTOM = ALLOC_VECTOR(NCON2+2);
          rate_adj = ALLOC_VECTOR(NCON2+2);
        }

for (i = 0; i < num_rates+1; i++)
        rate_B[i] = reset_Slope(rate_B,i);

IITOM = 0;
j = i = 0;
BEST_REL = 10000;


//*************************************************************************
//***Write Depths to an output file which will be used to calculate *********
//***Probability Density Function ***************************************
//*************************************************************************

if (MTOM == 4)
        {
            for (ii = 0; ii < NDAT; ii++)
```

```
                       {
                         if (ii == 0)
                             {
                                  fprintf (out_time,"                              ");
                                  fprintf (out_time,"\ni j ");
                             }
                         fprintf (out_time,"%6.3lf ",100*ZN[ii]);
                       }
                  fprintf (out_time,"\n");
                }

//****************************************************************************
//***Linear search with one or two variable parameter(s)              ****
//****************************************************************************
    do
        {
          if (strcmp(strupr(PARA_NAME[0]),"RATE")== 0)
                ATOM[0] = chg_R_to_B(i);
          else
                ATOM[0] = (i)/(double)(NCON1);
          if (strcmp(strupr(PARA_NAME[0]),"SURFACE")== 0)
                {
                  DENOMO (0,&PBMAX);
                  P[0] = PBMAX;
                  calc_E_O();
                }

//****************************************************************************
//***One parameter when NNN = 1
//****************************************************************************
  if (NNN == 1)
    {
      IITOM = LOB;

       if (i == 0)
      chk_i_j = 0;
      else chk_i_j = 1;

      LEAD(chk_i_j);
      if (i == 0)
       SS00 = SSSS;

      LLL = 1;
      XTOM[i] = BTOM[0][KTOM[LOB]];
      YTOM[i] = LIN_CHI;

      if (strcmp(strupr(PARA_NAME[0]),"RATE")== 0)
        rate_adj[i] = reset_Slope(XTOM,i);

      if (BEST_FLAG)
        {
          SBEST = SSSS;
          ABEST = ATOM[LOB];
          IBEST = i+1;
```

```
      KBEST = K;
      if (strcmp(strupr(PARA_NAME[0]),"RATE")== 0)
             XBEST = rate_adj[i];
      else
             XBEST = XTOM[i];
      ZBEST = YTOM[i];
      PBMAX_BST = PBMAX;
      WGT_BEST = WEIGHT;
      BEST_FLAG = 0;
   }

  if (strcmp(strupr(GRAPHICS),"ON") != 0)
   {
gotoxy(20,8);
printf ("THIS IS A LINEAR SEARCH");
gotoxy(20,10);
printf ("THERE ARE %ld ITERATIONS FOR %s.",NCON1, strupr(PARA_NAME[0]));
gotoxy(20,12);
printf ("THIS IS THE %ld TH STEP.    ", i+1);
gotoxy(20,14);
printf ("THE MSR IN THIS STEP IS %6.2lf.    ",YTOM[i]);
gotoxy(20,16);
printf ("TO TERMINATE PRIOR TO %ld ITERATIONS -- TAP A KEY:", NCON1);
gotoxy(15,18);
printf("***************************************************");
   }


  if (i == 0)
   {
     fprintf (out1," I   PARAMETERS   M.S.R. \n");
     fprintf(OUT_B, "SURFACE RATE CHI DEPTH_TIME\n");
   }

  if (strcmp(strupr(PARA_NAME[0]),"RATE")== 0)
    {
       fprintf (out1," %ld %14.4lf %14.4lf\n", i+1, rate_adj[i], YTOM[i]);
       NCON1 = num_rates-1;
    }
  else
     fprintf (out1," %ld %14.4lf %14.4lf\n", i+1, XTOM[i], YTOM[i]);

  fprintf (out1, "A(n) = ");

  for (k = 0; k < K; k++)
    fprintf (out1, "%18.5lf",tmp_COEF[k]);
  fprintf (out1, "\nB(n) = ");

  for (k = 0; k < K; k++)
    fprintf (out1, "%18.5lf",tmp_COEF[k+K]);
   fprintf(out1, "\n\n");


  fprintf(OUT_B, "%5.3lf %1.6E %5.5lf", PARA[0][0], XTOM[i], YTOM[i]);
```

```
    for (ii = 0; ii < NTIME; ii++)
       fprintf(OUT_B, "%5.3lf ", DEPTH_TIME[ii]);
    fprintf(OUT_B, '\n");

  }
//*****************************************************************************
//***Second parameter when NNN = 2
//*****************************************************************************
       else
         do
           {
                if (strcmp(strupr(PARA_NAME[1]),"RATE")== 0)
                  ATOM[1] = chg_R_to_B(j);
                else
                  ATOM[1] = (j)/(float)(NCON2);

                BEST_FLAG =         chk_local = 0;
                if (i == 0 && j == 0)
                  chk_i_j = 0;
                else chk_i_j = 1;

//*****************************************************************************
//**Primary Operation within linear search
//*****************************************************************************
                LEAD(chk_i_j);

                XTOM[i] = BTOM[0][KTOM[0]];
                YTOM[j] = BTOM[0][KTOM[1]];
                ZZZZ = SSSS;
                if (strcmp(strupr(PARA_NAME[1]),"RATE")== 0)
                  rate_adj[j] = reset_Slope(YTOM,j);

//*****************************************************************************
//**Save the global best if Best_flag is set in Lead
//*****************************************************************************
                if (BEST_FLAG)
                  {
                        SBEST = SSSS;
                        ABEST = ATOM[0];
                        BBEST = ATOM[1];
                        IBEST = i+1;
                        JBEST = j+1;
                        KBEST = K;
                        XBEST = XTOM[i];
                        if (strcmp(strupr(PARA_NAME[1]),"RATE")== 0)
                          YBEST = rate_adj[j];
                        else
                          YBEST = YTOM[j];
                        ZBEST = ZZZZ;
                  }
//*****************************************************************************
//**Visual display indicating how the model is doing without Graphic on
//*****************************************************************************
         if (strcmp(strupr(GRAPHICS),"ON") != 0)
```

```
      {
        gotoxy(20,8);
        printf ("THIS IS A LINEAR SEARCH FOR TWO PARAMETERS.");
        gotoxy(20,10);
        printf ("THERE ARE %ld ITERATIONS FOR %s,",NCON1+1, PARA_NAME[0]);
        gotoxy(20,12);
        printf ("AND %ld ITERATIONS FOR %s.",NCON2+1, PARA_NAME[1]);
        gotoxy(20,14);
        printf ("THIS IS THE %ld TH AND %ld STEP.", i+1, j+1);
        gotoxy(20,16);
        printf ("THE MSR IN THIS STEP IS %6.2lf.",LIN_CHI);
        gotoxy(20,18);
        tot = (NCON1*NCON2);
        printf ("TO TERMINATE PRIOR TO %ld ITERATIONS -- TAP A KEY:",tot);
        gotoxy(15,20);
        printf("*************************************************");
      }

    if (j == 0 && i == 0)
      {
            fprintf (out1,"I  J    PARAMETERS    M.S.R \n");
            fprintf(OUT_B, "SURFACE    RATE    CHI   DEPTH_TIME\n");
      }

  if (chk_local)
      {
        if (strcmp(strupr(PARA_NAME[1]),"RATE")== 0)
          {
            fprintf (out1, "%ld %ld %14.4lf %14.4lf %14.4lf\n",(i+1),(j+1),
                                       XTOM[i],rate_adj[j],LIN_CHI);
            NCON2 = num_rates-1;
          }
        else
            fprintf (out1, "%ld %ld %14.4lf %14.4lf %14.4lf\n",i+1,j+1,
                                       XTOM[i],YTOM[j],LIN_CHI);

        fprintf (out1, "A(n) = ");
        for (k = 0; k < K; k++)
            fprintf (out1, "%10.5lf",tmp_COEF[k]);

        fprintf (out1, "\nB(n) = ");
        for (k = 0; k < K; k++)
            fprintf (out1, "%10.5lf",tmp_COEF[k+K]);

        if (MTOM != 4)
            fprintf(out1, "\n\n");
        else
            fprintf(out1, "\n");

        if (LIN_CHI > MAXCHI)
            LIN_CHI = MAXCHI;
        fprintf(OUT_B,"%5.3lf %1.6E %5.5lf ",XTOM[i],rate_adj[j],LIN_CHI);
        for (ii = 0; ii < NTIME; ii++)
            fprintf(OUT_B, "%5.3lf ", DEPTH_TIME[ii]);
```

```
        fprintf(OUT_B, "\n");

//*********************************************************************************
//**If MTOM = 4 then it indicates that the user is restricting the data
//**by time, fit or relative fit
//*********************************************************************************
      if (MTOM == 4)
        {
//*********************************************************************************
//**This section is for relative fit
//*********************************************************************************
        CHI_PB_TIME = CURRENT_PB = CURRENT_TIME = 0;
        k = 0;
        P[0] = LOC_P[0];
        if (cnt_bst == 0)
          {
           INITIAL_TIME = INITIAL_PB =0;
           for (ii = 0; ii < NDAT; ii++)
             {
              CURRENT_PB = INITIAL_PB +=
                                    pow((P[ii] - LOC_P[ii]),2)/VAR_DATA[ii];
              if (fabs(SZV[ii]-DEP_TIM_RAW[0][k]) < 0.001 && NTIME)
                  CURRENT_TIME = INITIAL_TIME += pow((TIME_VEC[ii] -
                                          DEP_TIM_RAW[1][k]),2)/VAR_TIME[k++];
             }
          }
       else
         for (ii = 0; ii < NDAT; ii++)
            {
             CURRENT_PB  += pow((P[ii] - LOC_P[ii]),2)/VAR_DATA[ii];
             if (fabs(SZV[ii]-DEP_TIM_RAW[0][k]) < 0.001 && NTIME)
                 CURRENT_TIME += pow((TIME_VEC[ii]-
                                         DEP_TIM_RAW[1][k]),2)/VAR_TIME[k++];
            }

       if (INITIAL_PB == 0)
        INITIAL_PB = 1;

       if (INITIAL_TIME == 0)
        INITIAL_TIME = 1;

       if (SZV[NDAT-1] >= DEP_TIM_RAW[0][0] && NTIME)
        CHI_PB_TIME=(CURRENT_PB/INITIAL_PB)*(NTIME/((double)(NDAT+NTIME)))+
                    (CURRENT_TIME/INITIAL_TIME)*(NDAT/((double)(NDAT+NTIME)));
       else
        CHI_PB_TIME = (CURRENT_PB/INITIAL_PB);
//*********************************************************************************
       fprintf(out1, 'RELATIVE CHI = %5.6lf\n\n", CHI_PB_TIME);

       if (BEST_REL > CHI_PB_TIME)
           {
                   BEST_REL = CHI_PB_TIME;
                   for (ii = 0; ii < N; ii++)
                     BEST_COEF[ii] = tmp_COEF[ii];
```

```
                    for (ii = 0; ii < NDAT; ii++)
                      {
                              PBST[ii] = LOC_P[ii];
                              TBST[ii] = TIME_VEC[ii];
                              CBST[ii] = LOC_SRC[ii];
                              IBEST = i+1;
                              JBEST = j+1;
                              XBEST = PBST[0];
                              YBEST = rate_adj[j];
                      }
              for (ii = 0; ii < NDAT-1; ii++)
                    RBST[ii+1] = (SZV[ii+1] - SZV[ii])/(TBST[ii+1] - TBST[ii]);
            }
//*********************************************************************
//**Output information for Probabilty Density Function
//*********************************************************************
                  FILE_PROB_TYPES(i,j);
                  cnt_bst++;
            }
      }
//*********************************************************************
//**Visual display indicating how the model is doing with Graphic on
//*********************************************************************
            if (strcmp(strupr(GRAPHICS),"ON") == 0 && BEST_FLAG)
             draw_updated_data_graph(IJK);
            if (NNN == 0)
              ATOM[0] = ATOM[1] = 1;
            j++;
      }while (fabs(1.0-ATOM[1]) > 0.001 && !kbhit());
   if (NCON2 == 0)
    if (strcmp(strupr(GRAPHICS),"ON") == 0 && BEST_FLAG)
     draw_updated_data_graph(IJK);
   j = 0;
   i++;
    } while (fabs(1.0-ATOM[0]) > 0.001  && !kbhit() );

//*********************************************************************
//  Also for Probability Density
//*********************************************************************
   rewind(out_time);
   SET_PROB_TYPES();


//*********************************************************************
//  Calculate the fit and the FLUX
//*********************************************************************
   CHI = 0;
   P[0] = PBST[0];
        for (j = 0; j < NDAT; j++)
          {
            CHI += pow(PBST[j] - P[j], 2);
            FLUX[j] = (PBST[j]*exp(FLAMD*TBST[j]))*FLUX_COEF[j]*RBST[j];
          }
   CHI /=NDAT;
```

```
   if (MTOM == 4)
        BST_CHI = CHI;

//*****************************************************************************
//  This section outputs the bottom of the *.out file
//*****************************************************************************
    fprintf(out1, "\nTHE BEST FIT IS GIVEN BY: \n\n");

    fprintf (out1, "  A(n)   B(n)\n");
    for (k = 0; k < K; k++)
      fprintf (out1, "%10.5lf %10.5lf\n",BEST_COEF[k],BEST_COEF[k+K]);
    fprintf (out1,"\n");

    fprintf (out1,"THE CHI-SQUARED FOR DATA TO MODEL IS %6.4lf.\n",CHI);

    if (END_CORE_TIME > 1000)
      {
             for (i = 0; i < NDAT; i++)
               {
                     TBST[i] /= 1000;
                     RBST[i] *= 1000;
                     FLUX[i] *= 1000;
               }
             fprintf (out1, "NOTE: The time is listed as k-yrs.\n");
      }

    SET_CORE_DATE();
    if (NNN == 1)
    {
      fprintf (out1, "\nThe Parameter adjusted: %s.\n\n", PARA_NAME[0]);
      fprintf (out1, " I   PARAMETER  M.S.R.\n");
      fprintf (out1, " %ld %10.4lf %10.4lf\n\n", IBEST,XBEST,BST_CHI);
    }
    else
    {
      fprintf (out1, "\nThe Parameters adjusted: %s and %s.\n\n",
                                        PARA_NAME[0],PARA_NAME[1]);

      if (MTOM == 4)
          {
           fprintf (out1, " I J PARAMETERS CHI RELATIVE CHI\n");
           fprintf (out1, "%ld %ld %10.4lf %10.4lf %10.4lf %10.6lf\n",
                            IBEST,JBEST,XBEST,YBEST,BST_CHI,BEST_REL);
          }
      else
          {
           fprintf (out1, " I J PARAMETERS M.S.R.\n");
           fprintf (out1, "%ld %ld %10.4lf %10.4lf %10.4lf\n",
                                       IBEST,JBEST,XBEST,YBEST,BST_CHI);
          }
    }
    ATOM[LOB] = ABEST;

    if (NNN != 1)
```

```
      ATOM[1] = BBEST;

      if (END_CORE_TIME < 1000)
        {
          fprintf(out1,"\n DEPTH AGE TIME RATE SRCS FLUX DATA MODEL\n");
          for (i = 0; i < NDAT; i++)
          fprintf (out1, "%6.2lf %6.2lf %6.2lf %6.4lf %6.2lf %6.2lf %7.3lf
                                                                    %8.3lf\n",
            SZV[i],CORE_DATE[i],TBST[i],RBST[i],DBST[i],FLUX[i],P[i],PBST[i]);
          write_xyz_file_w(); // write an XYZ file W/ time values > 1000
        }
      else
        {
          fprintf(out1,"\n DEPTH TIME RATE SRCS FLUX DATA MODEL\n");
          for (i = 0; i < NDAT; i++)
          fprintf (out1, "%6.2lf %6.2lf %6.4lf %6.2lf %6.2lf %7.3lf %8.3lf\n",
                          SZV[i],TBST[i],RBST[i],DBST[i],FLUX[i],P[i],PBST[i]);
          write_xyz_file_wo(); // write an XYZ file W/ time values < 1000
        }
      VAR_DATA[0] = 1;
//********************************************************************************
//********************************************************************************


          fcloseall();

//********************************************************************************
// Display a graphical representation to the screen for "BEST"
//********************************************************************************
          draw_data_graph();
          grph_fin_vals(NNN);

          delay(10000);
          closegraph();
          restorecrtmode();
//********************************************************************************
//********************************************************************************


          exit(0);


}
//********************************************************************************
//***This procedure calculates the zero term Tomography.  Finds the       *****
//***predicted values for Pb.  Simply put a regression line through       *****
//***the raw data. The user may vary the surface value                    *****
//********************************************************************************
void CONST_SEDM_LEAD()
{
   long JI, k, sign_y;
   double CHKA1, CHKA2, ABSA1, ABSA2, FM, SRS, CS14, PIM, SR1,
          SR2, CS12, CS13, TIMECON, TIME1ON, SQ11, SQ12, CHECK,
          CHI, PPDIF, EPD, EXPSED, EXPSRC, slope, inv_slope,
          X_subn, Y_subn, y_dist, SUM_XY, SUM_XX, SUM_X, SUM_Y;
   double  *Y;
```

```
    Y = ALLOC_VECTOR(int_NDAT+2);

    W = -1;
    BST_CHI = 10000;
    CHI =  SUM_X = SUM_Y = SUM_XY = SUM_XX = 0;

    fprintf (out1,"*******************************************************\n");
    fprintf (out1,"**                                                   **\n");
    fprintf (out1,"**            ZERO TERM TOMOGRAPHY OUTPUT             **\n");
    fprintf (out1,"**                            .                      **\n");
    fprintf (out1,"*******************************************************\n\n");

    DENOMO (0,&PBMAX);
    DENOMO (1,&SLOPE);
    DENOMO (2,&WEIGHT);

    NCON1 = (long)NUM_SPACE[0]+1;

    k = 0;
    Z[0] = 0;
    DLTAZ[0] = 0;
    for (i = 0; i < NDAT; i++)
            {
            DLTAZ[i+1] = Z[i+1]-Z[i];
            Y[i] = P[i];
            }
// check graphics and then set them if graphics is ON
    if (strcmp(strupr(GRAPHICS),"ON") == 0)
            {
            detectgraph(&gdriver, &gmode);
             initgraph(&gdriver, &gmode, "");

            // read result of initialization
            errorcode = graphresult();

            if (errorcode != grOk)  // an error occurred
            {
              printf("Graphics error: %s\n", grapherrormsg(errorcode));
              printf("Press any key to halt:");
              getch();
              closegraph();
              restorecrtmode();

              exit(1);            // return with error code
            }
            SET_GRAPH();
            }

//***********************************************************************
//   calculate the Regression Line
//***********************************************************************

    while (k < NCON1)
```

```
      {
        if (MTOM == 0) Y[0] = PARA[1][0] + k*DELTA_PARA[0];
        else k = NCON1;

        CHI = SUM_X = SUM_Y = SUM_XX = SUM_XY = 0;

        for (i = 0; i < NDAT; i++)
        {
                SUM_X += SZV[i];
                SUM_Y += log(Y[i]);
                SUM_XY += SZV[i]*log(Y[i]);
                SUM_XX += SZV[i]*SZV[i];
        }
        slope = (NDAT*SUM_XY-SUM_X*SUM_Y)/(NDAT*SUM_XX-pow(SUM_X,2));

        for (i = 0; i < NDAT; i++)
         Pred_P[i] = Y[0]*exp(slope*SZV[i]);
//**********************************************************************
//**********************************************************************
        for (i = 0; i < NDAT; i++)
         CHI += pow(Pred_P[i]-Y[i],2);

        N = 2*K;
        TIME1[0] = 0;

        CHI /= NDAT;   // final CHI Squared

        if (MTOM == 0)
        {
         fprintf (out1, "\nSTEP %ld", k+1);
         fprintf (out1, "\nSURFACE = %10.4lf MSR = %10.4lf.\n",Y[0],CHI);
        }
//**********************************************************************
// calculate the "BEST"
//**********************************************************************
      if (CHI < BST_CHI)
      {
        for (i = 0; i < NDAT; i++)
          TIME1[i] = fabs(slope)*SZV[i]/FLAMD;//(log(Pred_P[i]/PBMAX))/-FLAMD;

        RBST[0] = 0;
        for (i = 0; i < NDAT; i++)
             { P[0] = Pred_P[0];
              SZ[i] = SZV[i+1] - SZV[i];
              DLTAT[i] = TIME1[i+1] - TIME1[i];
              RBST[i+1] = SZ[i]/DLTAT[i];
              DBST[i] = 0;// SEDM[i];
              PBST[i] =Pred_P[i];
              TBST[i] = TIME1[i];
              PBMAX_BST = P[0];
              WGT_BEST = WEIGHT;
               }
```

```
        IJKBST = k+1;
        BST_SLP = slope;
        CHSQL1 = BST_CHI = CHI;
        SBEST = BST_CHI;
      }

    if (strcmp(strupr(GRAPHICS),"ON") == 0)
      draw_updated_data_graph(IJK);
    k++;
  }
 free(Y);
}

//*************************************************************************
//**Constant Flux Model and Constant Specific Activity Model*************
//*************************************************************************
//*************************************************************************

void Current_Models()
{
    double INT_TIME, *part_sum;
    part_sum = ALLOC_VECTOR(int_NDAT+2);

    if (MTOM == 5)
        {
//   CONSTANT SEDIMENTATION -- VARIABLE SOURCE
        CONS_SRCM = 0;
        for (i = 0; i < NDAT; i++)
              {
                SZ[i+1] = SZV[i+1] - SZV[i];
                CONS_SRCM += P[i]*SZ[i];
              }
        for (i = 0; i < NDAT; i++)
            for (j = i+1; j < NDAT; j++)
              {
                SZ[j+1] = SZV[j+1] - SZV[j];
                part_sum[i] += P[j]*SZ[j];
              }

        for (i = 0; i < NDAT-1; i++)
              {
                TIME1[i] = (1/FLAMD)*log(CONS_SRCM/part_sum[i]);
                SRATE[i+1] = FLAMD*part_sum[i]/P[i];
              }

        fprintf (out1,"\n\n**********************************************\n");
        fprintf (out1,"**                                          **\n");
        fprintf (out1,"**         CONSTANT FLUX MODEL OUTPUT        **\n");
        fprintf (out1,"**                                          **\n");
        fprintf (out1,"**********************************************\n\n");

        gotoxy(5,10);
        printf ("******************************************************\n");
        gotoxy(5,11);
```

```
       printf ("**                                              **\n");
       gotoxy(5,12);
       printf ("**                   CONSTANT FLUX MODEL          **\n");
       gotoxy(5,13);
       printf ("**                                              **\n");
       gotoxy(5,14);
       printf ("*****************************************************\n\n");

       fprintf(out1,"\n DEPTH  TIME  RATE  DATA\n");
       for (i = 0; i < NDAT; i++)
              fprintf (out1, "%6.2lf %10.4lf %10.4lf %8.2lf\n",
                                     SZV[i], TIME1[i], SRATE[i],  P[i]);
    }
  else
    {
    for (i = 0; i < NDAT-1; i++)
           {
            TIME1[i+1] = (1/FLAMD)*log(P[0]/P[i+1]);
            SRATE[i+1] = SZV[i+1]/TIME1[i+1];
           }

    fprintf (out1,"\n\n***********************************************\n");
    fprintf (out1,"**                                              **\n");
    fprintf (out1,"**      CONSTANT SPECIFIC ACTIVITY MODEL OUTPUT  **\n");
    fprintf (out1,"**                                              **\n");
    fprintf (out1,"***********************************************\n\n");

    gotoxy(5,10);
    printf ("*****************************************************\n");
    gotoxy(5,11);
    printf ("**                                              **\n");
    gotoxy(5,12);
    printf ("**            CONSTANT SPECIFIC ACTIVITY MODEL    **\n");
    gotoxy(5,13);
    printf ("**                                              **\n");
    gotoxy(5,14);
    printf ("*****************************************************\n\n");

    fprintf(out1,"\n DEPTH  TIME  RATE  DATA\n");
    for (i = 0; i < NDAT; i++)
           fprintf (out1, "%6.2lf %10.4lf %10.4lf %8.2lf\n",
                                     SZV[i], TIME1[i], SRATE[i],  P[i]);
    }
  fcloseall();
  closegraph();
  restorecrtmode();

  exit(0);
}


//***********************************************************************
//**The following procedures are designated to the Cubic-B Spline routine****
//***********************************************************************
```

```
void SET_BLENDING(long t)
{
  double u, u_cube, u_square, uMinus_1_cube;

  for (i = 0; i < t; i++)
        {
         u = (i+1.0)/t;
         u_square = u*u;
         u_cube = u*u_square;
         uMinus_1_cube =(1-u)*pow(1-u,2);

         BLEND[0][i] = uMinus_1_cube/6;
         BLEND[3][i] = u_cube/6;
         BLEND[2][i] = -u_cube/2 + u_square/2 + u/2 +1.0/6;
         BLEND[1][i] = u_cube/2 - u_square + 2.0/3;
        }
}

void put_in_sm(double x, double y)
{
  Xsm[3] = x;
  Ysm[3] = y;
}

void make_curve(double huge **b)
{
  double x,y;

  for (i = 0; i < LINES_PER_SEC; i++)
        {
         x = y = 0;
         for (j = 0; j < 4; j++)
        {
          x += Xsm[j]*b[j][i];
          y += Ysm[j]*b[j][i];
        }
         X0 = x;
         Y0 = y;
        }
}

void next_section()
{
 for (i = 0; i < 3; i++)
   {
         Xsm[i] = Xsm[i+1];
         Ysm[i] = Ysm[i+1];
   }
}

void curve_abs_2(double x, double y, long k)
{
  put_in_sm(x,y);
  make_curve(BLEND);
```

```
  P[k] = X0;
  next_section();
}

void start_spline(double *ax, double *ay)
{
  for (i = 0; i < 3; i++)
        {
         Xsm[i] = ax[0];
         Ysm[i] = ay[0];
        }
  Xsm[3] = ax[1];
  Ysm[3] = ay[1];
  make_curve(BLEND);
  next_section();
}

void end_SPLINE()
{
  put_in_sm(Ax[NDAT-1],Ay[NDAT-1]);
  make_curve(BLEND);
  next_section();
  put_in_sm(Ax[NDAT-1],Ay[NDAT-1]);
  make_curve(BLEND);
}


void INITIALIZE()
{
 for (i = 0; i < NDAT; i++)
  {
          Ax[i] = P[i];
          Ay[i] = SZV[i];
  }
  X0 = Ax[0];
  Y0 = Ay[0];
}


void B_SPLINE()
{
  double tmpx, tmpy;

        LINES_PER_SEC = 10;

  for (M = 0; M < NUM_SPLINE; M++)
        {
           INITIALIZE();

           if (M == 0)
           for (i = 0; i < NDAT; i++)
             Pre_SPLINE_P[i] = Ax[i];

           SET_BLENDING(LINES_PER_SEC);
```

```
          start_spline(Ax,Ay);
           for (k = 2; k < NDAT; k++)
         curve_abs_2(P[k], SZV[k],k-1);

           end_SPLINE();
         }
}

//*****************************************************************************
//*****************************************************************************


//*****************************************************************************
//**CORE CUTTING USING A LINEAR SEARCH ROUTINE TO CALCULATE THE VALUES*******
//**AT EACH SPECIFIED DEPTH****************************************************
//*****************************************************************************
void LIN_SRCH_CORE()
{
        int dec, sign, chk_key;
        long i, j, tot, NDAT_INIT,NDAT_INIT_LOOP,ij,j_loc_bst[100],MIN_TERM,
              MAX_TERM, chk_time_values, ii, CNT_NDAT, chk_i_j;
        double *XTOM, *YTOM, ONE_TERM_CHI, TWO_TERM_CHI, LIN_CHI,
                      THREE_TERM_CHI, sum_dat[20];

        char str_grph[80], *tmpstr;


   LOB = 0;
   LIN_CHI = SBEST = 1.0E7;
   ONE_TERM_CHI = TWO_TERM_CHI = THREE_TERM_CHI = 1.0e5;

        NDAT_INIT = NDAT;

 if (strcmp(strupr(GRAPHICS),"ON") == 0)
         {
         detectgraph(&gdriver, &gmode);
         initgraph(&gdriver, &gmode, "");

         // read result of initialization
         errorcode = graphresult();

         if (errorcode != grOk)  // an error occurred
           {
          printf("Graphics error: %s\n", grapherrormsg(errorcode));
          printf("Press any key to halt:");
          getch();
          closegraph();
          restorecrtmode();
          exit(1);              // return with error code
           }
         SET_GRAPH();
         }

   if (TERMS == 10)
```

```
            {
             MIN_TERM = 1;
             MAX_TERM = 4;
             }
   else if (TERMS == 1)
            {
             MIN_TERM = 1;
             MAX_TERM = 2;
            }
   else if (TERMS == 2)
             {
              MIN_TERM = 2;
              MAX_TERM = 3;
             }
   else
             {
              MIN_TERM = 3;
              MAX_TERM = 4;
              }

   CNT_NDAT = NUM_BREAKS;

   fprintf (out_time, "%s\n", DATFL);
//****************************************************************************
//  Minimum number of terms is defined in the parameter file
//****************************************************************************

   for (NUM_CORE_TERM = MIN_TERM; NUM_CORE_TERM < MAX_TERM; NUM_CORE_TERM++)
           {
              if (NUM_CORE_TERM == 1)
                     TERMS = 1;
              else if (NUM_CORE_TERM == 2)
                     TERMS = 2;
              else TERMS = 3;

              SET_BTOM();
              NORMALIZE();

              DENOMO (0,&PBMAX);
              DENOMO (1,&SLOPE);
              DENOMO (2,&WEIGHT);

              K = TERMS;
              int_K = K;
              N = 2*K;
              ALLOC_int_K_DATA();

              loc_chi_time = 100000;

              SBEST = 1.0E5;
              if (NUM_CORE_TERM == MIN_TERM)
                {
                 fprintf (out1,"*******************************************\n");
                 fprintf (out1,"**                                      **\n");
```

```
                fprintf (out1,"***      LINEAR SEARCH TOMOGRAPHY OUTPUT      **\n");
                fprintf (out1,"***                                          **\n");
                fprintf (out1,"*****************************************\n\n");
              }
            NCON1 = (long)NUM_SPACE[0];
            if (NNN == 2)
               NCON2 = (long)NUM_SPACE[1];

            if (NNN == 1) NCON2 = 0;

//*************************************************************************
//  Allocate the amount of memory needed to run this procedure
//*************************************************************************
            if (NCON1 > NCON2)
              {
              XTOM = ALLOC_VECTOR(NCON1+2);
              YTOM = ALLOC_VECTOR(NCON1+2);
              rate_adj = ALLOC_VECTOR(NCON1+2);

              }
            else
              {
              XTOM = ALLOC_VECTOR(NCON2+2);
              YTOM = ALLOC_VECTOR(NCON2+2);
              rate_adj = ALLOC_VECTOR(NCON2+2);
              }


              if (NUM_CORE_TERM == 1)
              {
                ONE_TERM = ALLOC_MATRIX(15, NDAT+2);
                NDAT_INIT_LOOP = MIN_NUM_1_TERMS;
                ONE_CHI = ALLOC_VECTOR(NDAT-MIN_NUM_1_TERMS);
              }
              else if (NUM_CORE_TERM == 2)
              {
                TWO_TERM = ALLOC_MATRIX(20, NDAT+2);
                NDAT_INIT_LOOP = MIN_NUM_2_TERMS;
                TWO_CHI = ALLOC_VECTOR(NDAT-MIN_NUM_2_TERMS);
              }
            else
              {
                THREE_TERM = ALLOC_MATRIX(25, NDAT+2);
                NDAT_INIT_LOOP =  MIN_NUM_3_TERMS;
                THREE_CHI = ALLOC_VECTOR(NDAT-MIN_NUM_3_TERMS);;
              }


//*************************************************************************
//*************************************************************************

//*************************************************************************
//  Cut the core  and the step by CNT_NDAT (ie. 1,2,...,n) general n < 10
//*************************************************************************
            for (NDAT = NDAT_INIT_LOOP; NDAT < NDAT_INIT+1; NDAT+=CNT_NDAT)
```

```
      {
        for (i = 0; i < num_rates+1; i++)
              rate_B[i] = PARA[1][1] + i*DELTA_PARA[1];

        for (i = 0; i < num_rates; i++)
              sum_dat[i] = 0;
        ZMAX = ZN[NDAT-1];
        PBMIN = P[NDAT-1];

        for (i = 0; i < num_rates+1; i++)
              rate_B[i] = reset_Slope(rate_B,i);

        SET_BTOM();
        NORMALIZE();
        for (i=0; i < NDAT; i++)
              Z[i] = ZN[i]/ZMAX;
        for (i=0; i < NDAT; i++)
              SZV[i] = Z[i]*ZMAX*100; // getting the depth back
        for (i = 0; i < NDAT; i++)
              DLTAZ[i+1] = Z[i+1]-Z[i];

        if (MTOM == 3)
              fprintf (out_time, "%4.3lf  ", (ZMAX*100));

        IITOM = 0;
        j = i = 0;
        CHSQL = 1.0E5;

//*****************************************************************************
//    This is the linear search routine-- It is very similar to LINEAR_SEARCH
//Except without the Probability Density Function (Modified for CORE CUTTING.
//*****************************************************************************
        do
        {
          if (strcmp(strupr(PARA_NAME[0]),"RATE")== 0)
            ATOM[0] = chg_R_to_B(i);
          else
          ATOM[0] = (i)/(double)(NCON1);
          if (strcmp(strupr(PARA_NAME[0]),"SURFACE")== 0)
            {
              DENOMO (0,&PBMAX);
              P[0] = PBMAX;
              calc_E_O();
            }

          CHSQL = 1.0E5;
          chk_time_values = 1;
          if (NNN == 1)
          {
            IITOM = LOB;

            LEAD(0+i);
            if (i == 0)
            SS00 = SSSS;
```

```
                LLL = 1;
                XTOM[i] = BTOM[0][KTOM[LOB]];
                YTOM[i] = SSSS;
                if (strcmp(strupr(PARA_NAME[0]),"RATE")== 0)
                        rate_adj[i] = reset_Slope(XTOM,i);

                if (BEST_FLAG)
                {
                 SBEST = SSSS;
                 ABEST = ATOM[LOB];
                 IBEST = i+1;
                 KBEST = K;
                 BST_SLP = XBEST = rate_adj[i] ;
                 ZBEST = YTOM[i];
                 PBMAX_BST = PBMAX;
                 WGT_BEST = WEIGHT;
                }

                if (strcmp(strupr(PARA_NAME[0]),"RATE")== 0)
                  NCON1 = 20;
                if (strcmp(strupr(GRAPHICS),"ON") != 0)
                {
                 gotoxy(20,8);
                 printf ("THIS IS A LINEAR SEARCH");
                 gotoxy(20,10);
                 printf ("THERE ARE %ld ITERATIONS FOR %s.",NCON1+1,
                         strupr(PARA_NAME[0]));
                 gotoxy(20,12);
                 printf ("THIS IS THE %ld TH STEP. ", i+1);
                 gotoxy(20,14);
                 printf ("THE MSR IN THIS STEP IS %6.2lf. ",YTOM[i]);
                 gotoxy(15,18);
printf("***********************************************");
                }
                }
                else
              do
                {
                   BEST_FLAG = 0;
                   if (strcmp(strupr(PARA_NAME[1]),"RATE")== 0)
                     ATOM[1] = chg_R_to_B(j);
                   else
                   ATOM[1] = (j)/(float)(NCON2);
//                 IITOM = 2;
//                 if (j == 0 && LLL == 0)
//                 if (KTOM[0] >= 31) IITOM = 1;
                   if (i == 0)
                     chk_i_j = 0;
                   else chk_i_j = 1;

                   LEAD(chk_i_j );

//                 LLL = 1;
                   XTOM[i] = BTOM[0][KTOM[0]];
```

```
              YTOM[j] = BTOM[0][KTOM[1]];
              ZZZZ = SSSS;
         if (strcmp(strupr(PARA_NAME[1]),"RATE")== 0)
                   rate_adj[j] = reset_Slope(YTOM,j);

         if (BEST_FLAG)
          {
                   SBEST = SSSS;
                   ABEST = ATOM[0];
                   BBEST = ATOM[1];
                   IBEST = i+1;
                   JBEST = j+1;
                   KBEST = K;
                   XBEST = XTOM[i];
                   BST_SLP = YBEST = rate_adj[j];
                   ZBEST = ZZZZ;
                   PBMAX_BST = PBMAX;
                   WGT_BEST = WEIGHT;
          }

         if (strcmp(strupr(PARA_NAME[1]),"RATE")== 0)
           NCON2 = 20;

         if (strcmp(strupr(GRAPHICS),"ON") != 0)
          {
          gotoxy(20,8);
          printf ("THIS IS A LINEAR SEARCH FOR TWO PARAMETERS.");
          gotoxy(20,10);
          printf ("THERE ARE %ld ITERATIONS FOR %s,",NCON1+1,
                   strupr(PARA_NAME[0]));
          gotoxy(20,12);
          printf ("AND %ld ITERATIONS FOR %s.",NCON2+1,
                   strupr(PARA_NAME[1]));
          gotoxy(20,14);
          printf ("THIS IS THE %ld TH AND %ld STEP. ", i+1, j+1);
          gotoxy(20,16);
          printf ("THE MSR IN THIS STEP IS %6.2lf. ",ZZZZ);
//        gotoxy(20,18);
//        tot = (NCON1*NCON2);
//        printf ("TO TERMINATE PRIOR TO %ld ITERATIONS--TAP A KEY:",tot);
          gotoxy(15,20);
printf("**********************************************");
          }
         if (strcmp(strupr(GRAPHICS),"ON") == 0 && BEST_FLAG)
           draw_updated_data_graph(IJK);

         if (loc_time_min[DEP_LOCO[0]] < DEP_TIM_RAW[1][0])
           {
             if (NUM_CORE_TERM == 1)
               {
                   for (ij = 0; ij < NTIME; ij++)
                   if (DEP_TIM_RAW[1][ij] <  TIME1[loc_time[ij]])
                     continue;
               }
```

```
                    else if (NUM_CORE_TERM == 2)
                       {
                            for (ij = 0; ij < NTIME; ij++)
                             if (DEP_TIM_RAW[1][ij] <  TIME1[loc_time[ij]])
                               continue;
                       }
                    else
                       {
                            for (ij = 0; ij < NTIME; ij++)
                             if (DEP_TIM_RAW[1][ij] <  TIME1[loc_time[ij]])
                               continue;

                       }
                 }

            j_loc_bst[NDAT] = JBEST;
            j++;
          }while (fabs(1-ATOM[1]) > 0.001); //
        if (NCCN2 == 0)
        if (strcmp(strupr(GRAPHICS),"ON") == 0 && BEST_FLAG)
         draw_updated_data_graph(IJK);

//*******************************************************************
//    Maintain the necessary value for the model (e.g. time)
//    for specified number of terms
//*******************************************************************
                 if (NUM_CORE_TERM == 1)
                    {
                     if (ONE_TERM_CHI > CHSQL)
                        {
                          final_1_ndat = NDAT;
                          ONE_TERM_CHI = CHSQL;
                          for (ij = 0; ij < N; ij++)
                                ONE_TERM[0][ij] = BEST_COEF[ij];

                          ONE_TERM[1][0] =   PBMAX_BST;
                          ONE_TERM[1][1] =   BST_SLP;
                          ONE_TERM[1][2] =   WGT_BEST;

                          for (ij = 0; ij < NDAT; ij++)
                             {
                               ONE_TERM[2][ij] =   TBST[ij];
                               ONE_TERM[3][ij] =   RBST[ij];
                               ONE_TERM[4][ij] =   PBST[ij];
                               ONE_TERM[5][ij] =   DBST[ij];
                               ONE_TERM[6][ij] =   CBST[ij];
                             }
                        }
                    }
                 else if (NUM_CORE_TERM == 2)
                    {
                          if (TWO_TERM_CHI > CHSQL)
                          {
                            final_2_ndat = NDAT;
```

```
                          TWO_TERM_CHI = CHSQL;
                          for (ij = 0; ij < N; ij++)
                                TWO_TERM[0][ij] = BEST_COEF[ij];

                          TWO_TERM[1][0] =   PBMAX_BST;
                          TWO_TERM[1][1] =   BST_SLP;
                          TWO_TERM[1][2] =   WGT_BEST;

                          for (ij = 0; ij < NDAT; ij++)
                                {
                                    TWO_TERM[2][ij] =   TBST[ij];
                                    TWO_TERM[3][ij] =   RBST[ij];
                                    TWO_TERM[4][ij] =   PBST[ij];
                                    TWO_TERM[5][ij] =   DBST[ij];
                                    TWO_TERM[6][ij] =   CBST[ij];
                                }
                        }
                    }
            else
                    {
                    THREE_CHI[NDAT-MIN_NUM_3_TERMS] = CHSQL;
                    if (THREE_TERM_CHI > CHSQL)
                        {
                            final_3_ndat = NDAT;
                            THREE_TERM_CHI = CHSQL;
                            for (ij = 0; ij < N; ij++)
                                THREE_TERM[0][ij] = BEST_COEF[ij];

                            THREE_TERM[1][0] =   PBMAX_BST;
                            THREE_TERM[1][1] =   BST_SLP;
                            THREE_TERM[1][2] =   WGT_BEST;

                            for (ij = 0; ij < NDAT; ij++)
                                {
                                    THREE_TERM[2][ij] =   TBST[ij];
                                    THREE_TERM[3][ij] =   RBST[ij];
                                    THREE_TERM[4][ij] =   PBST[ij];
                                    THREE_TERM[5][ij] =   DBST[ij];
                                    THREE_TERM[6][ij] =   CBST[ij];
                                }
                        }
                    }
//**************************************************************************
//**************************************************************************
            j = 0;
            i++;

            if (kbhit())
              {
                    int chk_key = getch();
                    if (!chk_key)
                      chk_key = getch();
                    if (chk_key == '\x1b')
                        {
```

```
                                    ATOM[0] = 1;
                                    NDAT = NDAT_INIT+1;
                                    NUM_CORE_TERM = MAX_TERM;
                                }
                        }

                } while (fabs(1-ATOM[0]) > 0.001);
            if (abs(NDAT-NDAT_INIT+1) < CNT_NDAT)
//**********************************************************************
//   end of the linear search
//**********************************************************************
                CNT_NDAT = NDAT_INIT - NDAT;
            if (CNT_NDAT <=0)
                    CNT_NDAT = NDAT_INIT;

        }// end NDAT FOR loop
        } // END FOR LOOP ""
//**********************************************************************
//   FIND THE BEST
//**********************************************************************
  for (i = 0; i < NDAT_INIT+1; i++)
    TBST[i] = RBST[i] = PBST[i] = DBST[i] = 0.0;

  CHSQL = 1.0e5;

    if ((ONE_TERM_CHI < TWO_TERM_CHI) && (ONE_TERM_CHI < THREE_TERM_CHI))
    {
            NDAT = NDAT_INIT = final_1_ndat;
            for (j = 0; j < N; j++)
                    BEST_COEF[j] = ONE_TERM[0][j]; //ONE_TERM[i][0][j];

            PBMAX_BST = ONE_TERM[1][0];  //ONE_TERM[i][1][0];
            BST_SLP   = ONE_TERM[1][1];  //ONE_TERM[i][2][0];
            WGT_BEST = ONE_TERM[1][2];   //ONE_TERM[i][3][0];

            for (j = 0; j < NDAT_INIT; j++)
                    {
                      TBST[j] = ONE_TERM[2][j];   //ONE_TERM[i][4][j];
                      RBST[j] = ONE_TERM[3][j];   //ONE_TERM[i][5][j];
                      PBST[j] = ONE_TERM[4][j];   //ONE_TERM[i][6][j];
                      DBST[j] = ONE_TERM[5][j];   //ONE_TERM[i][7][j];
                      CBST[j] = ONE_TERM[6][j];   //ONE_TERM[i][8][j];
                    }
            if (NNN == 1)
                    if (strcmp(strupr(PARA_NAME[i]),"SURFACE")== 0)
                      XBEST = PBMAX_BST;
                    else
                      XBEST = BST_SLP;
            if (NNN == 2)
                    {
                      XBEST = PBMAX_BST;
                      YBEST = BST_SLP;
                    }
    }
```

```
else if ((TWO_TERM_CHI < ONE_TERM_CHI) && (TWO_TERM_CHI < THREE_TERM_CHI))
{
                NDAT = NDAT_INIT = final_2_ndat;
                for (j = 0; j < N; j++)
                  BEST_COEF[j] = TWO_TERM[0][j];

                PBMAX_BST = TWO_TERM[1][0];
                BST_SLP   = TWO_TERM[1][1];
                WGT_BEST = TWO_TERM[1][2];

                for (j = 0; j < NDAT_INIT; j++)
                   {
                           TBST[j] = TWO_TERM[2][j];
                           RBST[j] = TWO_TERM[3][j];
                           PBST[j] = TWO_TERM[4][j];
                           DBST[j] = TWO_TERM[5][j];
                           CBST[j] = TWO_TERM[6][j];
                   }
         if (NNN == 1)
                 if (strcmp(strupr(PARA_NAME[0]),"SURFACE")== 0)
                         XBEST = PBMAX_BST;
                 else
                         XBEST = BST_SLP;
             if (NNN == 2)
                 {
                         XBEST = PBMAX_BST;
                         YBEST = BST_SLP;
                 }
}
else
{
                NDAT = NDAT_INIT = final_3_ndat;
                for (j = 0; j < N; j++)
                  BEST_COEF[j] = THREE_TERM[0][j];

                PBMAX_BST = THREE_TERM[1][0];
                BST_SLP   = THREE_TERM[1][1];
                WGT_BEST = THREE_TERM[1][2];

                for (j = 0; j < NDAT_INIT; j++)
                   {
                           TBST[j] = THREE_TERM[2][j];
                           RBST[j] = THREE_TERM[3][j];
                           PBST[j] = THREE_TERM[4][j];
                           DBST[j] = THREE_TERM[5][j];
                           CBST[j] = THREE_TERM[6][j];
                   }

         if (NNN == 1)
                 if (strcmp(strupr(PARA_NAME[0]),"SURFACE")== 0)
                         XBEST = PBMAX_BST;
                 else
                         XBEST = BST_SLP;
```

```
                  if (NNN == 2)
                    {
                            XBEST = PBMAX_BST;
                            YBEST = BST_SLP;
                    }
   }


//***********************************************************************
//***********************************************************************
//***********************************************************************

        P[0] = P3ST[0];

//***********************************************************************
// Print the best model to a file.
//***********************************************************************
    CHI = 0;
    for (j = 0; j < NDAT; j++)
     CHI += pow(P3ST[j] - P[j], 2);
    BST_CHI = CHI /=NDAT;

    fprintf(out1, "\nTHE BEST FIT IS GIVEN BY: \n\n");

    fprintf (out1, "    A(n)          B(n)\n");
    for (k = 0; k < K; k++)
      fprintf (out1, "%10.5lf %10.5lf\n",BEST_COEF[k],BEST_COEF[k+K]);
    fprintf (out1, "\n");

    fprintf (out1, "THE CHI-SQUARED FOR DATA TO MODEL IS %6.4lf.\n",CHI);

    if (END_CORE_TIME > 1000)
      {
              for (i = 0; i < NDAT; i++)
               {
                      TBST[i] /= 1000;
                      RBST[i] *= 1000;
               }
              fprintf (out1, "NOTE:  The time is listed as k-yrs.\n");
      }

    SET_CORE_DATE();
    if (NNN == 1)
    {
     fprintf (out1, '\nTHE PARAMETER ADJUSTED: %s.\n\n", PARA_NAME[0]);
     fprintf (out1, '%s = %5.4lf WITH AN M.S.R. = %5.4lf\n\n",PARA_NAME[0],
                                                 XBEST,BST_CHI);
    }
    else
    {
     fprintf (out1, '\nThe Parameters adjusted: %s and %s.\n\n",
                     PARA_NAME[0],PARA_NAME[1]);
     fprintf (out1, '%s = %5.4lf AND %s = %5.4lf WITH AN M.S.R.= %5.4lf\n\n",
                          PARA_NAME[0],XBEST,PARA_NAME[1],YBEST,BST_CHI);
    }
```

```
    ATOM[LOB] = ABEST;

    if (NNN != 1)
    ATOM[1] = BBEST;

    fprintf(out1,"\n  DEPTH  AGE TIME RATE SRCS DATA MODEL\n");
    for (i = 0; i < NDAT; i++)
     fprintf (out1, "%6.2lf %6.2lf %6.2lf %6.4lf %6.2lf %7.2lf %8.2lf\n",
             SZV[i], CORE_DATE[i], TBST[i], RBST[i], DBST[i], P[i], PBST[i]);

    write_xyz_file_w();
    fcloseall();

    draw_data_graph();
    grph_fin_vals(NNN);

    delay(10000);
    closegraph();
    restorecrtmode();

    exit(0);
}


//*******************************************************
//*******************************************************
//****M*********M*****A******IIIIIIIII**N******N*****
//****MM*******MM****A*A********I******NN*****N*****
//****M*M*****M*M***A***A*******I*****N*N****N*****
//****M**M***M**M**A*****A******I*****N**N***N*****
//****M***M*M***M**A******A*****I*****N***N**N*****
//****M***M****M**AAAAAAAA*****I*****N***N**N*****
//****M*********M**A******A*****I*****N****N*N*****
//***M*********M**A*******A*****I*****N*****NN****
//****M*********M**A*******A**IIIIIIIII**N******N*****
//*******************************************************
//*******************************************************
void main()
{
//*******************************************************************
// This section of code reads the configuration file which should have two
// information sections.
// First -- the name in the configuration is the last data file used
// Second -- the number indicates an alpha/numeric code which will be
// placed in the first character location of the output file.
//      1 represents "a"; 2 represents "b"; ... 34 represents "8";
//      35 represents "9"; and 36 represents "0".
// Therefore the limits on the number of runs (different model settings)
// completed are 1 through 36.
//*******************************************************************


// out_tmp =  fopen("testcoef.out","w");
  int dec, sign;
  char *tmpstr, str_grph[80];
```

```
//*******************************************************************
//          Check the configuration file for the last data file used
//*******************************************************************

  CFG = fopen(CFG_FILE, "r");

  if (CFG == NULL)
          {
          fclose(CFG);
          CFG = fopen(CFG_FILE,"w");
          fl_incr = 1;
          FILE_FIND_FIRST();
          fclose(CFG);
          }

  fgets(tempstr, 30,CFG);

  if (strlen(tempstr) == 0)
          {
          fl_incr = 1;
          FILE_FIND_FIRST();
          fclose(CFG);
          }
  else
          {
          sscanf(tempstr,"%s %ld", DATFL, &fl_incr);
          fclose (CFG);
          }

  CFG = fopen(CFG_FILE, "w");

  CHSQL = 1.E5;

  clrscr();
  UPDATE_DATA_FILE();

  if (fl_incr >36 || fl_incr < 1)
          fl_incr = 0;

  fprintf(CFG, "%s %ld", DATFL, fl_incr+1);
  fclose (CFG);

//*******************************************************************
//*******************************************************************
// This section of code sets up the output file with an alpha/numeric
// value in the first location and set the file extension to OUT.
//*******************************************************************

  ptr = strchr(DATFL,point);
  if (ptr == NULL)
          {
          OUTFL[0] = alpha_num[fl_incr-1];
          for (j = 0; j < strlen(DATFL); j++)
          OUTFL[j+1] = DATFL[j];
```

```
            strcpy(XYZFL,OUTFL);
            strcpy(linear,OUTFL);
            strcpy(probfile,OUTFL);

            strcat(XYZFL,XYZ2);
            strcat(OUTFL,OUT2);
            }
    else
            {
            OUTFL[0] = alpha_num[fl_incr-1];
            for (j = 0; j < ptr-DATFL+1; j++)
                    OUTFL[j+1] = DATFL[j];

            strcpy(XYZFL,OUTFL);
            strcpy(linear,OUTFL);
            strcpy(probfile,OUTFL);

            strcat(XYZFL,XYZ);
            strcat(OUTFL,OUT);
            }
//********************************************************************

    in2 = fopen(PARAFILE, "r");
    if (in2 == NULL)
            {
            printf ("\n\a\a\a\a\a\a Parameter file was not found in current
                                                             directory");
            exit(0);
            }

    out1 = fopen (OUTFL, "w");

    clrscr();

    _setcursortype(_NOCURSOR);

    gotoxy(15,1);
    printf("****************************************************");
    gotoxy(15,3);
    printf("          Program is running, please wait.          ");
    gotoxy(15,5);
    printf("****************************************************");

    input();
    NDAT++;


    j = 0;
    ZMAX = ZN[NDAT-1];
    PBMIN = P[NDAT-1];
    for (i=0; i < NDAT; i++)
            Z[i] = ZN[i]/ZMAX;
    for (i=1; i < NDAT; i++)
```

```
          {
            SZV[i] = Z[i]*ZMAX*100;   // getting the depth back
            if (fabs(SZV[i] - DEP_TIM_RAW[0][j]) < 0.0001)
                  loc_time[j++] = i;
          }

if (real_syn_data == 2)
        {
          AVG_ERR_DATA=0;
          for (i=1; i< NDAT; i++)
                AVG_ERR_DATA += VAR_DATA[i]/P[i];
          AVG_ERR_DATA /= NDAT;

          for (i=1; i < NDAT; i++)
                VAR_DATA[i] = AVG_ERR_DATA*P[i];
        }

if (strcmp(SPLINE, "SPLINE") == 0)
        B_SPLINE();


for (i = 0; i < NDAT; i++)
        DLTAZ[i+1] = Z[i+1]-Z[i];


SET_BTOM();
NORMALIZE();
NUM_PARA = NNN = i;
if (NNN == 0)
        {
          KTOM[0] = 0;
          KTOM[1] = 1;
        }


DENOMO (0,&PBMAX);
DENOMO (1,&SLOPE);
DENOMO (2,&WEIGHT);

if (TERMS != 10)
        {
          K = TERMS;
          int_K = K;
          N = 2*K;

          if (MTCM != 5 && MTOM != 6)
            {
                  ALLOC_int_K_DATA();
                  calc_E_O();
            }
        }
else K = 1;

loc_chi_time = 100000;
```

```
for (i = 0; i < NNN; i++)

        switch (KTOM[i])
        {
          case 0: {  strcpy(PARA_NAME[i],"Surface"); QN_LOCO[i] = 0; break;}
          case 1: {  strcpy(PARA_NAME[i],"RATE"); QN_LOCO[i] = 1;   break;}
          case 2: {  strcpy(PARA_NAME[i],"Weight"); QN_LOCO[i] = 2; break;}
          default:  printf("Bug!\n");
        }

  if ((MTOM == 0 || MTOM == 3 || MTOM == 4) && K != 0)
        {
          if (NNN > 2)
            {
            gotoxy(15,10);
            printf("\a\a\a***** TOO MANY PARAMETERS FOR LINEAR SEARCH *****");
            gotoxy(15,12);
            printf("************* CHECK THE PARAMETER FILE ************");
            fprintf(out1,"\n*** TOO MANY PARAMETERS FOR LINEAR SEARCH ***\n");
            fcloseall();
            exit(0);
             }
          if (MTOM == 3)
           LIN_SRCH_CORE();
          else
           LINEAR_SEARCH();
        }

  if (MTOM == 5 || MTOM == 6)
    Current_Models();

  IJK = 1;
  set_min_flag = LLL = 0;


  if (K == 0 && NNN != 0)
        CONST_SEDM_LEAD();
//  else if (strcmp(BREAK_CORE,"CUT_CORE") == 0)
//          CORE();


  if (NNN)
          {
          fprintf (out1, "\nTHE PARAMETER(S) ADJUSTED: ");
          for (i = 0; i < NNN; i++)
            fprintf(out1, "%s ", strupr(PARA_NAME[i]));
          fprintf (out1, "\n\n");
           }

  _setcursortype(_NORMALCURSOR);
  fprintf (out1,"THE BEST PARAMETERS ARE : \n\n");

  CHI = 0;
  for (j = 0; j < NDAT; j++)
```

```
        {
          CHI += pow((PBST[j] - P[j])/P[j], 2);
          FLUX[j] = (PBST[j]*exp(FLAMD*TBST[j]))*FLUX_COEF[j]*RBST[j];
        }

CHI /=NDAT;

if (K == 0)
        fprintf (out1,"CHI-SQUARED = %12.4lf WHICH WAS OBTAINED AT %dTH
                                        ITERATION\n",BST_CHI,IJKBST);
else
    {
      fprintf (out1, "THE CHI-SQUARED FOR DATA TO MODEL IS %6.4lf.\n",CHI);
      fprintf (out1,"CHI-SQUARED = %12.4lf WHICH WAS OBTAINED AT %dTH
                                        ITERATION\n",BST_CHI,IJKBST);
      fprintf (out1, "\n   A(n)  B(n)\n");
      for (i = 0; i < K; i++)
        fprintf (out1, "%12.4lf  %12.4lf\n", BEST_COEF[i], BEST_COEF[i+K]);
        fprintf(out1, "\n");
    }

fprintf (out1, "\nPBMAX  = %12.4lf\n", PBMAX_BST);
fprintf (out1, "SLOPE  = %12.4lf\n", BST_SLP);
fprintf (out1, "WEIGHT = %12.4lf\n\n", WGT_BEST);

if (END_CORE_TIME > 1000)
        {
                for (i = 0; i < NDAT; i++)

                {
                        TBST[i] /= 1000;
                        RBST[i] *= 1000;
                }
                fprintf (out1, "NOTE: The time is listed as k-yrs.\n");
        }

SET_CORE_DATE();

if (strcmp(SPLINE, "SPLINE") != 0)
    {
      fprintf(out1,'\n DEPTH AGE TIME RATE SRCS FLUX DATA MODEL\n");
      for (i = 0; i < NDAT; i++)
       fprintf (out1, "%6.2lf %6.2lf %6.2lf %6.4lf %6.2lf %6.2lf %7.3lf
                                        %8.3lf\n",
         SZV[i],CORE_DATE[i],TBST[i],RBST[i],DBST[i],FLUX[i],P[i].PBST[i]);
    }
else
    {
      Pre_SPLINE_P[0] = P[0];
      fprintf(out1,'\n ORG.  SPLINE");
      fprintf(out1,'\n DEPTH AGE TIME RATE SRCS FLUX DATA DATA MODEL\n");
      for (i = 0; i < NDAT; i++)
       fprintf (out1, "%6.2lf %6.2lf %6.2lf %6.4lf %6.2lf %6.2lf
                                        %7.3lf %7.3lf %8.3lf\n",
```

```
            SZV[i], CORE_DATE[i], TBST[i], RBST[i], DBST[i], FLUX[i],
                                    Pre_SPLINE_P[i], P[i], PBST[i]);
      }
  write_xyz_file_w();
  fcloseall();

  draw_data_graph();
  grph_fin_vals(NNN);

  delay(10000);
  closegraph();
  restorecrtmode();

  exit(0);

//   FREE_int_K_DATA();
//   free(CHISQM);
//   FREE_HUGE_MATRIX(ANM,int_NMPT+1);
//   FREE_DATA_VARS();
}
//******************************************************************************
```

References

Aller, R. C. (1980a). Quantifying solute distributions on the bioturbated zone of marine sediments by defining an average microenvironment. *Geochim. Cosmochim. Acta, 44*, 1955–1965.

Aller, R. C. (1980b). Relationships of tube-dwelling benthos with sediment and overlying water chemistry. In Tenore, K. R. & Coull, B. C. (Eds.). *Marine Benthic Dynamics* (pp. 285–308). Columbia: University of South Carolina Press.

Aller, R. C. & Cochran, J. K. (1976). ^{234}Th/^{238}U disequilibrium in near-shore sediment: particle reworking and diagenetic time scales. *Earth Planet. Sci. Lett., 29*, 37–50.

Anderson, R. F., Bopp, R. F., Buesseler, K. O. & Biscaye, P. E. (1988). Mixing of particles and organic constituents in sediments from the continental shelf and slope off Cape Cod:SEEP-1 results. *Cont. Shelf Res., 8* (5–7), 925–946.

Appleby, P. G. & Oldfield, F. (1978a). Application of ^{210}Pb to sedimentation studies. In Ivanovich, M. & Harmon, R. S. (Eds.). *Uranium-series Disequilibrium: Applications to Earth, Marine, and Environment Sciences* (2nd edn, 1992). (pp. 730–760). Oxford: Oxford Science Publications.

Appleby, P. G. & Oldfield, F. (1978b). The calculation of Lead-210 dates assuming a constant rate of supply of unsupported ^{210}Pb to the sediment, *Catena, 5*, 1–8.

AREC (Arkhangelsk Regional Environmental Committee) (1998). *1997, Status and Protection of the Environment in Arkhangelsk Region* (88 pp). Arkhangelsk (in Russian).

Balistrieri, L. S. & Murray, J. W. (1984). Marine scavenging: trace metal adsorption by interfacial sediment from MANOP Site H. *Geochemica et Cosmochemica Acta, 48*, 921–929.

Bendat, J. S. & Piersol, A. G. (1971). *Random Data: Analysis and Measurement Procedures*. New York: Wiley Interscience.

Benninger, L. K. (1978). ^{210}Pb balance in Long Island sound. *Geochemica et Cosmochemica Acta, 42*, 1165–1174.

Benninger, L. K., Aller, R. C., Cochran, J. K. & Turekian, K. K. (1979). Effects of biological mixing on the ^{210}Pb chronology and trace metal distribution in a Long Island sound sediment core. *Earth Planetary Sciences Letter, 43*, 241–259.

Berger, W. H. & Heath, G. R. (1968). Vertical mixing in pelagic sediments. *J. Mar. Res., 26*, 134–143.

Berner, R. A. (1980). *Early Diagenesis: A Theoretical Approach*. (241 pp.). Princeton: Princeton University Press.

Boehm, P. D., Page, D. S., Gilfillan, E. S., Bence, A. E., Burns, W. A. & Mankiewicz, P. J. (1998). Study of the fates and effects of the Exxon Valdez oil spill on benthic sediments in two bays in Prince William Sound, Alaska. I. Study design, chemistry, and source fingerprinting. *Environ. Sci. Technol., 32*, 567–576.

Boudreau, B. P. (1986). Mathematics of tracer mixing in sediments. I. Spatially-dependent, diffusive mixing (Modele mathematique du mixage des traceurs dans les sediments. I. Mixage par diffusion avec variation spatiale). *American Journal of Science, 286* (3), 161–198.

Boudreau, B. P. (1986). Mathematics of tracer mixing in sediments. II. Non-local mixing and biological conveyor-belt phenomena. *American Journal of Science, 286*, 199–238.

Boudreau, B. P. (1998). Mean mixed depth of sediment: the wherefore and the why. *Limnology and Oceanography, 43* (3), 524–526.

Boudreau, B. P. & Imboden, D. M. (1987). Mathematics of tracer mixing in sediments. III. The theory of nonlocal mixing within sediments (Les calculs mathematiques du melange des traceurs dans les sediments. III La theorie du melange non local dans les sediments). *American Journal of Science, 287* (7), 693–719.

Broecker, W. S. & Peng, T. H. (1982). *Tracers in the Sea*. Palisades, NY: Lamont Doherty Geological Observatory, Columbia University.

Burke, W. H., Denison, R. E., Hertherington, E. A., Koepnick, R. B., Nelson, H. F. & Otto, J. B. (1982). Variation of seawater $^{87}Sr/^{86}Sr$ throughout Phanerozoic time. *Geology, 10,* 516–519.

Buscail, R., Ambatsian, P., Monaco, A. & Bernat, M. (1997). Pb-210, manganese and carbon: Indicators of focusing processes on the northwestern Mediterranean continental margin. *Marine Geology, 137,* 271–286.

Calvert, S. E. (1987). Oceanographic controls on the accumulation of organic matter in marine sediments. In Brooks, J. & Fleet, A. J. (Eds.). *Marine Petroleum Source Rocks* (pp. 137–151). Geological Society Special Publication, no. 26.

Calvert, S. E., et al. (1991). Low organic carbon accumulation in Black Sea sediments. *Nature, 350,* 692–695.

Carroll, J., Boisson, F., Teyssie, J.-L., King, S. E., Krosshavn, M., Carroll, M. L., Fowler, S. W., Povinec, P. P. & Baxter, M. S. (1999). Distribution coefficients for use in risk assessment models of the Kara Sea. *Applied Radiation and Isotopes, 51,* 121–129.

Carroll, J., Boisson, F., Fowler, S. W. & Teyssie, J.-L. (1999). Radionuclide adsorption to sediments from nuclear waste dumping sites in the Kara Sea. *Marine Pollution Bulletin, 35,* 296–304.

Carroll, J. & Harms, I. H. (1999). Uncertainty analysis of partition coefficients in a radionuclide transport model. *Water Research, 33,* 2617–2626.

Carroll, J. & Lerche, I. (1990). A cautionary note on the use of $^{87}Sr/^{86}Sr$ sediment ages in stratigraphy studies. *Nucl. Geophys., 4,* 461–466.

Carroll, J., Lerche, I., Abraham, J. A. & Cisar, D. J. (1995). Model-determined sediment ages from depth profiles of radioisotopes: Theory and examples using synthetic data. *Nucl. Geophys., 9,* 553–565.

Carroll, J., Lerche, I., Abraham, J. D. & Cisar, D. J. (1999). Sediment age and flux variations from depth profiles of radioisotopes: Lake and marine examples. *Appl. Rad. Isotopes, 50,* 793–804.

Chanton, J. P., Martens, C. S. & Kipphut, G. W. (1989). Lead-210 sediment geochronology in a changing coastal environment. *Geochim. Cosmochim. Acta, 47,* 1791–1804.

Christensen, E. R. (1982). A model for radionuclides in sediments influenced by mixing and compaction. *J. Geophys. Res., 87,* 566–572.

Cochran, J. K. (1982). The oceanic chemistry of the U- and Th-series nuclides. In Ivanovich, M. & Harmon, R. S. (Eds.). *Uranium Series Disequilibrium: Applications to Earth, Marine and Environmental Problems* (571 pp.). Oxford: Clarendon Press.

Cooper, L. W., Larsen, I. L., Beasley, T. M., Dolvin, S. S., Grebmeier, J. M., Kelley, J. M., Scott, M. & Johnson-Pyrtle, A. (1998). The distribution of Radiocesium and Plutonium in Sea ice-entrained Arctic sediments in relation to potential sources and sinks. *Journal of Environmental Radioactivity, 39,* 279–303.

Crusius, J. & Anderson, R. F. (1995). Sediment focussing in six small lakes inferred from radionuclide profiles. *Journal of Paleolimnology, 13,* 143–155.

Cutshall, N. H., Larsen, I. L. & Olsen, C. R. (1983). Direct analysis of ^{210}Pb in sediment samples: self adsorption corrections. *Nuclear Instruments & Methods, 206,* 309–312.

Dellapenna, T. M., Kuehl, S. A. & Schaffner, L. C. (1998). Sea-bed mixing and particle residence times in biologically and physically dominated estuarine systems: a comparison of lower Chesapeake Bay and the York River subestuary. *Estuarine, Coastal and Shelf Science, 46,* 777–795.

Demaison, G. J. & Moore, G. T. (1980). Anoxic environments and oil source bed genesis. *American Association of Petroleum Geologists Bulletin, 64,* 1179–1209.

DeMaster, D. J. & Cochran, J. K. (1982). Particle mixing rates in deep-sea sediments determined from excess ^{210}Pb and ^{32}Si profiles. *Earth Planet. Sci. Lett., 61,* 257–271.

DeMaster, D. J., Kuehl, S. A. & Nittrouer, C. A. (1986). Effects of suspended sediments on geochemical processes near the mouth of the Amazon River: examination of biological silica uptake and the fate of particle-reactive elements. *Cont. Shelf Res., 6,* 107–125.

DePaolo, D. J. (1986). Detailed record of the Neogene Sr isotopic evolution of seawater from DSDP Site 590B. *Geology, 14,* 103–106.

DePaolo, D. J. & Ingram, B. L. (1985). High-resolution stratigraphy with strontium isotopes. *Oceanography, 42,* 1517–1529; *Science, 227,* 938–940.

Dickin, A. P. (1997). *Radiogenic Isotope Geology* (490 pp.). Cambridge: Cambridge University Press.

Dobrovolsky, A. D. & Zalogin, B. S. (1982). *The Seas of USSR* (192 pp.). Moscow State University Publishing House (in Russian).

Dukat, D. A. (1993). Investigation of non-steady state ^{210}Pb flux and the use of $^{228}Ra/^{226}Ra$ as a geochronological tool on the Amazon continental shelf. Masters Thesis. University of South Carolina.

Duursma, E. K. & Carroll, J. (1996). *Environmental Compartments* (277 pp.). Berlin: Springer Verlag.

Edgington, D. N., Klump, J. Val., Robbins, J. A., Kusner, Y. S., Pampurav, D. & Saudimirov, I. V. (1991). Sedimentation rates, residence times and radionuclide inventories in Lake Baikal from ^{137}Cs and ^{210}Pb in sediment cores. *Nature, 350,* 601–604.

Feller, W. (1957). *An Introduction to Probability Theory and its Applications*, Vol. 1 (461 pp.). New York: John Wiley and Sons.

Figueiredo, A. G., Gamboa, L. A. P., Gorini, M. & Costa Alves, E. (1972). Naturaza da sedimentacao atual do Rio Amazonas – testemunhos e geomorfologia submarina (canyn) Amazonas testemunhos submarinos. *Congresso Brasilliero de Geologia* (26), 2.51–2.56.

Fuglseth, T. J. (1991). Organic carbon preservation in deep-sea environments: a comparison between the Sulu and South China seas. M.S. Thesis. University of South Carolina.

Gardner, L. R., Sharma, P. & Moore, W. S. (1987). A regeneration model for the effect of bioturbation by fiddler crabs on ^{210}Pb profiles in salt marsh sediments. *J. Environ. Radioactivity, 5,* 25–36.

Gibbs, R. J. (1967). The geochemistry of the Amazon river system: Part I. The factors that control the salinity and the composition and concentration of the suspended solids. *Geol. Soc. Am. Bull., 78,* 1203–1232.

Girkurov, G. E. (1993). Assessment of marine contamination in the Eurasian Arctic shelf by NFO Sevmorgeologia. *Proceedings of the Workshop on Arctic Contamination, 2/7 May 1993, Anchorage, Alaska* (pp. 246–256). Interagency Arctic Research Policy Committee, Arctic Research of the United States.

Goldberg, E. D. (1963). Geochronology with ^{210}Pb. In *Radioactive Dating* (pp. 121–131). Vienna: IAEA.

Goldberg, E. D. & Bruland, K. (1974). Radioactive geochronologies. In Goldberg, E. D. (Ed.). *The Sea, Vol. 5: Marine Chemistry.* New York: Wiley-Interscience.

Goldberg, E. D., Hodge, V., Koide, M., Griffin, J., Bricker, O. P., Matisoff, G., Holdren, G. R. & Braun, R. (1978). A pollution history of Chesapeake bay. *Geochim. Cosmochim. Acta, 42,* 1413–1425.

Goldberg, E. D. & Koide, M. (1962). Geochronological studies of deep-sea sediments by the Io/Th method. *Geochim. Cosmochim. Acta, 26,* 417–450.

Goreau, T. J. (1977). Quantitative effects of sediment mixing on stratigraphy and biogeochemistry: signal theory approach. *Nature, 265,* 525–526.

Gradshteyn, I. S. & Ryzhik, I. M. (1965). *Table of integrals, Series and Products* (1086 pp.). New York: Academic Press.

Gradstein, F. M. (1993). Presentation at a meeting at Saga Petroleum. Oslo.

Gradstein, F. M., Agterberg, F. P., Aubry, M.-P., Berggren, W. A., Flynn, J. J., Hewitt, R., Kent, D. V., Klitgord, K. D., Miller, K. G., Obradovich, J. D., Ogg, J. G., Prothero, D. R. & Westermann, G. E. G. (1998). Sea level history. *Science, 214,* 599–601.

Guinasso, N. L. J. & Schink, D. R. (1975). Quantitative estimates of biological mixing rates in Abyssal sediments. *Journal of Geophysical Research, 80,* 3032–3043.

Hahn, C. L. (1955). Reservoir sedimentation in Ohio (87 pp.). Ohio Dept. Nat. Res. Div. Water Bull. No. 24.

Hamilton, T. F., Ballestra, S., Baxter, M. S., Gastaud, J., Osvath, I., Parsi, P., Povinec, P. P. & Scott, E. M. (1994). Radiometric investigations of Kara Sea sediments and preliminary radiological assessment related to dumping of radioactive wastes in the Arctic seas. *Journal of Environmental Radioactivity, 25,* 113–134.

Hancock, G. J. & Hunter, J. R. (1999). Use of excess ^{210}Pb and ^{228}Th to estimate rates of sediment accumulation and bioturbation in Port Phillip Bay, Australia. *Marine and Freshwater Research, 50* (6), 533–545.

Hansen, J. R., Hansson, R. & Norris, S. (1996). The state of the European Arctic environment. European Environment Agency. EEA Environmental Monograph No. 3 (135 pp.). Oslo, Norway: Norwegian Polar Institute Meddelelser No. 141.

Harland, W. B., Armstrong, R. L., Cox, A. V., Craig, L. E., Smith, A. G. & Smith, D. G. (1990). *A Geological Time Scale 1989* (265 pp.). Cambridge: Cambridge University Press.

Harms, I. H. & Karcher, M. J. (1999). Modelling the seasonal variability of circulation and hydrography in the Kara Sea. *Journal of Geophysical Research, 104* (C6), 13 431–13 488.

Henderson, G. M., Lindsay, F. N. & Slowey, N. C. (1999). Variation in bioturbation with water depth on marine slopes: a study on the Little Bahamas Bank. *Marine Geology, 160* (1–2), 105–118.

Hess, J., Berder, M. L. & Schilling, J. G. (1986). Evolution of the ratio of strontium-87 to strontium-86 in seawater from Cretaceous to present. *Science, 231,* 979–984.

Hewitt, R., Kent, D. V., Klitgord, K. D., Miller, K. G., Obradovich, J. D., Ogg, J. G., Prothero, D. R. & Westermann, G. E. G. (1988). Sea level history. *Science, 214,* 599–601.

Hollander, D. J. & McKenzie, J. M. (1991). CO_2 control on carbon-isotope fractionation during aqueous photosynthesis: A paleo-pCO_2 barometer. *Geology, 19*, 929–932.

Honeyman, B. D., Baliestrieri, L. S. & Murray, J. W. (1988). Oceanic trace metal scavenging: the importance of particle concentration. *Deep-Sea Research, 35*, 227–246.

Hurst, R. W. (1986). Strontium isotopic chronostratigraphy and correlation of the Miocene Monterey Formation in the Ventura and Santa Maria basins of California. *Geology, 14*, 459–462.

IAEA (1985). Sediment Kds and concentration factors for radionuclides in the marine environment. IAEA Tech. Report No. 247 (73 pp.). Vienna: IAEA.

IARC (1987). *IARC Monographs on the Evaluation of the Carcinogenic Risk of Chemicals to Humans*. Overall Evaluation of Carcinogenicity: An Updating of IAPC Monographs. Vols. 1–42 (Suppl. 7). Lyon, France: International Agency for Research on Cancer.

IASAP (1998). Radiological conditions of the Western Kara Sea: Assessment of the radiological impact of the dumping of radioactive waste in the Arctic Seas. Report on the International Arctic Seas Assessment Project (IASAP) (124 pp.). Vienna: IAEA.

Klump, J. V. & Martens, C. S. (1981). Biogeochemical cycling in an organic rich coastal marine basin. 2. Nutrient sediment–water exchange processes. *Geochim. Cosmochim. Acta, 43*, 101–121.

Koide, M., Bruland, K. W. & Goldberg, E. D. (1973). Th-228/Th-232 and Pb-210 geochronologies in marine and lake sediments. *Geochim. Cosmochim. Acta, 37*, 1171–1187.

Koide, M., Soutar, A. & Goldberg, E. D. (1972). Marine geochronology with [210]Pb. *Earth Planet. Sci. Lett., 14*, 442–446.

Krishnaswami, S., Benninger, L. K., Aller, R. C. & Von Damm, K. L. (1980). Atmospherically derived radionuclides as tracers of sediment mixing and accumulation in near shore marine and lake sediments; Evidence for [7]Be, ^{210}Pb and [239, 240]Pu. *Earth Planet. Sci. Lett., 47*, 307–318.

Krishnaswami, S., Lal, D., Martin, J. M. & Meybeck, M. (1971). Geochronology of lake sediments. *Earth Planet. Sci. Lett., 11*, 407–414.

Kuehl, S. A. (1985). Sediment accumulation and the formation of sedimentary structure on the Amazon continental shelf. Ph.D. Thesis (213 pp.). North Carolina State University.

Kuehl, S. A., DeMaster, D. J. & Nittrouer, C. A. (1986). Nature of sediment accumulation on the Amazon continental shelf. *J. Cont. Shelf Res., 6*, 209–225.

Kuehl, S. A., Fuglseth, T. J. & Thunell, R. C. (1993). Sediment mixing and accumulation rates in the Sulu and South China Seas: Implications for organic carbon preservation in deep-sea sediments. *Marine Geology, 111*, 15–35.

Kuehl, S. A., Hariu, T. M. & Moore, W. S. (1989). Shelf Sedimentation off the Ganges-Brahmaputra river system: Evidence for sediment bypassing the Bengal Fan. *Geology, 17*, 1132–1135.

Lauerman, L. M. L., Smoak, J. M., Shaw, T. J., Moore, W. S. & Smith Jr., K. L. (1997). [234]Th and ^{210}Pb evidence for rapid ingestion of settling particles by mobile epibenthic megafauna in the abyssal NE Pacific. *Limnol. Oceanogr., 42*, 589–595.

Lerche, I. (1992). *Oil Exploration: Basin Analysis and Economics* (178 pp.). San Diego: Academic Press.

Liu, J., Carroll, J. & Lerche, I. (1991). A technique for disentangling temporal source and sediment variations from radioactive isotope measurements with depth. *Nucl. Geophys., 5*, 31–45.

Martens, C. S. & Klump, J. V. (1980). Biogeochemical cycling in an organic-rich coastal marine basin. I. Methane sediment–water exchange processes. *Geochim. Cosmochim. Acta, 44*, 471–490.

Martin, P. & Grachev, M. (1994). Lake Baikal. *Arch. Hydrobiol. Beih. Ergebn. Limnol. Stuttgart, 44*, 1–9.

Martin, P., Goddeeris, B. & Martens, K. (1993). Sediment oxygen distribution in ancient lakes. *Verh. Internat. Verein. Limnol. Stuttgart, 25*, 793–794.

Martinson, D. G., Menke, W. & Stoffa, P. (1982). An inverse approach to signal correlation. *J. Geophys. Res., 87*, 4807–4818.

Matisoff, G. (1982). Mathematical models of bioturbation. In McCall, P. L. & Tevesz, M. J. (Eds.). *Animal-Sediment Relations, The Biogenic Alteration of Sediments* (pp. 289–330). New York: Plenum.

Matisoff, G. (1994). Private communication.

Matthews, R. K. (1984). *Dynamic Stratigraphy: An Introduction to Sedimentation and Stratigraphy* (267 pp.). Englewood Cliffs, NJ: Prentice-Hall.

McCall, P. L., Robbins, J. A. & Matisoff, G. (1984). [137]Cs and ^{210}Pb transport and geochronologies in urbanized reservoirs with rapidly increasing sedimentation rates. *Chemical Geology, 44*, 33–65.

McClimans, T. A., Johnson, D. R., Krosshavn, M., King, S. E., Carroll, J. & Grenness, Ø. (2000). Transport processes in the Kara Sea. *Journal of Geophysical Research, 105*, 14 121–14 139.

McElroy, A. E., Farrington, J. W. & Teal, J. M. (1989). Bioavailability of polycyclic aromatic hydrocarbons in the aquatic environment. In Varanasi, U. (Ed.). *Metabolism of Polycyclic Aromatic Hydrocarbons in the Aquatic Environment* (pp. 1–39). Boca Raton, FL: CRC Press Inc.

Meade, R. H. (1985). Suspended sediment in the Amazon River and its Tributaries in Brazil during 1982–84. U.S. Geological Survey Open-File Report 85-492.

Meade, R. H., Dunne, T., Richey, J. E., Santos, V. D. & Salati, E. (1985). Storage and remobilization of suspended sediment in the lower Amazon River of Brazil. *Science, 228*, 488–490.

Mélières, M.-A., Pourchet, M. & Pinglot, J.-F. (1991). Comment on deconvolution profiles in sediment cores. *Water, Air and Soil Pollution, 60*, 35–42.

Melnikov, S. A., Vlasov, C. V., Rizhov, O. V., Gorshkov, A. N. & Kuzin, A. I. (1993). Zones of relatively enhanced contamination levels in the Russian Arctic Seas. *Proceedings of the Workshop on Arctic Contamination, Interagency Arctic Research Policy Committee, 2/7 May 1993, Anchorage, Alaska* (8 pp.). Arctic Research of the United States.

Menke, W. (1984). *Geophysical Data Analysis: Discrete Inverse Methods*. Orlando: Academic Press.

Mikhailov, V. N. (1997). River Mouths of Russia and Adjacent Countries: Past, Present and Future (413 pp.). Moscow: GEOS (in Russian).

Milliman, J. D., Summerhayes, C. P. & Barretto, H. T. (1975). Quaternary sedimentation on the Amazon continental margin: a model. *Geological Society of America Bulletin, 86*, 610–614.

Mitra, S., Dellapenna, T. M. & Dickhut, R. M. (1999). Polycyclic aromatic hydrocarbon distribution within lower Hudson River estuarine sediments: Physical mixing vs. sediment geochemistry. *Estuarine, Coastal and Shelf Science, 49* (3), 311–326.

Moore, W. S. (1984). Radium isotope measurements using Germanium detectors. *Nuclear Instruments and Methods, 223*, 407–411.

Moore, W. S. & Dymond, J. (1988). Correlation of ^{210}Pb removal with organic carbon fluxes in the Pacific Ocean. *Nature, 331*, 339–341.

Morrison, S. J. & Cahn, L. S. (1991). Mineralogical residence of alpha-emitting contamination and implications for mobilization from uranium mill tailings. *Journal of Contaminant Hydrology, 8*, 1–21.

Muller, P. J. & Suess, E. (1979). Productivity, sedimentation rate, and sedimentary organic matter in the oceans. I. Organic carbon preservation. *Deep-Sea Research, 26A*, 1347–1362.

Mulsow, S., Boudreau, B. P. & Smith, J. N. (1998). Bioturbation and porosity gradients. *Limnol. Oceanogr., 43*, 1–9.

Nittrouer, C. A., Curtin, T. B. & Demaster, D. J. (1986). Concentration and flux of suspended sediments on the Amazon Continental Shelf. *Continental Shelf Research, 6*, 151–174.

Nittrouer, C. A., DeMaster, D J., McKee, B. A., Cutshall, N. H. & Larsen, I. L. (1983/84). The effect of sediment mixing on Pb-210 accumulation rates for the Washington continental shelf. *Mar. Geol., 54*, 201–221.

Nittrouer, C. A., Kuehl, S. A., Demaster, D. J. & Kowsman, R. O. (1986). The deltaic nature of Amazon Shelf sedimentation. *Geological Society of America Bulletin, 97*, 444–458.

Officer, C. B. (1982). Mixing, sedimentation rates and age dating for sediment cores. *Mar. Geol., 46*, 261–278.

Olsen, C. R., Simpson, H. J., Peng, T. H., Bopp, R. T. & Trier, R. M. (1981). Sediment mixing and accumulation rate effects on radionuclide depth profiles in Hudson estuary sediments. *J. Geophys. Res., 86*, 11 020–11 028.

Osvath, I., Ballestra, S., Baxter, M. S., Gastaud, J., Hamilton, T., Harms, I., Liong Wee Kwong, L., Parsi, P. & Povinec, P. P. (1995). IAEA programmes related to the radioactive waste dumped in the Artic seas. Part 2. IAEA–MEL's contribution to the investigation of the Kara Sea dumping sites. *Abstract IASAP Meeting Vienna, Austria, May 1995*.

Page, D. S., Boehm, P. D., Douglas, G. S., Bence, A. E., Burns, W. A. & Mankiewicz, P. J. (1999). Pyrogenic polycyclic aromatic hydrocarbons in sediments record past human activity: a case study in Prince William Sound, Alaska. *Mar. Poll. Bull., 38*, 247–260.

Palmer, M. R. & Elderfield, H. (1985). Sr isotope composition of seawater over the past 75 Myr. *Nature 314*, 526–528.

Pampoura, V. D. & Sandimirov, I. V. (1993). Total organic carbon, uranium and thorium in modern and Holocene deposits of Baikal, International Project of Paleolimnology and Late Cenozoic Climate. *Newsletter, 7*, 40–42.

Paulsen, S. C., List, E. J. & Santschi, P. H. (1999). Modeling variability in ^{210}Pb and sediment fluxes near the Whites Point outfalls, Palos Verdes Shelf, California. *Environmental Science and Technology, 33*, 3077–3085.

Pedersen, T. F. & Calvert, S. E. (1990). Anoxia vs. productivity: What controls the formation of organic carbon-rich sediments and sedimentary rocks? *American Association of Petroleum Geologists Bulletin, 74*, 454–466.

Peng, T. H., Broecker, W. S. & Berger, W. H. (1979). Rates of benthic mixing in deep-sea sediment as determined by radioactive tracers. *Quat. Res., 11*, 141–149.

Pheiffer-Madsen, P. & Sørensen, J. (1979). Validation of the Lead-210 method. *J. Radioanal. Chem., 54*, 39–48.

Punning, J.-M., Ilomets, M., Karofeld, E., Koff, T., Kozlova, M., Laugusta, R., Taure, I., Rajamäe, R. & Varvus, M. (1989). Technogeensed muutused biogeokeemilises aineringes. In Ilomets, M. (Ed.). Kurtna Järvestika Looduslik Seisund ja Selle Areng II (pp. 14–28) (in Estonian).

Radakovitch, O. & Heussner, S. (1999). Fluxes and budget of Pb-210 on the continental margin of the Bay of Biscay (northeastern Atlantic): Deep-sea research. Part II. *Topical Studies in Oceanography, 46*, 2175–2203.

Ritchie, J. C. & McHenry, R. (1990). Application of radioactive fallout Cesium-137 for measuring soil erosion and sediment accumulation rates and patterns: a review. *J. Environ. Qual., 19*, 215–233.

Robbins, J. A. (1978). Geochemical and geophysical applications of radioactive lead isotopes. In Nriago, J. P. (Ed.). *Biogeochemistry of Lead* (pp. 285–393). North-Holland: Elsevier.

Robbins, J. A. (1986). A model for particle-selective transport of tracers in sediments with conveyor belt deposit feeders (Modele de transport selectif des sediments des traceurs dans des sediments par distributeur de depot a courroie transporteuse). *Journal of Geophysical Research, 91* (7), 8542–8558.

Robbins, J. A. & Edgington, D. N. (1975). Determination of recent sedimentation rates in Lake Michigan using Pb-210 and Cs-137. *Geochim. Cosmochim. Acta, 39*, 285–304.

Robertson, A. (1998). Petroleum Hydrocarbons. In *AMAP Assessment Report: Arctic Pollution Issues. Arctic Monitoring and Assessment Programme (AMAP), Oslo, Norway* (pp. 661–716).

Rogozin, A. A. (1993). *Coast zones of Lakes Baikal and Xubsugul* (167 pp.). Novosibirsk: Nauka.

Rundberg, Y. & Smalley, P. C. (1989). High-resolution dating of Cenozoic sediments from Northern North Sea using $^{87}Sr/^{86}Sr$ stratigraphy. *AAPG Bulletin, 73*, 298–308.

Sagris, A. (1987). Môningate Kurtna järvestiku järvede veebilanisist. In Ilomets, M. (Ed.). *Kurtna Järvestiku Looduslik Seisund ja Selle Areng* (pp. 144–147) (in Estonian).

Savinov, V. M., Savinova, T. N., Carroll, J., Matishov, G. G. & Næs, K. (2000). Polycyclic aromatic hydrocarbons (PAHs) in sediments of the White Sea, Russia. *Marine Pollution Bulletin, 40*, 807–818.

Scarlato, O. A. (1991). Oceanographic conditions and biological productivity in the White Sea (annotated atlas) (115 pp.). Murmansk: PINRO.

Schell, W. T. & Sibley, T. H. (1982). Distribution coefficients for radionuclides in aquatic environments. Final Summary Report to U.S. Nuclear Regulatory Comm., NUREG/CR-1869.

Schlüter, M., Sauter, E., Hansen, H.-P. & Suess, E. (2000). Seasonal variations of bio-irrigation in coastal sediments: modeling of field data. *Geochimica et Cosmochimica Acta, 64*, 821–834.

Schulz, M. & Schaefer-Neth, C. (1997). Translating Milankovitch climate forcing into eustatic fluctuations via deep thermal water expansion: a conceptual link. *Terra Nova, 9*, 228–231.

Sholkovitz, E. R. & Mann, D. R. (1984). The pore water chemistry of $^{239,240}Pu$ and ^{137}Cs in sediments of Buzzards Bay, Massachusetts. *Geochimica et Cosmochimica Acta, 48*, 1107–1114.

Sholkovitz, E. R., Cochran, J. K. & Carey, A. E. (1983). Laboratory studies of the diagenesis and mobility of $^{239,240}Pu$ and ^{137}Cs in nearshore sediments. *Geochimica et Cosmochimica Acta, 47*, 1369–1379.

Sivintsev, Y. (1994). Study of Nuclides Composition and Characteristics of Fuel in Dumped Submarine Reactors and Atomic Icebreaker "LENIN" Part 1 and Part 2. Working Material of the International Arctic Seas Assessment Project. Vienna, Austria: IAEA.

Smith, C. R., Jumars, P. A. & DeMaster, D. J. (1986). In situ studies of megafaunal mounds indicate rapid sediment turnover and community response at the deep-sea floor. *Nature, 323*, 251–253.

Smith, C. R., Pope, R. H., Demaster, D. J. & Magaard, L. (1993). Age-dependent mixing of deep-sea sediments. *Geochim. Cosmochim. Acta, 57* (7), 1473–1488.

Smith, J. N., Boudreau, B. P. & Noshkin, V. (1986). Plutonium and ^{210}Pb distributions in northeast Atlantic sediments: subsurface anomalies caused by non-local mixing. *Earth and Planetary Science Letters, 81*, 15–28.

Smith, J. T. & Comans, R. N. J. (1996). Modeling the diffusive transport and remobilisation of ^{137}Cs in sediments: The effects of sorption kinetics and reversibility. *Geochimica et Cosmochimica Acta, 60*, 995–1004.

Smith, J. N., Ellis, K. M., Næs, K., Dahle, S. & Matishov, D., 1995, Sedimentation and mixing rates of radionuclides in Barents Sea sediments off Novaya Zemlya. *Deep-Sea Research, 42*, 1471–1493.

Smith, J. N. & Walton, A. (1980). Sediment accumulation rates and geochronologies measured in the Saguenay Fjord using the Pb-210 dating method. *Geochim. Cosmochim. Acta, 44*, 225–240.

Soetaert, K., Herman, P. M. J. & Middelburg, J. J. (1996). A model of early diagenetic processes from the shelf to abyssal depths. *Geochimica et Cosmochimica Acta, 60* (6), 1019–1040.

Soetaert, K., Herman, P. M. J. Middelburg, J. J., Help, C., deStigter, H. S., van Weering, T. C. E., Epping, E. & Helder, W. (1996). Modelling Pb-210-derived mixing activity in ocean margin sediments: Diffusive versus nonlocal mixing. *J. Mar. Res., 54*, 1207–1227.

Somayajulu, B. L. K., Bhushan, R., Sarkar, A., Burr, G. S. & Jull, A. J. T. (1999). Sediment deposition rates on the continental margins of the eastern Arabian Sea using ^{210}Pb, ^{137}Cs and ^{14}C. *The Science of the Total Environment*, 237–238, 429–439.

Statistical Book (1998). Moscow: State Committee of the Russian Federation on Statistics.

Tedesco, L. P. & Aller, R. C. (1997). Pb-210 chronology of sequences affected by burrow excavation and infilling: Examples from shallow marine carbonate sediment sequences, Holocene South Florida and Caicos Platform, British West Indies. *Journal of Sedimentary Research, 67* (1), 36–46.

Tolstov, G. P. (1962). *Fourier Series*. New York: Dover Publications.

Turekian, K. K., Nozaki, Y. & Benninger, L. K. (1977). Geochemistry of atmospheric radon and radon products. *Ann. Rev. Earth Planet. Sci., 5*, 227–255.

Varvus, M. & Punning, J.-M. (1993). Use of the ^{210}Pb method in studies of the development and human-impact history of some Estonian lakes. *The Holocene, 3*, 34–44.

Von Damm, K. L., Benninger, L. K. & Turekian, K. K. (1979). The ^{210}Pb geochronology of a core from Mirror Lake, New Hampshire. *Limnol. Oceanog., 24*, 434–439.

Votinsev, K. K. & Popovskaya G. I. (1979). The peculiarity of the biotic cycle in Lake Baikal. *Doklady Akademii Nauk SSSR, 216*, 666–669 (in Russian).

Waugh, W. J., Carroll, J., Abraham, J. D., Thiede, M. E. & Landeen. D. S. (1998). Use of dendrochronology and sediment geochronology to establish reference episodes for evaluations of environmental radioactivity. *Journal of Environmental Radioactivity, 41*, 269–286.

Weiss, R. F., Carmack, E. C. & Koropalov, V. M. (1991). Deep-water renewal and biological production in Lake Baikal. *Nature, 349*, 665–669.

Wepener, V., van Vuren, J. H. J. & duPreez, H. H. (2000). Application of the equilibrium partitioning method to derive copper and zinc quality criteria for water and sediment: A South African perspective. *Water SA, 26* (1), 97–104.

Weyhenmeyer, G. A., Håkanson, L. & Meili, M. (1997). A validated model for daily variations in the flux, origin and distribution of settling particles within lakes. *Limnology and Oceanography, 42*, 1517–1529.

Wheatcroft, R. A. (1992). Experimental tests for particle size dependent bioturbation in the deep ocean. *Limnology and Oceanography, 37*, 90–104.

White Book 3 (1993). Facts and problems related to radioactive waste disposal in seas adjacent to the territory of the Russian Federation. Moscow: Office of the President of the Russian Federation.

Williams, D. F., Lerche, I. & Full, W. E. (1988). *Isotope Chronostratigraphy: Theory and Methods* (3–4 pp.). San Diego: Academic Press.

Williams, D. F. , Qui, L., Karabanov, E. & Gvozdkov, A. (1993). Goechemical indicators of productivity and sources of organic matter in surficial sediments of Lake Baikal. *Russian Geologu and Geophisics, 33*, 111–125.

Yefimov, E. (1994). Radionuclides composition, characteristics of shielding barriers and analyses of weak points of the dumped reactors of submarine N 601. Working Material of the International Arctic Seas Assessment Project. Vienna, Austria: IAEA.

Subject Index